# Green Energy and Technology

Climate change, environmental impact and the limited natural resources urge scientific research and novel technical solutions. The monograph series Green Energy and Technology serves as a publishing platform for scientific and technological approaches to "green"—i.e. environmentally friendly and sustainable—technologies. While a focus lies on energy and power supply, it also covers "green" solutions in industrial engineering and engineering design. Green Energy and Technology addresses researchers, advanced students, technical consultants as well as decision makers in industries and politics. Hence, the level of presentation spans from instructional to highly technical.

**Indexed in Scopus**.

More information about this series at http://www.springer.com/series/8059

Malti Goel · T. Satyanarayana ·
Maruthadu Sudhakar · D. P. Agrawal
Editors

# Climate Change and Green Chemistry of $CO_2$ Sequestration

Springer

*Editors*
Malti Goel
Climate Change Research Institute
New Delhi, India

T. Satyanarayana
Netaji Subhas University of Technology
New Delhi, India

Maruthadu Sudhakar
Centre for Marine Living Resources
and Ecology
Kochi, Kerala, India

D. P. Agrawal
Climate Change Research Institute
New Delhi, India

ISSN 1865-3529 ISSN 1865-3537 (electronic)
Green Energy and Technology
ISBN 978-981-16-0028-9 ISBN 978-981-16-0029-6 (eBook)
https://doi.org/10.1007/978-981-16-0029-6



This Springer imprint is published by the registered company Springer Nature Singapore Pte Ltd.
The registered company address is: 152 Beach Road, #21-01/04 Gateway East, Singapore 189721, Singapore

# Foreword

In the twenty-first century, climate change has become an ecological and technological challenge, and a serious concern for health and sustainability.

As forests, soil, oceans, the atmosphere, living beings and fossil fuels are vital storehouses of carbon; the burning fossil fuels is a primary source of carbon dioxide. Disproportionate emissions of $CO_2$ and other greenhouse gases in the global atmosphere from anthropogenic activities are causing concerns for global warming and climate change. India has ratified the Paris Agreement on Climate Change 2015 and moves ahead toward its compliance by 2030. We anticipate strictor emission standards in an effort to counteract climate change through the adoption of not one, but all available options in our energy systems.

According to the International Energy Agency, carbon capture, utilization and storage (CCUS)—$CO_2$ sequestration is a technology set of options that can significantly reduce fossil fuel emissions. Yet, there are challenges in its application; commercialized $CO_2$ sequestration technology is not yet fully developed. Capturing carbon dioxide in the air is a vast technological issue and is a costly proposition. Carbon storage is both technical as well as a safety issue. Stored carbon dioxide may leak into the atmosphere due to some accidents or acts of nature. That should be properly addressed so that ecological consequences are taken into consideration for its sustenance.

I am happy that the book on 'Climate Change and Green Chemistry of $CO_2$ Sequestration,' edited by Malti Goel, T. Satyanarayana, Maruthadu Sudhakar, and D. P. Agrawal, has a knowledge-sharing perspective on R&D in green chemistry of carbon management. I compliment the Climate Change Research Institute and Springer Nature, as editors and publishers for broad coverage of topics in the seventeen chapters of the book. It amply illustrates the importance of $CO_2$ sequestration research and the directions in which this field is advancing in India and abroad.

I am sure the book will motivate and inspire the readers.

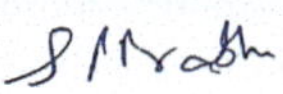

Shri Suresh Prabhu
G20 Sherpa; Former Union Minister, Ministry of Commerce, Industry, Civil Aviation and Railways, New Delhi, India

# Preface

*The time is past when humankind thought it could selfishly draw on exhaustible resources, we know now that world is not a commodity.*

—Francois Hollande on Climate Change, President of French Republic (2015)

A vast area of multi-disciplinary research is unfolding and awaiting the initiatives to improve our understanding to address the complex relationships between environment, human health and development. From industries and other developmental activities, increased greenhouse gas emissions and their accumulation in the atmosphere have led to global warming and climate change risks. The International Protocol as Paris Agreement on Climate Change was evoked in 2015 to bring nations together to reduce greenhouse gas emissions to keep average global temperatures from rising less than 2 °C above the pre-industrial levels and to make efforts to limit the increase to 1.5 °C.

In 2020, COVID-19 pandemic took the world by surprise. It became evident with the onset of this public health emergency that people's participation in environmental and climate change issues is almost a mandatory requirement for sustainable growth and social development.

The global $CO_2$ concentrations in the atmosphere hit a record high as 414 ppm in 2019 and 416 ppm in 2020. With the increasing demand for clean energy and our reliance on fossil fuels in the medium term, the carbon dioxide ($CO_2$) in the atmosphere presents us with an unprecedented situation of a further increase in greenhouse gas emissions. Carbon sequestration, i.e., capturing, utilizing and storing carbon emitted from the global energy systems, has emerged as a proposed strategy for reducing atmospheric $CO_2$ concentrations. But for it to be a viable option, it needs to be cost-effective, reliable and safe compared to the other strategies. These concerns are demanding scientific and technological innovations at the national and international levels.

At present, $CO_2$ sequestration and utilization are in the research domain. National governments are investing heavily in carbon capture and utilization research, and there are important ongoing research programs worldwide. India is among the few developing countries, which began investing in the national program on $CO_2$ sequestration research starting in 2006. Carbon capture innovations has got renewed thrust with the increase in R&D investments under the Mission Innovation international program launched in 2015 in which India is a partner country.

The book on *Climate Change and Green Chemistry of $CO_2$ Sequestration* provides a multi-disciplinary outlook for academic exchange of current research in the field of $CO_2$ sequestration and describes challenges as well as opportunities. Information from different scientific disciplines concerning $CO_2$ capture and utilization is put together to provide the reader a wider perspective about the directions in which this field is advancing in India and abroad. Climate change mitigation through bio-route is discussed, and a chapter on modeling of $CO_2$ fluxes over the Asian region has been included.

## Structure of the Book

This book features contributions of leading scientists and technocrats across 17 chapters of particular interest to the readers. State-of-the-art scientific reviews and research papers on advances in green chemistry for carbon management strategies in response to climate change are covered. It exposes readers to a variety of new approaches and methods to recycle and reuse $CO_2$, and minimize emissions. The book is organized into three parts, namely

1. *$CO_2$ Sequestration Opportunities for India's Energy Security,*
2. *Carbon Dioxide Capture and Green Chemistry of Its Conversion,*
3. *Climate Change Mitigation Through Bio-route and Assimilation of $CO_2$ Fluxes.*

The first part ponders on $CO_2$ sequestration opportunities for India's energy security in various sectors of the economy. The R&D experience from the Indian oil industry for reducing carbon footprints is shared in chapter "Carbon Capture and Sequestration: Implications and Opportunities for India" with policy recommendations. Opportunities arising from de-carbonization initiatives in India and the potential pathways for implementation of $CO_2$ sequestration are the highlights. Chapter "An Assessment of $CO_2$ Reduction Potential from Carbon Sequestration Versus Renewable Energy Targets in India" describes the potential of carbon sequestration technologies in India's energy industry for the long term reduction of emissions vis-à-vis high renewable energy targets. Chapter "Challenges for $CO_2$ Management Through Renewable Energy, Conventional Power and System Reliability in Indian Power Sector" evaluates the impact of increased renewable energy penetration into the grid and its impact on coal-based power generation. Adapting advanced coal combustion technologies for thermal power plants can facilitate $CO_2$ capture. The possibility of using advanced technology of chemical looping combustion for India's high ash coal is examined in chapter "CFD Analysis of Chemical Looping Combustion with Special Emphasis on Indian Coal: A Review." Approach to integration of $CO_2$ sequestration technologies in large-scale demonstration projects for controlling emissions from the automobile sector is the theme of chapter "Feasibility Study of $CO_2$ Control in Automobile Sources." Technology development at a place relies on the progress made in patenting the innovations. Patent analysis of the trends in $CO_2$ sequestration technology: global and Indian, is described in chapter "Patent Landscape Analysis in Carbon Dioxide Capture Technologies."

The second part on carbon dioxide capture and green chemistry of its conversion delves deeper into sustainability pathways for $CO_2$ mitigation in the next five chapters. Chapter "$CO_2$ Conversion into Chemicals and Fuel: India's Perspective" describes the progress made in system design for electrocatalytic and photocatalytic processes to convert $CO_2$ into chemicals and challenges in developing a commercially viable catalyst. Chapter "Thermodynamic Analysis of Ionic Liquids for $CO_2$ Capture, Regeneration and Conversion" highlights thermodynamic considerations of using ionic liquids (ILs) for $CO_2$ capture and its conversation. Using predictive models of thermodynamic properties, higher absorption and improved energy efficiency solvents become feasible. Chapter "Heterogeneous Catalysis for Chemical Fixation of $CO_2$ via Carbonylation Reactions," discusses design studies for heterogeneous catalysts for chemical fixation of $CO_2$ and the generation of useful chemicals for prospective uses in the chemical industry and the petroleum refining industry. The next chapter "Application $CO_2$ as Refrigerant for HVAC, Supermarket, and Cold Chain—A Review" explains the potential of $CO_2$ as a replacement to conventional refrigerants in HVAC sectors like automobiles, air conditioning in supermarkets and cold chain applications in the agriculture sector. Chapter "Research & Development Pathways/Challenges in Direct Air Capture of $CO_2$" deals with direct air capture (DAC), as a negative emission technology and challenges in its commercial-scale development.

The third part of the book explores climate change mitigation through bio-route and also describes the assimilation of $CO_2$ fluxes. Among the various biological methods available for carbon capture from industrial point sources, carbonic anhydrase (CA)-based $CO_2$ capture is gaining immense attention in recent years. Chapter "Mitigating Global Warming Through Carbonic Anhydrase-Mediated Carbon Sequestration" describes innovations in biomimetic carbon sequestration using thermo-alkali-stable CAs as a promising tool for climate change mitigation. The CA is a zinc metalloenzyme, which is an essential biocatalyst for all living beings. Chapter "Biomass-Based $CO_2$ Adsorbents for Biogas Upgradation with Pressure Swing Adsorption" identifies the research gaps in biogas production from biomass using pressure swing adsorption and challenges in the selection of low-cost adsorbents to maximize utilization of $CO_2$ into useful biofuels in transport. The next chapter "$CO_2$ Sequestration Through Phytoremediation Techniques with Special Emphasis on Urban Forestry to Mitigate Climate Change Impact" is about applying the phyto-remediation techniques and developing a green belt model in urban forestry by planting more $CO_2$ assimilating plants. How forest management practices like harvesting, thinning and burning activities affect the $CO_2$ soil fluxes and minimize GHG emissions, is described in chapter "Effect of Anthropogenic Activities on the Rate of Soil $CO_2$ Flux in the Dipterocarpus Forest Ecosystem of Northeast India." Chapter "Estimation of Carbon Dioxide Fluxes Between Land, Ocean and Atmosphere During 2006–2011 with a 4D Variational Assimilation Scheme and Special Reference to Asia" presents the assimilation of $CO_2$ fluxes between land, ocean and atmosphere into the LMDZ transport model using carbon dioxide data from over a hundred global stations. The $CO_2$ observed fluxes from three stations in India have been integrated into models.

We hope that the book will be of great interest to the researchers, postgraduate students, professors, scientists, government officers, industry technocrats, and policymakers.

New Delhi, India
January 2021

Malti Goel
T. Satyanarayana
Maruthadu Sudhakar
D. P. Agrawal

# Acknowledgements

The editors acknowledge all renowned authors and contributors for their strong engagement, hard work and patience. The book's genesis is the awareness and capacity building workshop on Recent Advances in $CO_2$ Capture Technologies and its Sectoral Applications. The present book includes the lectures delivered in the workshop and invited papers from the eminent experts. Our special thanks are due to Prof. Anjan Ray, Shri V. S. Verma, Dr. Ajit Kumar Parwani, Dr. P. Senthilkumar, Dr. H. Purushotham, Prof. Sudhasatwa Basu, Prof. Ramesh L. Gardas, Dr. Ganpati V. Shanbhag, Dr. Dileep Kumar Gupta, Shri Gautam Sen, Dr. Ranjan R. Pradhan, Dr. M. Govindaraju, Prof. P. S. Yadava and Prof. P. S. Swathi for their insightful contributions. We also take this opportunity to thank the Peer Review team members, especially, Prof. S. Basu, Director, CSIR-IMMT, and Prof. V. S. Verma, Ex-CERC, for their timely and valuable help.

We are deeply indebted to honorable Shri Suresh Prabhu, India's G20 Sherpa and former Union Minister, Commerce and Industry, Civil Aviation and Railways, to grace the workshop and for his guidance and blessings. We also convey our sincere thanks to Prof. Anjan Ray, Director, CSIR-Indian Institute of Petroleum, Dehradun, for his unstinted support and delivering the keynote address despite being unwell. We are obliged to distinguished speakers Dr. S. S. V. Ramakumar, Director (R&D) and Board Member, Indian Oil Corporation, Faridabad, and Dr. Anupam Agnihotri, Director, Jawaharlal Nehru Research, Design and Development Centre, Nagpur, for taking out valuable time from their busy schedules and for inspiring the deliberations.

We record our appreciation to the Ministry of Earth Sciences, Government of India, for providing partial financial support for the workshop.

We express our most sincere thanks and appreciation to Ms. Swati Meherishi, Executive Editor, Applied Science & Engineering, Springer Nature. She and her team deserve praise for their dynamism and constant help. We thank CCRI team, especially Shri Anish Tripathi, Shri L. K. Bansal, and other members for their immense contribution.

New Delhi, India
January 2021

Malti Goel
T. Satyanarayana
Maruthadu Sudhakar
D. P. Agrawal

# Annexure

Contributing Institutions

- Anna University, Chennai, Tamil Nadu
- Bharathidasan University, Tiruchirappalli, Tamil Nadu
- Climate Change Research Institute, New Delhi, Delhi
- CSIR-Institute of Minerals and Materials Technology, Bhubaneswar, Odisha
- CSIR Fourth Paradigm Institute, Bengaluru, Karnataka
- CSIR-Indian Institute of Petroleum, Dehradun, Uttarakhand
- Indian Institute of Technology, Kharagpur, West Bengal
- Indian Institute of Technology Madras, Chennai, Tamil Nadu
- Institute of Infrastructure Technology, Research and Management, Ahmedabad, Gujarat
- Manipur University, Imphal, Manipur
- Netaji Subhas University of Technology, New Delhi
- Poornaprajna Institute of Scientific Research, Bengaluru, Karnataka
- University of Guleph, Canada

# Contents

# Editors and Contributors

## About the Editors

**Prof. Dr. (Mrs.) Malti Goel** is a physicist and former senior Advisor and Scientist 'G', Ministry of Science & Technology, Government of India. She headed the Inter-sectoral Science & Technology Advisory Committee division from 1998-2008, and coordinated with the socio-economic Ministries. She showed exemplary leadership in spearheading multi-institutional research and technology projects in new and emerging areas of the national importance. She took several initiatives in promoting research at the national level, particularly in carbon srequestration, physical sciences thrust areas, and atmospheric science programmes on MONTBLEX90 and U-PROBE. Dr. Malti Goel is a recipient of Gold Medal in M.Sc Physics from Birla Institute of Technology and Science, Pilani, India. She did Post M.Sc DIIT in Solid State Physics and Ph.D in Physics from Indian Institute of Technology Delhi. She was Research Associate and Scientific Pools Officer at IIT Delhi, conducted pioneering post-doctoral research on polymers and composites. As Emeritus Scientist, she got affiliated with the Indian National Science Academy and the Jawaharlal Nahru University, New Delhi. She became Professor Adjunct, Jamia Hamdard University. Recipient of several awards and honours for her academic involvements, Dr. Malti Goel received the '*Lifetime Achievement Award 2016*' to recognise her outstanding contributions and achievement in climate change research in India. She is author and editor of eleven scientific books, thirty book chapters and has published nearly three hundred papers in international repute journals and conferences. Dr. Malti Goel is a Fellow of National Environment Science Academy and a member of many prestigious professional bodies. As Chief Executive, Climate Change Research Institute, she is engaged in promoting climate change education and research.

**Prof. T. Satyanarayana** is Professor Emeritus at the Department of Biological Sciences & Engineering, Netaji Subhas University of Technology, New Delhi, India after superannuating from the Department of Microbiology, University of Delhi, New Delhi, as Professor and Head, in 2016. He has over 280 scientific papers and reviews, 11 edited books and 2 patents to his credit. He is a fellow of the National Academy of Agricultural Sciences (NAAS), Association of Microbiologists of India (AMI), Biotech Research Society (India), Mycological Society of India (MSI) and Telengana Academy of Sciences. He is a recipient of Dr. Manjrekar award of AMI, Dr. Agnihotrudu award of MSI and Malaviya Memorial award of BRSI. He has over 40 years of research and teaching experience and has mentored 30 scholars for Ph.D. He was president of AMI and MSI during 2014–2015 and 2015–2016, respectively. He has successfully executed several major research projects from government agencies like, DST, DBT, UGC, CSIR and the Ministry of Environment & Forests. His research efforts have been focused on understanding the diversity and applications of yeasts, thermophilic fungi and bacteria and their enzymes, metagenomics, carbon sequestration employing extremophilic bacterial carbonic anhydrases and bioethanol production from lignocellulosic substrates using enzyme cocktails.

**Dr. Maruthadu Sudhakar** was the Director, Centre for Marine Living Resources & Ecology (CMLRE) at Kochi (2015–2019) and held the position of Advisor to the Government of India, Ministry of Earth Sciences (2009–2015). Between 1982 and 2009, he served premier national institutes of the country as Scientist in various grades at National Centre for Polar and Ocean Research (NCPOR), and National Institute of Oceanography (NIO), Goa. Dr. Sudhakar contributed significantly to India's deep-sea programme 'Surveys for Polymetallic Nodules in Indian Ocean' and participated in negotiations for the allocation of Pioneer Area in Central Indian Ocean Basin by the Preparatory Commission (PrepCom) to the Law of the Sea in 1987. He was instrumental to launch and lead a pilot expedition to the Southern Ocean and first expedition to Larsemann Hills, Eastern Antarctica, carried out swath bathymetric mapping of seafloor. He was an elected member of Legal and Technical Commission of International Seabed Authority (ISA), Jamaica for two terms. He was also the Commissioner of India and led Indian delegation to the annual sessions of Convention on Conservation of Antarctic Living Resources (CCAMLR), Hobart, Australia, and Commissioner of Yeosu World Expo in Republic of Korea (2012). He is a Fellow of Geological Society of India, Mining Engineers Association of India, Honorary Fellow of ISAG, IGU, IGC and AP Akademi of Sciences. He published more than 150 papers in peer-reviewed journals, equal number in seminar/symposia/conferences and 7 books/chapters.

**Prof. D. P. Agrawal** is Ex-Chairman, Union Public Service Commission, New Delhi. Before joining this prestigious position Prof. Agrawal was founder Director of ABV Indian Institute of Information Technology and Management (IIITM), Gwalior. IIITM was set up by Ministry of Human Resource Development in 1998 and declared as a Deemed University by the Government of India in 2001. Under his leadership IIITM, has entered into MOUs with several international institutions/universities for joint collaboration in the areas of research, exchange programmes for faculty and students, and consultancy to companies in India and abroad. Many initiatives started by him at IIITM, Gwalior have been replicated in the country. He was Professor and Dean at IIT-Delhi from where, in 1994, he took over as Joint Educational Adviser

(T) in the Ministry of Human Resources Development, Government of India. He has planned and developed a number of initiatives at national level in technical education, particularly in the areas of information technology and computer education. He has made substantial contribution for the development of vocational education through advice and guidance.

## Contributors

**D. P. Agrawal** Ex-Union Public Service Commission, New Delhi, India

**Suddhasatwa Basu** Materials Chemistry Department, CSIR-Institute of Minerals and Materials Technology, Bhubaneswar, Odisha, India

**Himadri Bose** Department of Biotechnology, Indian Institute of Technology Kharagpur, Kharagpur, India

**Yatendra S. Chaudhary** Materials Chemistry Department, CSIR-Institute of Minerals and Materials Technology, Bhubaneswar, Odisha, India

**Yi Wai Chiang** School of Engineering, University of Guelph, Guelph, Canada

**Indrajit Das** Department of Chemistry, Indian Institute of Technology Madras, Chennai, India

**Animesh Dutta** School of Engineering, University of Guelph, Guelph, Canada

**J. Fowmitha Banu** Department of Environmental Biotechnology, School of Environmental Science, Centre for Climate Change Research, Bharathidasan University, Tiruchirappalli, Tamil Nadu, India

**Ramesh L. Gardas** Department of Chemistry, Indian Institute of Technology Madras, Chennai, India

**Pragyan P. Garnaik** Milton, Canada

**Malti Goel** Ex-Ministry of Science & Technology, New Delhi, India

**M. Govindaraju** Department of Environmental Biotechnology, School of Environmental Science, Centre for Climate Change Research, Bharathidasan University, Tiruchirappalli, Tamil Nadu, India

**Dileep Kumar Gupta** Mechanical Engineering Department, Institute of Infrastructure, Technology, Research and Management, Ahmedabad, India

**N. K. Indira** CSIR Fourth Paradigm Institute, NAL Belur Campus, Bengaluru, India

**Prashrita Kaushal** National Research Development Corporation, New Delhi, India

**Nagendra Kulal** Materials Science and Catalysis Division, Poornaprajna Institute of Scientific Research (PPISR), Bengaluru, India

**Pulkit Kumar** Institute of Infrastructure Technology Research and Management, Ahmedabad, Gujarat, India

**Sanjeev Majumdar** National Research Development Corporation, New Delhi, India

**Niharika** Materials Chemistry Department, CSIR-Institute of Minerals and Materials Technology, Bhubaneswar, Odisha, India

**Manju Lata Pal** Mechanical Engineering Department, Institute of Infrastructure, Technology, Research and Management, Ahmedabad, India

**Rupali Pal** Ex-Climate Change Research Institute, New Delhi, India

**Ajit Kumar Parwani** School of Engineering, University of Guelph, Guelph, Canada

**Ranjan R. Pradhan** Institute of Infrastructure Technology Research and Management, Ahmedabad, Gujarat, India

**Hanumanthu Purushotham** National Research Development Corporation, New Delhi, India

**V. Ramkumar** Department of Chemistry, Indian Institute of Technology Madras, Chennai, India

**M. Ramonet** LSCE, Gif sur Yvette, France

**Anjan Ray** CSIR-Indian Institute of Petroleum, Dehradun, India

**T. Satyanarayana** Division of Biological Sciences and Engineering, Netaji Subhas University of Technology, Dwarka, New Delhi, India

**Gautam Sen** Reliance Industries Ltd. (E&P), New Delhi, India;
ONGC, New Delhi, India

**S. Senthamil Selvi** Department of Botany, School of Life Science, Bharathidasan University, Tiruchirappalli, Tamil Nadu, India

**P. Senthilkumar** Department of Automobile Engineering, Madras Institute of Technology Campus, Anna University, Chromepet, Chennai, Tamil Nadu, India

**Ganapati V. Shanbhag** Materials Science and Catalysis Division, Poornaprajna Institute of Scientific Research (PPISR), Bengaluru, India

**Aditya Sharma** Research Scholar, Central University of Rajasthan, Kishangarh, India

**Maruthadu Sudhakar** Ex-Ministry of Earth Sciences, New Delhi, India

**P. S. Swathi** CSIR Fourth Paradigm Institute, NAL Belur Campus, Bengaluru, India

**Amrabati Thokchom** Department of Life Sciences, Manipur University, Imphal, India

**B. J. Vaishnavi** Materials Science and Catalysis Division, Poornaprajna Institute of Scientific Research (PPISR), Bengaluru, India

**V. S. Verma** Ex-Central Electricity Regulatory Commission, New Delhi, India

**Ashish Kumar Yadav** Mechanical Engineering Department, Institute of Infrastructure, Technology, Research and Management, Ahmedabad, India

**P. S. Yadava** Department of Life Sciences, Manipur University, Imphal, India

# Introduction

**Malti Goel, T. Satyanarayana, Maruthadu Sudhakar, and D. P. Agrawal**

**Abstract** Environment and climate change impacts due to global warming have been engaging the attention of scientists and policymakers for quite some time. The use of fossil fuel combustion-based energy systems gives rise to increasing emissions of carbon dioxide ($CO_2$), a greenhouse gas that gets accumulated in the atmosphere. This is contributing to global warming and climate change. Paris Agreement on Climate Change is an International Protocol evoked in 2015 for nations to take actions collectively to fulfill their commitments. This chapter introduces the importance of $CO_2$ sequestration and how advances in green chemistry can be an opportunity for climate change mitigation. The $CO_2$ is an inert and non-toxic gas, and a green chemical pathway for its utilization from the flue gases of an industry as a resource, complements the green chemistry. Green chemistry with its many advantages has a vast scope for waste elimination through the conversion of captured $CO_2$. Interdisciplinary approaches for the application of green chemistry to $CO_2$ sequestration are summarized.

**Keywords** Carbon sequestration · Green chemistry · Greenhouse gases · Global warming · Climate change mitigation

M. Goel (✉)
Science & Technology Solutions for Sustainable Energy Future, Climate Change Research Institute, C-85 Shivalik, New Delhi, India
e-mail: maltigoel2008@gmail.com

T. Satyanarayana
A-113, Produogiki Apartments, Plot no. 11, Sector 3, Dwarka, New Delhi 110078, India

M. Sudhakar
F-105, Devashri Garden Povorim, Goa 403501, India

D. P. Agrawal
B-25, 3rd Floor, Sarvodaya Enclave, New Delhi 10017, India

 
M. Goel et al. (eds.), *Climate Change and Green Chemistry of $CO_2$ Sequestration*, Green Energy and Technology, https://doi.org/10.1007/978-981-16-0029-6_1

# 1 Climate Change Actions

> Pour ce qui est de l'avenir, il ne s'agit pas de le prévoir, mais de le rendre possible.
> —Antoine de Saint Exupéry, Citadelle, 1948.

The above quote about future predictions is very apt in the context of climate change mitigation. Translated in English it means '*As far as the future is concerned, it is not a question of predicting it, but of making it possible*' said in 1948.

## *1.1 Paris Agreement on Climate Change*

In the 21st Conference of Parties meeting of the United Nations Framework Convention on Climate Change (UNFCCC) held during December 2015, the Paris Agreement on Climate Change was adopted by all 197 countries. This International Protocol came into being, as the Post Kyoto Protocol phase was coming to an end. It entered into force on November 4, 2016, and is a framework to bring nations together to take their predetermined actions to reduce greenhouse gas emissions to keep average global temperatures from rising less than 2 °C above the pre-industrial levels and to make efforts to limit the increase to 1.5 °C. Unlike the Kyoto Protocol which aimed at stabilization of greenhouses gases according to the targets given to Annex I countries and no targets assigned to non-Annex I countries, the Paris Climate Agreement was seen as a truly global deal because it allowed each country to set its own emission reduction targets as nationally determined contributions (NDCs) and then in their compliance to adopt its own strategies.

Intergovernmental Panel on Climate Change (IPCC) Special Report [1] on Impact of Global Warming of 1.5 °C above pre-industrial levels (SR1.5) had issued a warning that recent trends in emissions and the level of international commitments of the nations indicated by their nationally determined contributions (NDCs) of Paris Agreement on Climate Change are inadequate and a sharp decline in greenhouse gas emissions by 2030 can only make global warming to limit to 1.5 °C in the following decades. Earlier, the IPCC report (2007) had predicted that $CO_2$ emissions may have to be reduced by 15% of the year 2000 value by the year 2050, with the peak emissions occurring somewhere between 2000 and 2015. As the peaking of emissions from different nations has yet to reach, mitigating climate change will evidently require much larger reductions to be made rapidly.

The climate change threat, which is arising from increasing anthropogenic greenhouse gas emissions and their accumulation in the atmosphere, suggests that about 77% of the climatic radiative forcing of anthropogenic greenhouse gas emissions from all sources comes from carbon dioxide. The cumulative total of anthropogenic $CO_2$ emissions from 1870 onward is increasing. The global temperature has already risen by 1 °C in 2019 since the pre-industrial age. Climate change is yet to be fully understood, and our knowledge of its science is on the rise but many uncertainties remain. On the other hand, climate change impacts are being manifested in several ways and are visible in terms of increased frequency of disasters, frequent extremes of weather events, snow cover changes, sea-level changes, migration of species and adverse health impacts, and others. If the emissions continue to increase at the current rate, global warming is likely to reach 1.5 °C between 2030 and 2052 according to SR1.5. By reaching the tipping point early would mean a devastating impact on human health and well-being. According to the World Health Organization, it was estimated that between 2030 and 2050, the effects of climate change will lead to approximately 250,000 additional deaths [2].[1] Finding strategies for accelerated reduction of $CO_2$ emissions in the atmosphere has become our goal toward sustainable development.

## 1.2 Carbon Sequestration

Carbon sequestration is a science and technology solution to reduce $CO_2$ concentrations in the atmosphere and to mitigate the environmental impacts of fossil fuel combustion and industrial processes. Carbon dioxide can be sequestered by capturing carbon dioxide from its industrial point sources or directly from air [3] and utilizing it into products through value addition or storing it away from the atmosphere. For the utilization of $CO_2$ it needs to be activated. Once activated, it becomes a versatile chemical feedstock that has proven utility across a variety of industries. Carbon dioxide, an inert and non-toxic gas, becomes a renewable resource when anthropogenic $CO_2$ is used for producing chemicals and fuels. It creates the application of carbon dioxide sequestration within the scope of green chemistry, which has vast coverage in multidisciplinary research.

Carbon capture and storage (CCS) and carbon capture and utilization (CCU) combine into carbon capture, utilization, and storage (CCUS) or $CO_2$ sequestration.[2] The $CO_2$ sequestration technologies are a set of technologies classified into $CO_2$ capture, $CO_2$ passive geological storage, $CO_2$ geological utilization into enhanced oil ($CO_2$-EOR) or gas ($CO_2$-EOG) recovery, $CO_2$ chemical utilization, and $CO_2$ biological utilization etc. Particularly in the field of $CO_2$-EOR, several demonstration projects have been undertaken worldwide. Innovations are taking place in the $CO_2$ utilization technologies by using chemical and biological means thus making carbon dioxide a feedstock for the production of value-added chemicals, fuels and products on scale. Through *$CO_2$Chem*—a network of academics from different disciplines—in the Chemistry Department of the University of Sheffield, the UK is aiming to develop leadership in $CO_2$ utilization chemistry [4, 5]. A large number of useful chemicals

[1] The global onset of pandemic COVID-19 has demonstrated that the health tipping point is already approaching in 2020.

[2] CCUS and CCS are used to describe carbon capture, utilization and/or storage at various places in the book. For the sake of uniformity, the term $CO_2$ sequestration has been used here.

and products such as hydrocarbons, fuels, formic acid, urea, cyclic carbonates, cyclic carbamates, methanol, and carboxylic acid can be obtained. The $CO_2$ utilization chemistry is thus signifies as green or sustainable chemistry.

## 2 Green Chemistry

Green chemistry is a subset of chemistry as the name implies. Chemistry is at the heart of understanding the climate system and also in finding solutions through new emission reduction technologies as well as through clean energy alternatives. Chemistry is also a very prominent part of our daily lives—from food and drinks made safe to consume, the development of a range of cosmetics, the use of gadgets for automation, other equipment, and the pharmaceuticals as new cures for diseases, all have origin in chemistry [6]. Chemistry has improved our quality of life and also extended the life span. The chemical processes, however, have also contributed to new environmental problems, and some of the products of the industry have turned out to be more harmful than unexpected. In the early 1990s, the need was felt to develop 'greener' chemical products through clean processes that minimize environmental pollution. 'Green chemistry' soon emerged as a theme to develop solutions aimed at pollution prevention at the molecular level, i.e., chemical reactions. Due to its immense contribution to preventing adverse implications to the environment, water bodies, and human health, it has become an extremely important area of chemistry.

Green chemistry is the chemistry of products, and processes designed for reducing or eliminating the use and/or generation of environmentally harmful substances [7, 8]. The green chemistry approach is adopted in both the synthesis and the use of chemicals in material production in a sustainable manner as well as the processing of the waste in a more environmentally benign way. If technology reduces or eliminates hazardous chemicals from the environment or cleans up environmental contaminants, this technology would qualify as a green chemistry technology. Quoted with different names such as environmentally benign chemistry, clean chemistry, atom economy, and benign by design chemistry; 'green chemistry or sustainable chemistry' is more commonly used.

Green chemistry applications are emerging and are getting extended in many industry areas [9]. Pharma industry is ahead in benchmarking and developing green chemistry protocols [10]. The main driver toward finding the solution for green chemistry options by industry has been the implementation of the US Pollution Prevention Act of 1990. It aimed to promote the development of new chemistries for industrial applications that are less hazardous to the environment and human health. This could be achieved by reducing waste, employing efficient material resource and energy use, and minimizing risks/hazards. The US EPA in 1993 launched the 'US Green Chemistry Program,' which promoted companies and individuals that supported the environmental and economic benefits of using greener chemistry and instituted the 'Presidential Green Chemistry Challenge Award' [11].

The American Chemical Society (ACS) Committee on Environmental Improvement (CEI) stated the green chemistry approach as 'a sustainable world enabled through the sustainable practice and use of chemistry' [12].

## 2.1 Its Twelve Principles

The '12 Principles of Green Chemistry' have been declared as a philosophy to provide guidelines for developing green chemistry technologies to motivate academic and industrial scientists. These provide variety of ways to reduce the environmental and health impacts of chemical production and are stated below [13].

(1) *Prevention*: Preventing waste is better than treating or cleaning up waste after it is created.
(2) *Atom economy*: Synthetic methods should try to maximize the incorporation of all materials used in the process into the final product.
(3) *Less hazardous chemical syntheses*: Wherever practicable, synthetic methods should be designed to use and generate substances that possess little or no toxicity to human health and the environment.
(4) *Designing safer chemicals*: Chemical products should be designed to preserve the efficacy of function while reducing toxicity.
(5) *Safer solvents and auxiliaries*: The use of auxiliary substances (e.g., solvents, separation agents, etc.) should be made unnecessary wherever possible and innocuous when used.
(6) *Design for energy efficiency*: Energy requirements should be recognized for their environmental and economic impacts and should be minimized. Synthetic methods should be conducted at ambient temperature and pressure.
(7) *Use of renewable feedstocks*: A raw material or feedstock should be renewable rather than depleting whenever technically and economically practicable.
(8) *Reduce derivatives*: Unnecessary derivatization (use of blocking groups, protection/deprotection, and temporary modification of physical/chemical processes) should be minimized or avoided if possible because such steps require additional reagents and can generate waste.
(9) *Catalysis*: Catalytic reagents (as selective as possible) are superior to stoichiometric reagents.
(10) *Design for degradation*: Chemical products to be designed so that at the end of their function, they breakdown into innocuous degradation products and do not persist in the environment.
(11) *Real-time analysis for pollution prevention*: Analytical methodologies need to be further developed, to allow for real-time, in-process monitoring, and controlled before the formation of hazardous substances.
(12) *Inherently safer chemistry for accident prevention*: Substances and the form of a substance used in a chemical process should be chosen to minimize the potential for chemical accidents, including releases, explosions, and fires.

Green chemistry principles apply across the life cycle of a chemical product, including its design, manufacture, use, and its ultimate disposal [14]. Two main aspects of green chemistry are dealing with the process step, right from the design stage; and the product step including material and energy efficiencies and waste elimination.

(a) *Green synthetic chemistry*—Green synthetic chemistry aims at pollution prevention by using safer solvent and from the molecular reactions designed for synthesis. The pollution at the source can be reduced by minimizing or eliminating the hazards of chemical feedstocks, reagents, and solvents. It is becoming an extremely important area of chemistry due to its implications in improving the atom efficiency [15] at the product level.
(b) *Green chemistry of waste recycling*—Green chemistry eliminates hazards and develops the reuse of waste by giving innovative scientific solutions. It reduces the negative impacts of chemical products use on humans and environmental health. The industries are evolving green protocols taking into account advancements in waste minimization and are recycling waste to make circular economy feasible.

According to the US Environment Protection Act, it has been estimated that the 'green chemistry' innovations could save over 21 billion gallons of water and eliminate 7.8 billion pounds of carbon dioxide releases to air each year. Indian Chemical Industry worth 60 billion US$ has taken several initiatives for reducing its environmental footprints by using green chemistry [16]. The pharma industry faces environmental challenges from very high toxicity and has taken the first steps toward benchmarking green chemistry adoption. In the academic sector, Delhi University has taken an initiative to create a green chemistry network centre for the exchange of expertise, discussion, and knowledge between industrialists and academicians as well as between chemists and engineers with interests and expertise relevant to green chemistry.

## 3 Green Chemistry for Climate Change Mitigation

For climate change mitigation, the potential role of green chemistry would require some new and stimulating chemistry innovations which would aim either at (i) reducing greenhouse gas emissions and/or (ii) removing (sequestering) greenhouse gases from the atmosphere. A brief overview of these with examples from industry is included here.

## 3.1 Reducing Greenhouse Gas Emissions (GHGs)

The main human-induced greenhouse gas emissions recognized by the IPCC, which contribute to the 'enhanced' greenhouse gas effect and global warming are as follows:

- Carbon dioxide ($CO_2$)
- Nitrous oxide ($N_2O$)
- Methane ($CH_4$)
- Hydrofluorocarbons (HFCs)
- Perfluorocarbons (PFCs)
- Sulfur hexafluoride ($SF_6$).

Atmospheric chemistry has an important role in the contribution of these gases to global warming. Their common sources, atmospheric lifetime after emission, and contribution to global warming potential are depicted in Table 1.

To stabilize or reduce atmospheric concentrations for climate change mitigation, global emissions of GHGs must be reduced by at least an order of magnitude. The aim would be to find ways to decarbonize energy systems to reduce their carbon footprints. The greenhouse gases, viz. the $CO_2$ and other air pollutants such as nitrogen oxides, methane, and CFCs can be partially reduced by chemical processes. Among the GHGs, the $CO_2$ being most potent, its concentration in the atmosphere was measured 412 ppm in 2018 (an increase of 46% from a pre-industrial value of 280 ppm). Use of low-carbon fuels, use of renewable sources, and improvement in energy efficiency so that either there are low or no emissions during operation or emissions are minimized through efficient processes are the proposed strategies to minimize $CO_2$ emissions. Approaches in green chemistry are making many such alternatives possible for reducing carbon footprints in industrial operations. Innovation such as making more efficient solar cells using organic dyes [17] having titanium dioxide covering with a layer of chlorophyll using green chemistry processes would help in reducing $CO_2$ footprints.

Recent developments in green chemistry have led to phasing out the use of chlorofluorocarbons (CFCs) in refrigerants. The CFCs are ozone-depleting substances and also contribute to global warming. The $CO_2$ is an inert gas that can be used as a green refrigerant. It can also replace other harmful ozone-depleting substances in the supercritical stage. Dow Chemical Company developed a green chemistry process to use $CO_2$ in making polystyrene for use in many applications. This could result in keeping 3.5 million pounds of CFCs away from the atmosphere and helped in reducing global levels of $CO_2$ [18]. Replacement of titanium dioxide an energy-intensive additive to paints by a polymeric additive that requires less titanium dioxide as well as reduces the carbon footprint of the wall paints by 22% and water consumption by 30% is possible through greener chemical reactions.

Replacement of oil or gasoline by biofuels helps in minimizing emissions in the transport sector. Microbial production of renewable petroleum fuels and chemicals is an important area of research. Using sugarcane, corn syrup, sweet sorghum syrup,

**Table 1** GHGs, their sources, lifetime, and global warming potential

| GHG | Name | Common sources | Atmospheric lifetime (years)[a] | Global warming potential |
|---|---|---|---|---|
| $CO_2$ | Carbon dioxide | Fossil fuel combustion, forest clearing, cement production, etc | 50–200 | 1 |
| $CH_4$ | Methane | Landfills, production, and distribution of natural gas and petroleum, fermentation from the digestive system of livestock, rice cultivation, fossil fuel combustion, etc | 12 | 21 |
| $N_2O$ | Nitrous oxide | Fossil fuel combustion, fertilizers, nylon production, manure, etc | 150 | 310 |
| HFC's | Hydrofluoro-carbons | Refrigeration gases, aluminum smelting, semiconductor manufacturing, etc | 264 | Up to 11,700 |
| PFC's | Perfluorocarbons | Aluminum production, semiconductor industry, etc | 10,000 | Up to 9200X |
| $SF_6$ | Sulfur hexafluoride | Electrical transmission and distribution systems, circuit breakers, magnesium production, etc | 3200 | Up to 23,900X |

[a] defined as the amount of the gas in the atmosphere divided by the rate at which it is removed from the atmosphere. For carbon dioxide life time is variable (Source: AR-5)

molasses, glycerin, and biomass feedstock, microbes have been developed to manufacture transport fuel for use as diesel or jet fuel [19]. It could also produce olefins for lubricants and polymers, fatty alcohols for use as surfactants, aldehydes for insulation and resins, fatty acids for soaps, and chemical intermediates in a single-unit operation. This resulted in 85% reduction in greenhouse gas emissions.

### 3.2 *Removing (Sequestering) Greenhouse Gases from the Atmosphere*

Green chemistry technologies for the removal of anthropogenic carbon dioxide in the atmosphere aim at pathways for its conversion into useful products through chemical fixation or fixation routes. Carbon dioxide sequestration strategies deals with research advances in capturing carbon dioxide from its point sources, its utilization, and storage. A key step for sequestration of $CO_2$ is its cost-effective capture. The low-cost absorbers and safer membranes that are less energy-intensive are under development. Novel nano-materials show the amazing property of high absorption of $CO_2$. Lithium silicate-based materials developed by Toshiba in 2003 showed high $CO_2$ absorption capacity and captured up to 400 times its own volume [20]. The captured carbon dioxide could be reused. The use of green catalysts such as liquid metals is emerging as a viable negative emission green chemistry technology.

Direct air capture of $CO_2$ is also receiving attention. Karlsruhe Institute of Technology, Germany, has launched NEgative Carbon diOxide to Carbon (NECOC) research project to build a unique test facility using green chemistry to convert $CO_2$ contained in ambient air into highly pure carbon black powder aiming at the active reduction of atmospheric carbon dioxide [21]. LanzaTech process makes it possible to utilize a gas fermentation processes based on biological catalysts to make fuel and chemicals by recycling of carbon dioxide from the waste gases in oil refineries and industries. The life cycle studies have shown that this fuel creates 70% fewer greenhouse gases than the traditional fossil fuel [22].

There are also innovations in mimicking photosynthesis in plants. Photosynthesis is the process in which plants use sunlight, carbon dioxide, and water for conversion to carbohydrates and oxygen. In addition to C3 and C4 plants, Crassulacean Acid Metabolism (CAM) in plants provides carbon fixation pathways with high $CO_2$ adsorption capacity at night [23] and removes $CO_2$ from the atmosphere. Artificial photosynthesis which can also act as a photovoltaic cell and generate electrons by absorbing light is based on the concept of a dye adopting green chemical pathways. Green chemistry innovations are not only safer but also add to sustainability.

## 4 Applications of Green Chemistry for $CO_2$ Sequestration

In the present book, we are concerned with the green chemistry technology of minimizing emissions and recycling waste carbon dioxide from the air aimed at climate change mitigation. The application of green or sustainable chemistry described in the book illustrates certain ways of managing $CO_2$ conversion and sequestration by adopting interdisciplinary approaches. The book has its origin in the capacity-building workshop on "Recent Advances in $CO_2$ Capture Technologies and its Sectoral Applications" organized during August 29–September 1, 2018, by the Climate Change Research Institute. The workshop discussed the sectoral applications of the $CO_2$ sequestration in the oil industry, transport, power, agriculture, aluminum, and others. The insightful lectures delivered in the workshop and invited papers from the experts shape the book.

The scope of the book 'Climate Change and Green Chemistry of $CO_2$ Sequestration' is in the introduction Chap. 1. It is providing a brief outline to topics covered in the following chapters. The Chap. 2 is by Anjan Ray (Director, CSIR-IIP, Dehradun) delivered as keynote address in the workshop. Potential pathways to carbon capture and sequestration and the opportunities of decarbonization initiatives in India that exist through the deployment of domestic carbon sources such as agricultural/forest residue, biogas, used cooking oil, and municipal solid waste through conversion into fuels using green chemistry advances needed to be explored as potential sinks for carbon. This has been demonstrated through indigenously developed technology for bio-jet fuel at the Indian Institute of Petroleum, Dehradun, for use in a civilian flight by Spice Jet and also by the Indian Air force. Novel processes for converting polyolefin wastes into polyethylene and polypropylene as well as diesel or gasoline have been developed for the oil industry. A long-term strategy would maximize this approach as a feasible carbon sequestration option for India.

Carbon management from the energy sector is a key challenge before the nations. India having dominance of coal in energy generation has been undergoing rapid economic and social transitions, and several policies that work toward climate change control by reducing or avoiding greenhouse emission have been introduced. India has set ambitious renewable energy targets to achieve 175 GW of installed capacity by 2022 and proposing 450 GW by 2030. Chapters 3 and 4 raise research issues in the light of the current policy framework. In Chap. 3, Malti Goel, R. Pal, and A. Sharma aim to assess the potential impact of $CO_2$ sequestration technologies in curtailing emissions from the energy industry. Using $CO_2$ emission factors in three scenarios of; high coal use, business-as-usual and high renewable energy; the emissions without and with the adoption of $CO_2$ sequestration are calculated for 2017–2030. It emphasizes the need for government guidelines and assessing the economic implications of $CO_2$ sequestration for coal-based energy systems to facilitate technology development and adoption. V. S. Verma in Chap. 4 evaluates the impact of high renewable energy integration in the Indian power sector on the electricity grid. The efficiency of generation and cost implications for the power plants, whenever ramping up or ramping down is required, needs to be addressed to study the overall economics of operation of the power system. In a high renewable energy use scenario, there is a cost attached in reducing the generation of coal-fired power stations below a certain limit. Therefore, an optimal operation of the total system would need to be resorted so that there could be smooth change over from the present scenario of low renewable contribution to a future higher share in the national grid.

Innovation in coal combustion technology for thermal power plants can facilitate low-cost $CO_2$ capture. The application of chemical looping combustion (CLC) technology is proving most efficient and economical for $CO_2$ capture and sequestration. Crucial use of CLC and numerical simulations for the indigenous coal resource in India are described by Praveen Kumar and A. K. Parwani in Chap. 5. Multi-modeling Eulerian and Langarian approaches are used by the authors in computational fluid dynamics simulations for the proposed CLC reactor using India's high ash coal. A review of CFD analysis and the need for carrying out 3D simulations to get accurate results are the highlights.

As a result of the increasing global vehicular population (reaching almost 1.5 billion), the $CO_2$ levels are increasing rapidly. Pathways for controlling $CO_2$ from the automobile sector are dealt with by P. Senthil Kumar in Chap. 6. It suggests government strategies to reduce $CO_2$ in the automobile sector. World scenario of sector-wise $CO_2$ emissions is presented, and data for vehicular emissions in major metropolitan cities in India are described. Indian scenario of carbon-based fuels, vehicles, and strategies adopted for mitigating $CO_2$ emission is discussed. The author has explained the techniques being used for $CO_2$ capture. The integration of capture, transport, storage, and utilization in full-scale projects is needed to gain the knowledge and experience required for more widespread deployment of $CO_2$ sequestration technologies.

The $CO_2$ sequestration opportunities for the industry at a place are dependent on the progress made in obtaining patents. Patent analysis for carbon dioxide capture technologies in the 15 years period (2003–2018) is reported in the next Chap. 7 by H. Purshotham, Sanjeev Majumdar, and Prashrita Kaushal for understanding the global and Indian technology trends. Geographic distribution analysis suggests that the maximum number of patent families exists in the USA, China, South Korea, European Union, Japan, France, Taiwan, Australia, and Canada in that order. Among the industry, General Electric and Alston are the top players in global CCS patents. Indian scenario of CCSU research and patents obtained is also analyzed with the top assignees.

Chemical transformation of $CO_2$ by an appropriate catalyst into useful components is an important step for making advances in green chemistry. Chemical fixation and conversion of $CO_2$ at room temperature to solid carbonaceous species using liquid metal electrocatalysts are emerging as a viable negative emission technology. Considerable progress has been made in systems designed in India using different metal catalysts for the conversion of $CO_2$. In the next Chap. 8, Niharika, Yatendra S. Chaudhary, and Suddhasatwa Basu describe the crucial importance of major electrocatalytic developments in India for the reduction of $CO_2$. The photocatalytic conversion has also been investigated using several catalysts and sunlight as a source of energy. The carbon monoxide dehydrogenase (CODH) derived $CO_2$ reduction to CO, and its nano-integrated system shows high selectivity. Scale-up of these processes and the challenges in developing a commercially viable catalyst for the reduction are explored.

$CO_2$ is inherently thermodynamically stable gas and requires activation through external means either by using catalysts or applying high temperature or pressure for its conversion. Indrajit Das, V. Ramkumar, and Ramesh L. Gardas in Chap. 9 present thermodynamic analysis considerations of using ionic liquids (IL) for $CO_2$ capture and its conversation. Amine-based and non-amine-based ionic liquids are deep eutectic solvents and as a combination can form good adsorbents and have the potential to replace conventional absorbents for $CO_2$ capture. Feasibility of high absorption and improved energy efficiency could be achieved by changing the constituents of ILs, by using detailed predictive models of thermodynamic properties.

Chapter 10 by Ganapati V. Shanbhag, Nagendra Kulal, and Vaishnavi B. J. considers the use of heterogeneous catalysis for chemical fixation of $CO_2$ and the generation of useful chemicals. Heterogeneous catalysts offer easy extraction from the reaction mixture in comparison with homogeneous catalysts and are preferred for commercial applications. Design studies for the production of various cyclic organic carbonates, carbamates, and polycarbonates are presented using carbonylation reactions of epoxides, amines, and alcohols as green catalysts with gaseous $CO_2$. New insights into the impact of Zr doping on the formation of dimethyl carbonate (DMC) synthesis for the design of highly efficient Ce-$O_2$-based catalysis for carbonylation of methanol and $CO_2$ are provided, which can lead to environmentally benign chemicals for prospective use in the chemical industry and petroleum refining industry.

Manju Lata Pal, Ashish Kumar Yadav, Dileep Kumar Gupta in Chap. 11 discuss the use of captured $CO_2$ as a refrigerant in various applications in HVAC sectors like automobile air-conditioning supermarkets and the cold chain for the agriculture products. A two-way strategy to minimize HCFC emissions and to maximize the use of anthropogenic $CO_2$ in refrigeration has been studied. Recent developments and advantages of $CO_2$ as refrigerant along with technical challenges and opportunities for mitigating harmful emissions from ozone-depleting substances for the supermarket and cold chain applications are presented.

Direct air capture (DAC) is receiving attention as negative emission technology. Gautam Sen in Chap. 12 describes the challenges like the requirement of high energy, high water consumption, and economic viability being faced in the adoption of DAC of $CO_2$ on a commercial scale. Techniques such as bio-sequestration, absorption, and adsorption are discussed along with the challenges involved in their implementation. Few companies have incorporated these techniques in their start-ups, and these have been deliberated along with the financial costs. Commercial outlook of start-ups and the possibilities for $CO_2$ reduction through investment in large-scale research and development projects as a future strategy for climate change mitigation are discussed.

Among the various methods available for carbon capture from industrial point sources, carbonic anhydrase (CA)-based biological methods of $CO_2$ capture are gaining immense attention in recent years. Chapter 13 by H. Bose and T. Satyanarayana describes carbonic anhydrase (CA)-mediated carbon sequestration. Carbonic anhydrase, a zinc metalloenzyme is an essential enzyme in living beings and plays a vital role in accelerating the hydration and dehydration of carbon dioxide in the body. Biomimetic carbon sequestration methods using it as a biocatalyst with thermo-alkali stable CAs are promising for $CO_2$ sequestration as they can be immobilized on various solid surfaces generating useful calcium carbonate. This extensive review describes various innovations in utilizing CAs of prokaryotes in carbon capture technologies which can lead to efficient bioreactors as a tool for climate change mitigation.

Research gaps in biogas production and its upgradation to deliver renewable natural gas for the transport sector to help to minimize greenhouse gas emission are the theme of Chap. 14. Using the pressure swing asorption technique for biomass conversion, Pragyan P. Garnaik, Ranjan R. Pradhan, Yi Wai Chiang, and Animesh Dutta explain the challenges in the production and selection of low-cost adsorbents to maximize $CO_2$ utilization into useful biofuels. Insights into the working principle of an energy-efficient bio-fixation technique are given. Conclusions are drawn to augment the recent developments in sustainable technologies for their large-scale applications.

Urban areas and cities are having a significant contribution to the anthropogenic accumulation of greenhouse gases in the atmosphere. To reduce the impact of climate change, bio-fixation of $CO_2$ through enhanced photosynthesis can be very effective as one of the most important goals of green chemistry. Chapter 15 by M. Govindaraju, J. FowmithaBanu, S. Senthamilselvi, and Malti Goel is about the application of the phytoremediation technique in urban forestry. Developments in the green belt model by planting more $CO_2$ assimilating plants are discussed to achieve bio-sequestration of $CO_2$. It also improves the microclimate of the region and adds to urban biodiversity. Case studies for best practices adopted for sustainable development in Tamil Nadu and Delhi highlight the need for multi-agency collaboration initiatives from urban institutions, and local bodies to mitigate the impact.

P. S. Yadava and Amravati Thokchom in Chap. 16 present a study of the effect of burning and logging of forests on $CO_2$ fluxes to determine the rate of bio-fixation of $CO_2$ in the soil for climate change mitigation. Physico-chemical characteristics of soil in the Dipterocarpus forest sites are discussed. In the Dipterocarpus forests of Northeast India, it is seen that due to anthropogenic activities rate of soil $CO_2$ flux was the highest in the burnt forest and the lowest in logged forest. Seasonal variations in soil $CO_2$ flux are also reported using linear regression methods. Forest management practices like harvesting, thinning, and burning can help in changing the $CO_2$ fluxes bio-fixation rate in the soil and minimizing GHG emissions.

Chapter 17 by P. S. Swathi, N. K. Indira, and M. Ramonet provides estimation of $CO_2$ fluxes between land, ocean, and atmosphere using observed carbon dioxide data of 2006–2011 from 103 stations across the globe and assimilated into the adjoint of the LMDZ transport model. This included carbon dioxide data from three stations in India namely, Hanle, Pondicherry, and Port Blair. The grid point-wise results are aggregated into larger regions representing India, China, USA, Western Europe, and the Rest of the World to reflect the quantification of sources and sinks in these regions and potential for $CO_2$ sequestration.

## 5 Concluding Remarks

Recent advances in green chemistry are aimed to efficiently minimize the use of material resources in processes, minimize waste generation, maximize the use of renewable resources, improved atom efficiency in the synthesis of new materials, and avoided use of toxic and/or hazardous reagents or solvents in the manufacture and application of chemical products. Also defined as green or sustainable chemistry toward environmental sustainability, it is transforming the chemical industry.

Carbon sequestration aims to reduce $CO_2$ emissions from energy systems and mitigate the impacts of fossil fuel combustion in power and industrial processes. Both climate change mitigation and $CO_2$ sequestration are among the most complex to understand scientifically and implement on the scale. Green chemistry is turning out to be a viable interdisciplinary approach to find scientific solutions for carbon dioxide mitigation to address the global environmental problems of global warming and climate change.

This chapter covers the scope of the book on "Climate change and green chemistry of $CO_2$ sequestration" and outlines the interdisciplinary approaches adopted in the application of green chemistry for carbon dioxide sequestration through chemical, biochemical and biological means. The carbon dioxide utilization by using green chemistry pathways is an emerging field and the progress would depend on the development of knowledge, know-how, and tools for the design of cost-effective catalysts and bioreactors. Increasing investment in research and scale-up studies are necessary to make it a profitable choice for the nations.

## References


1. IPCC (2018) Global Warming of 1.5 °C. An IPCC Special Report on the impacts of global warming of 1.5 °C above pre-industrial levels and related global greenhouse gas emission pathways, in the context of strengthening the global response to the threat of climate change, sustainable development, and efforts to eradicate poverty [Masson-Delmotte V, Zhai P, Pörtner H-O, Roberts D, Skea J, Shukla PR, Pirani A, Moufouma-Okia W, Péan C, Pidcock R, Connors S, Matthews JBR, Chen Y, Zhou X, Gomis MI, Lonnoy E, Maycock T, Tignor M, Waterfield T (eds)]
2. https://www.who.int/health-topics/climate-change#tab=tab_1
3. Lackner K, Bob van der Z, Allen W, Alissa P, Jürg M, Sarah B (2012) The urgency of the development of $CO_2$ capture from ambient air, PNAS August 14, 109(33):13156–13162. https://doi.org/10.1073/pnas.1108765109
4. Styring P et al (2011) Carbon capture and utilization in the green economy. Centre for Low Carbon Futures, https://bit.ly/2nlKsIO
5. Clarke J (2018) Carbon dioxide: greenhouse gas or useful chemical feedstock? $CO_2$ Chem Green Sustain Chem
6. Cann M, Connelly M (2000) Real-world cases in green chemistry. Am Chem Soc
7. Anastas P (2003) Meeting the challenges to sustainability through green chemistry. Green Chem 5(2):G29–G34
8. Dunn P (2012) The importance of green chemistry in-process research and development. Chem Soc Rev 41(4):1452–1461
9. EPA (2017) Basics of green chemistry. The United States Environmental Protection Agency
10. Singhal M, Singh A, Khan P, Khan S, Eram S, Sachan N (2012) Green chemistry potential for past present and future perspective. Int Res J Pharm 3(4)
11. Oristian J (2015) Green chemistry innovations for climate change. ACS Green Chemistry Institute
12. Noce A (2017) In 25 years of green chemistry. ACS comments, vol 95, Issue 28, p 33, Date: July 10, 2017
13. Anastas P, Warner J (1998) Green chemistry. Theory and practice. Oxford University Press, New York, p 30

14. Horváth I, Anastas P (2007) Introduction: green chemistry. Chem Rev 107(6):2167–2168. Publication Date: 13 June 2007. https://doi.org/10.1021/cr0783784
15. Clark J, Macquarrie D (2008) Handbook of green chemistry and technology. Wiley
16. Mehta N (2012) Industrial green chemistry initiatives in India. In: 16th annual green chemistry & engineering conference, 18 June 2012
17. Bissember A (2017) Green chemistry is key to reducing waste and improving sustainability. Down To Earth
18. https://www.weforum.org/agenda/2016/10/4-ways-green-chemistry-is-helping-make-the-world-a-better-place/
19. Esposito M, Soufani K, Tse T (2016) 4 ways green chemistry is helping make the world a better place. World Economic Forum
20. https://www.azom.com/article.aspx?ArticleID=2052
21. KIT Press Release 019/2020, From Greenhouse Gas to a High-tech Resource, https://www.kit.edu/kit/english/pi_2020_019_from-greenhouse-gas-to-a-high-tech-resource.php
22. https://energy.economictimes.indiatimes.com/
23. https://www.britannica.com/science/photosynthesis/Carbon-fixation-in-C4-plants

# $CO_2$ Sequestration Opportunities for India's Energy Security

# Carbon Capture and Sequestration: Implications and Opportunities for India

**Anjan Ray**

**Abstract** At the COP-21 UNFCCC Summit, India has made a challenging and significant commitment (INDC) to the goal of confining global warming to no more than 1.5 °C. This is particularly remarkable because India's economic growth trajectory requires a substantial increase in energy access to underpin the development and ensure the widespread benefits of such growth to all strata of society. At the same time, India depends significantly on fossil fuels imports and is projected to be primarily fossil-fuel-dependent for at least two decades going forward. This indicates that both demand-side interventions, i.e., improved energy efficiencies as well as reduced carbon-derived energy, and supply-side interventions involving greater use of renewable carbon energy and lower footprint of deployed non-renewable carbon energy are critical to achieving simultaneous targets of energy access, energy security, and GHG mitigation. In this chapter, an innovative approach to analyzing the implications of these elements at a level of carbon atoms in the supply chain, their energy conversion, implications on GHG emissions, and potential long-term strategies including carbon capture, sequestration, and utilization is presented.

**Keywords** Carbon capture · Carbon import · Waste carbon sources · Bio-jet fuel

## Abbreviations

| | |
|---|---|
| BTX | Benzene-Toluene-Xylene |
| COP21 | 21st Conference of Parties meeting |
| CSIR-IIP | CSIR—Indian Institute of Petroleum |
| FAME | Fatty Acid Methyl Ester |
| FSSAI | Food Safety and Standards Authority of India |
| GAIL | Gas Authority of India Limited |
| INDC | Intended nationally Determined Contribution |

A. Ray (✉)
CSIR-Indian Institute of Petroleum, Haridwar Road, Mohkampur, Dehradun 248005, India
e-mail: anjan.ray@iip.res.in



M. Goel et al. (eds.), *Climate Change and Green Chemistry of $CO_2$ Sequestration*, Green Energy and Technology, https://doi.org/10.1007/978-981-16-0029-6_2

| | |
|---|---|
| IPCC | Intergovernmental Panel on Climate Change |
| LPG | Liquefied Petroleum Gas |
| RUCO | Repurposing Used Cooking Oil |
| TBOs | Tree Borne Oils |
| UCO | Used Cooking Oil |
| UNFCCC | U.N. Framework Convention on Climate Change |

## 1 Introduction

India's Intended Nationally Determined Contribution (INDC) (INDC 2015) [4] underscores the country's commitment to the new international climate agreement under the aegis of the U.N. Framework Convention on Climate Change (UNFCCC) Conference of the Parties (COP21) in Paris in December 2015. Three key metrics outlined in the INDC [3] are highlighted in the context of actionable and measurable indicators:

- To reduce the emissions intensity of its GDP by 33–35% by 2030 from 2005 level.
- To achieve about 40% cumulative electric power installed capacity from non-fossil-fuel-based energy resources by 2030, with the help of the transfer of technology and low-cost international finance, including from Green Climate Fund.
- To create an additional carbon sink of 2.5–3 billion tonnes of $CO_2$ equivalent through additional forest and tree cover by 2030.

However, the importance of energy access as a likely causal factor in development and sustainability indicators cannot be understated. It has been argued [6] that access to clean cooking fuels and sanitation facilities consistently lag improved water and electricity access by a large margin. These two deprivations are more concentrated among poor countries and poor people in middle-income countries. However, some Asian countries, including India in recent years, have been able to achieve faster progress in electrification at lower income levels compared to industrialized countries' earlier efforts.

At the same time, increased energy access is accompanied by increasing GHG emissions in a fossil-fuel-dependent national energy portfolio. India's share of total global primary energy demand is set to roughly double to ~11% by 2040, underpinned by strong population growth and economic development. As per the BP Energy Outlook 2019, India would account for over 25% of the incremental global primary energy demand growth between 2017 and 2040. 42% of this additional energy requirement is met through coal, meaning $CO_2$ emissions could be doubled over this period if mitigation efforts fall short.

High coal usage in thermal power, open burning of crop residues, forest fires, and industrial as well as automotive emissions based on fossil fuels also contribute to air quality issues in India, especially in urban centres.

Against this backdrop, a host of decarbonization initiatives are visible in India. Widespread adoption of solar PV, wind power capacity augmentation, small hydro-electric projects, replacement of biomass-burning domestic cook stoves with LPG burners, city gas distribution, bio-fuel adoption, and a slew of energy efficiency regulations are moving in the right direction. It is the pace and scale of these programs, and effective metrics for tracking emerging carbon capture, sequestration, and utilization efforts, that will determine the eventual quantum of success.

## 2 The Case for Carbon Capture from Domestic Sources in India

To meet India's energy demand, fossil fuel imports are currently of the order of about 200 MMT each of petroleum and coal as well as 15 MMT of natural gas. In carbon atom terms, this means India brings approximately 330 MMT of carbon annually across its borders for fuel purposes (Table 1), not accounting for fuels exports from India's petroleum refineries which operate at capacities over national requirements to achieve economies of scale.

The reduction of such fossil fuels imports to enhance energy security, bring down the GHG emissions in transit between countries (which do not currently appear to be specifically assigned at a country level), and to reduce foreign exchange outflow can be achieved either by reducing demand-side requirements or augmenting the supply of carbon atoms into fuels production from domestic sources. The demand-side reduction can be achieved through non-carbon energy sources such as solar, wind, tidal, or geothermal, or through improved energy efficiency in generation, distribution, and utilization of energy. Supply-side replacement of fuel imports can be addressed through enhanced domestic primary fossil fuel production, and by maximizing indigenous carbon sources as elements of the fuel pool. These have largely remained untapped except for traditional informal use of biomass. Such domestic carbon sources are available substantially within India in different forms like agricultural residue, forest residue, bio-gas, used cooking oil, and municipal solid waste. These can potentially provide more than 500 MMT of carbon atoms every year (Table 2), which is more than the number of carbon atoms imported. However, deployment of carbon atoms from such sources requires their conversion to usable fuel forms

**Table 1** Indian carbon imports (approximate values)

| Commodity | Import, MMT/yr | %C | Imported C, MT/yr |
|---|---|---|---|
| Crude oil | 200 | 85 | 170 |
| Coal | 200 | 75 | 150 |
| Natural gas | 15 | 77 | 11.6 |
| **Total** | | | **331.6** |

*Source* [7]

**Table 2** Domestic carbon sources (estimated values)

| Commodity | Scope, MMT/yr | %C | Potential C, MT/yr |
|---|---|---|---|
| Agri-residue (surplus only) | 200 | 40 | 80 |
| Forest residue | 150 | 42 | 63 |
| Bio-gas sources, excl landfill | 800 | 45 | 360 |
| MSW | 50 | 25 | 12 |
| UCO | 5 | 85 | 4 |
| **Total** | | | **519** |

*Source* [7]

meeting standard specifications so that they may be included in the supply chain without major disruption.

This analysis throws up the evident need for realigning national carbon management approaches toward maximizing the use of domestic carbon resources. It may be noted that additional domestic carbon atoms over and above the Table 2 opportunities can be harvested from gaseous emissions of carbon monoxide (CO), methane (CH4), and carbon dioxide ($CO_2$). The first demonstration plant for industrial CO capture and its conversion to ethanol and 2, 3-butanediol is already under construction at Indian Oil Corporation near its Panipat refinery, using microbial gas fermentation technology from Lanzatech Inc, USA.

India's very substantial coal reserves [1], estimated at over 3.19 billion metric tons down to 1.2 km exploration depth, are also not accounted for in this analysis.

At current levels, about 90 million tons of diesel, 35 million tons of gasoline, 12 million tons of LPG, and 8 million tons of aviation fuels are available on the demand side for potential carbon replacement opportunities in India.

## 3 Capture, Conversion, and Utilization of Domestic Carbon

While there has been considerable technology focus around the world and in India or development of alternative fuels based on renewable carbon sources, limited success has been seen in India except for first-generation ethanol. With the new National Bio-fuels Policy issued in May 2018 providing impetus to an adoption based on performance (specification), cost, and scale-competitive bio-fuels, this is expected to change in the foreseeable near future. A few examples below will suffice to illustrate this point.

## 3.1 *Biodiesel (Fatty Acid Methyl Ester, FAME)*

Although policy support to biodiesel as well as technology for its production to scale has been available for over a decade in India, adoption remained somewhat limited because of the lack of established supply chains for non-edible oils and fats in India. According to the Biodiesel Association of India, installed capacity in India for FAME production is currently at about 1.1 million metric tons, but sales are a fraction of this due to mismatched feedstock supplies. Four key source categories of oils and fats offer scope to address the supply chain issue.

### 3.1.1 Used Cooking Oil (UCO)

- Edible oil refining by-products such as palm stearin and livestock-based industry by-products like dairy fat waste, fish oil, poultry fat, and tallow
- Tree-borne oils (TBOs), and
- Rotation crop oilseeds (such as camelina and brassica species).

Edible oils, explicitly prohibited for bio-fuels production on the sound logic that the country already has a severe deficit of domestic edible oil supply, can nonetheless contribute to the Used Cooking Oil (UCO) pool if such UCO can be collected and aggregated after primary use, and then suitably channelized into bio-fuel production. The Food Safety and Standards Authority of India (FSSAI) initiative for Repurposing Used Cooking Oil (RUCO) has estimated that upward of 2 million tons of UCO could be collected and redeployed in this manner.

### 3.1.2 Solvent Extraction and Meat Processing

Also, oil cakes from edible oil cold-expelling facilities can be further de-oiled by the solvent extraction route to augment the fuel-production pool. This and other by-products of industries such as dairy, meat, and fish processing could add up to several tens of thousands of incremental tons of feedstock for the purpose.

### 3.1.3 Tree-Borne Oilseeds (TBOs)

In a report [2] by environmental website downtoearth.org, there is a potential to collect 3.5 million tons of wild-growing tree-borne oilseeds as against 0.5 million tons today according to the Solvent Extractors' Association of India. The uncollected oilseeds would largely be mulched naturally, leading to methane emissions potential well over and above the natural carbon replenishment requirement of soil. Utilization of such oilseeds for biodiesel would thus not only reduce potent GHG emissions of methane from natural decomposition but also a part of the GHG emissions inherent in the use of diesel to the extent replaced by such biodiesel.

### 3.1.4 Short Rotation Oilseed Crops

Out of India's total agricultural land of about 179 million hectares, roughly 40% is irrigated [5]. Of the irrigated land, roughly half again is cultivated for a single crop cycle annually. This offers the opportunity to cultivate short-rotation oilseed crops like camelina and carinata species in 90 to 120-day periods on several million hectares.

## *3.2 Bio-Jet Fuels (Bio-Aviation Turbine Fuels, Bio-ATF)*

India's aviation industry has grown by leaps and bounds since the introduction of the UDAN-RCS scheme in 2016 to enhance regional connectivity and make flights affordable to citizens so that productivity and economic growth may be enabled in smaller towns across the country. The country is also surplus in kerosene and jet fuel production. However, de-carbonization of flights, especially to international destinations, is in global focus since the impact of aviation emissions in the atmosphere is assessed to be more significant per unit carbon dioxide produced than emissions at ground transportation levels. The same oils and fats available for FAME production, described above, can also be used for Bio-ATF production. For this, technologies are available globally and sustainable aviation fuels are already in commercial production at various sites in the USA, EU, and Singapore. Within India, pilot-scale production of bio-jet fuels from jatropha oil, Pongamia oil (both TBOs), and Used Cooking Oil has been demonstrated in flights by a civil airline operator (Spice Jet) and the Indian Air Force. The technology used has been indigenously developed by CSIR-Indian Institute of Petroleum (CSIR-IIP), a publicly funded research laboratory at Dehradun, India.

## *3.3 Fuels Derived from Waste Plastics*

In collaboration with GAIL (India) Ltd, a large gas-based polyolefin producer, CSIR-IIP has also developed a novel, patented process for converting polyolefin wastes like polyethylene and polypropylene, which account for about 70% of the total waste plastics, into useful products. The processing unit can swing production from such feed into gasoline, diesel, or petrochemicals (aromatics benzene-toluene-xylene, also referred to as BTX), depending on the selection of catalyst and process conditions. By-product propane and butane, the main constituents of liquefied petroleum gas (LPG), are used to provide energy for the process, which consists of a pyrolysis unit operation followed by catalytic conversion and subsequent condensation and fractionation.

1 kg of clean polyolefin waste can produce up to either 550–600 ml gasoline or 750–800 ml diesel, conforming to transportation fuel standards or 450–500 ml

aromatics. The fuel products have also been extensively tested on vehicles, and the emissions are at par with refinery-produced gasoline and diesel from petroleum crudes. A 1 metric ton per day pilot plant is operational in Dehradun for converting waste plastics to diesel using this process. The net GHG savings are more than simple calculations of diesel substitution would indicate, as dumped waste plastic is often burned in the open and also ingested by animals and birds leading to open putrefaction of carcasses. One study suggests that across their lifecycle, plastics account for 3.8% of global greenhouse gas emissions—nearly twice of the aviation sector [8].

## 4 Conclusions and Outlook

The conversion of waste carbon sources to fuels offers significant potential for GHG reduction while helping India reduce its total fuel imports. A focused strategy to maximize this approach may help the country become a very important global contributor in controlling global temperature increase above baseline to less than 1.5 °C as compared to 2 °C originally envisaged, as the Intergovernmental Panel on Climate Change (IPCC) identified that such temperature increase mitigation could go hand in hand with ensuring a more sustainable and equitable society.

## Reference


1. https://www.coal.nic.in/content/coal-reserves, Ministry of Coal, Government of India
2. https://www.downtoearth.org.in/news/forests/oil-grows-on-trees-60255, (2018)
3. https://www4.unfccc.int/sites/submissions/INDC/Published%20Documents/India/1/INDIA%20INDC%20TO%20UNFCCC.pdf
4. India's Intended Nationally Determined Contribution (2015)
5. Pocket Book of Agricultural Statistics (2017) Government of India—Ministry of Agriculture & Farmers, Directorate of Economics & Statistics
6. Rao N, Pachauri S (2017) Environ Res Lett 12(2)
7. Ray A, Kumar S, Atray N, Anil K (2019) J Gov 18(1)
8. Zheng J, Suh S (2019) Strategies to reduce the global carbon footprint of plastics. Nat Clim Change 9:374–378

# An Assessment of $CO_2$ Reduction Potential from Carbon Sequestration Versus Renewable Energy Targets in India

**Malti Goel, Rupali Pal, and Aditya Sharma**

**Abstract** Carbon management from the energy sector is a key challenge before the nations. India as a nation is undergoing rapid economic and social transitions requiring diversified energy resources to meet basic energy needs of the people. India is having a 7% share in the world coal resources, and coal has been the dominant fuel for energy. Several policies that work toward climate change control by reducing or avoiding greenhouse gas emissions, exist. This chapter describes India's energy scene, trends in coal consumption, and India's renewable energy targets. To assess the potential of $CO_2$ sequestration—carbon capture, utilization, and storage, in mitigating emissions from the Indian industry—$CO_2$ emission scenarios need to be generated. Using basic assumptions from Coal Vision 2030, the $CO_2$ emission projections in three scenarios of high coal, business-as-usual, and high renewable energy, without and with $CO_2$ sequestration are calculated for 2017–2030. The current national and international $CO_2$ sequestration research is highlighted.

**Keywords** Carbon management · Energy security · $CO_2$ reduction · Renewable energy · $CO_2$ sequestration

## Abbreviations

| | |
|---|---|
| ADB | Asia Development Bank |
| BAU | Business as Usual |
| BCS | Best case Scenario |

M. Goel (✉)
Science & Technology Solutions for Sustainable Energy Future, Climate Change Research Institute, New Delhi 110017, India
e-mail: maltigoel2008@gmail.com

R. Pal
Ex-Climate Change Research Institute, New Delhi 110017, India

A. Sharma
Science & Technology Solutions for Sustainable Energy Future, Climate Change Research Institute, New Delhi 110017, India



M. Goel et al. (eds.), *Climate Change and Green Chemistry of $CO_2$ Sequestration*, Green Energy and Technology, https://doi.org/10.1007/978-981-16-0029-6_3

| | |
|---|---|
| CCUS | Carbon capture, Utilization, and storage |
| CDIAC | Carbon Dioxide Information Analysis Center |
| CEA | Central Electricity Authority |
| CSMCRI | Central Salt & Marine Chemicals Research Institute |
| GISS | Goddard Institute of Space Studies |
| HELE | High-efficiency Low-emission |
| IEA | International Energy Agency |
| IES | Indian Energy Scenario |
| IESS | India Energy Security Scenario |
| IIASA | International Institute for Applied Systems Analysis |
| IITB | Indian Institute of Technology, Mumbai |
| IGCAR | Indira Gandhi Centre for Atomic Research |
| IGCC | Integrated Gasification Combined Cycle |
| IGFC | Integrated Gasification Fuel Cells |
| MANIT | Maulana Azad National Institute of Technology |
| MI | Mission Innovation |
| MNRE | Ministry of New and Renewable Sources of Energy |
| NAPCC | National Action Plan of Climate Change |
| NDC | National Determined Contribution |
| NETRA | NTPC Energy Technology Research Alliance |
| RCN | Research Council of Norway |
| SECI | Solar Energy Corporation of India |
| TREC-STEP | Tiruchi Regional Engineering College—Science and Technology Entrepreneurs Park |
| UNEP | United Nations Environment Programme |
| UNFCCC | U.N. Framework Convention on Climate Change |
| UNFPA | United Nation Population Fund |
| WUSTL | Washington University at ST. Louis |

## 1 Introduction

> The potential market for products made from carbon dioxide could be as much as 1.1 trillion by 2030.
>
> —I. Dainanieh, Global $CO_2$ Initiative.

The risk of global climate change in the long run is one of the central issues facing the world. The Earth's average temperature has been increasing since the first Industrial Revolution, which took place in 1750s. Between 1880 and 2015, the average global surface temperature rise was recorded as 0.8 °C [1]. In 2016, the Earth experienced its third consecutive hottest year of the decade [2]. There is a wide agreement that human activities are responsible for almost all of the increase in greenhouse gases in the atmosphere leading to a warmer planet [3]. The primary sources of GHG emissions are the burning of fossil fuels (coal, oil, and gas), followed by contributions from land-use changes and other industrial activities. The share of fossil fuels in world energy continues to dominate, giving rise to an increase of carbon dioxide ($CO_2$) emissions from anthropogenic yet developmental activities. In 2017, 80% of world energy was derived from fossil fuels, a trend that is expected to continue. Increasing the use of renewable sources of energy, increasing efficiency in use and the strategies to sequester $CO_2$ are suggested as clear pathways for mitigating global warming and climate change. Among these, the $CO_2$ sequestration can be achieved by neutralizing excess $CO_2$ in the atmosphere with the help of chemical, physical, or biological processes.

## 2 Global $CO_2$ Emissions

The $CO_2$ is a major greenhouse gas that absorbs and emits infrared solar radiation, creating the 'natural greenhouse effect.' Along with other natural greenhouse gases, such as nitrous oxide and methane, $CO_2$ has an important role in sustaining a habitable temperature for the planet. If there were no natural GHGs, our planet would simply be too cold. It has been estimated that without these gases, the average surface temperature of the Earth would be about ~(−)18 °C [4] and not ~(+)14 °C. Since the first industrial revolution, energy-driven consumption of fossil fuels has led to a rapid increase in anthropogenic $CO_2$ emissions and other pollutants causing changes in $CO_2$ fluxes between atmosphere, land, and oceans; and a disruption in the global carbon cycle. It is leading to a global warming effect also known as the 'enhanced greenhouse effect.'

Global warming and a changing climate have a range of potential ecological, physical, and health impacts, including extreme weather events (such as floods, droughts, storms, and heat waves), sea level rise, altered crop growth, and disrupted water systems. The most extensive analysis on the potential impacts of climatic change can be found in the $5^{th}$ Intergovernmental Panel on Climate Change (IPCC) report. Emissions from the number of growing economies have been increasing rapidly over the last few decades as shown in Fig. 1. Emissions vary widely across countries, and developing countries are predicted to drive future emission increases going forward. The temperature anomaly at the global level has risen sharply to approximately 0.9 °C (Fig. 2). The annual anomaly of the global average surface temperature in 2018 was +0.31 °C above the 1981–2010 average and was the fourth warmest year since 1891. On a longer time scale, global average surface temperatures have risen at a rate of about 0.73 °C in a century.

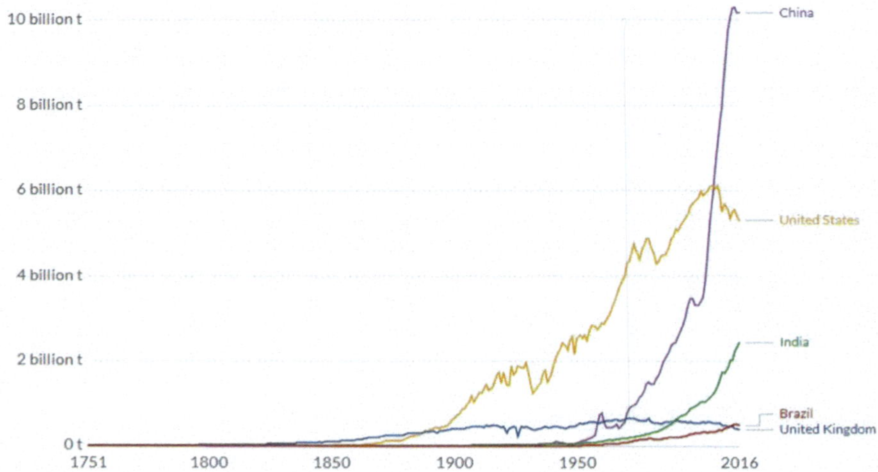


**Fig. 1** Annual $CO_2$ emissions in billion tones by country from 1751onward (*Source* [5])

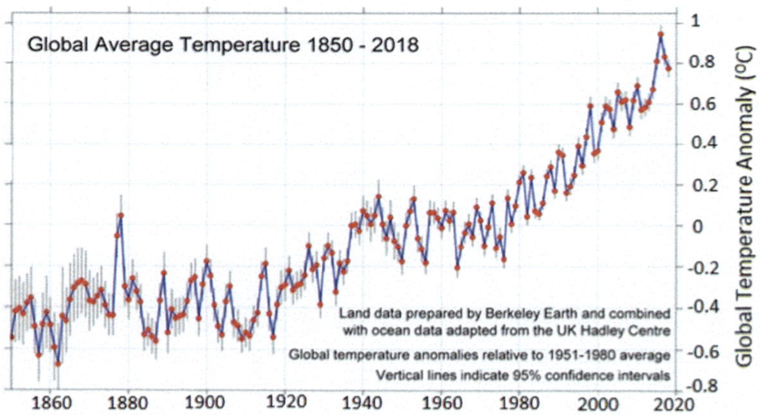


**Fig. 2** Global average land–sea temperature anomaly (°C) *Source* Global Temperature Report for 2018, Posted on January 24, 2019, by Robert Rohde, https://berkeleyearth.org/2018-temperatures/

## 3 India's Current Energy Scene

Energy is the backbone of the economy of a nation. India's total installed capacity of 329 GW in 2017 had a 59% share of coal as a dominant energy fuel contributing to 194 GW installed capacity (Fig. 3). The share of hydropower was 14%, natural gas 7%, nuclear 2%, and renewable energy 17%. To achieve decarbonization of the energy sector, a policy to close the old thermal plants (adding to approximately 30 GW) and achieve high renewable energy growth has been targeted. However, with

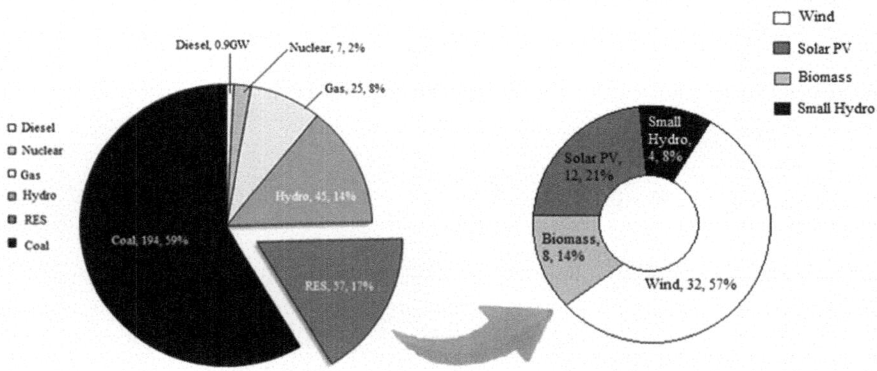


**Fig. 3** Share of different fuels in India's total installed capacity of 329 GW as of 31.03.2017 with a breakup of the renewable energy capacity of 57 GW

the country's demand for electricity expected to go up, coal-powered generation will continue to be the backbone in the energy mix for the next one or two decades. The bulk of India's future capacity would be based on supercritical and advanced ultra-supercritical technology. Because of the higher efficiency of new plants, less coal will be consumed per unit of electricity, and this would help in the reduction of $CO_2$ emission intensity.

Approximately, 7% of the global coal reserves exist in India [6]. Coal being an abundant resource, its production has bloomed for meeting the energy demand in comparison with other fossil fuels. The estimated reserves of coal in 2017 were 315.14 billion tones, an addition of 6.34 billion tones over the previous year. Coal deposits are mainly confined to the eastern and south-central parts of the country. Coal is the single largest source of India's total primary energy demand, and it is expected to be so even in 2030. Coal contributes 71% of India's total energy-related $CO_2$ emissions. The power sector had a share of 53% of total $CO_2$ emissions in 2017, which is expected to fall to 46% by 2030. In the total $CO_2$ emissions, approximately, 20–25% emissions are from industrial sources other than power—mainly cement production plants (~7% of the total emission), oil refineries ( ~6%), and iron and steel industry (~5%).

Coal production in the country during the year 2016–17 was 662.79 million tones (Mt) as compared to 639.23 Mt during 2015–16, registering a growth of 3.69%, whereas the total production of energy from commercial sources increased from 13,767.83 petajoules during 2015–16 to 14,149.29 petajoules during 2016–17, showing an increase of 2.77% [7]. In India, the power sector is the largest consumer of coal with electricity generation (utility sector) by source contributing to 986,591 GWh (75.9%) in 2017–2018, while the share of renewable energy was 15.3%. Fitch Solutions Macro Research in 2019 had predicted that India's coal output is expected to grow 4.3% per year up to 2028 [8] suggesting that India and China will have the largest impact on the global coal market in the coming years. The coal demand scenario at a place is influenced by the economic growth, energy efficiency, and emergence of alternate coal technologies. According to Coal Vision India (CVI), thermal coal demand is estimated to be 1150–1750 Mtpa (million tones per annum) with overall coal demand as 1300–1900 Mtpa by 2030 [9].

## 4 India's $CO_2$ Emissions

India is undergoing rapid economic and social transitions. By 2010, India accounted for 17.8% of the world population, and according to the State of the World Population 2018 report of United Nations Population Fund (UNFPA), India's population grew at 1.2% a year between 2010 and 2019, marginally higher than the global average of 1.1% a year in this period, but more than double of China's 0.5% a year [10]. India had a share of 6.6% of global emissions in 2018, and this will grow since India is starting at a low base of per capita emissions ~2.2 tons, which is 37% of the global average [11].

Several policies that work toward climate change control by reducing or avoiding greenhouse gas emissions exist. From a global perspective, India's total emissions are sufficiently high and is approaching to become the third-largest emitter in the world. However, from a national perspective, India's development future cannot be assured without increasing levels of energy use for millions, which would lead to an increase in emissions. As the final energy demand is projected to increase 4.3 times during 2010–2050, deep decarbonization pathways for India are required to be developed for the energy sector [12].

Pathways for $CO_2$ mitigation in the atmosphere include: (i) increasing use of renewable energy; (ii) improving energy efficiency in supply and demand sides; (iii) use of low-carbon fuels in place of coal, leading to a transition from coal to oil to natural gas to hydrogen; (iv) adoption of $CO_2$ capture, storage, and utilization. India's efforts to increase energy use are shaped by a climate mitigation context that is driving rapid changes in energy technology, as reflected in the commitments made to the Paris Agreement on climate change. The key highlights of India's Nationally Determined Contributions (NDCs) are:

(i) To reduce its carbon emission intensity, i.e., the emission per unit of GDP, by 33–35% from what it was in 2005, by 2030.
(ii) To have 40% of the total electricity capacity from sources other than fossil fuels.
(iii) To enhance $CO_2$ sinks by 2.5–3 billion tones per year by 2030. It may require increasing forest cover to 25% from 22% at present.

It is anticipated that with these ambitious targets, India could be moving ahead on the path of limiting global warming to a 1.5 °C rise [13] and become the main driver in the reduction of total energy-related $CO_2$ emissions during 2017–2030.

## *4.1 India's Policy Plan for Renewable Energy*

In India, several schemes, fiscal incentives, and targets have been launched to support the development of renewable energy sources toward sustainable development from time to time, ever since it is being promoted since the 1980s. The new targets stated in 2014 for renewable energy (RE) are to achieve 175 GW of installed capacity by 2022. This includes 100 GW of solar energy capacity, 60 GW of wind energy, 10 GW from biomass, and 5 GW from small hydro. Of the solar, 40 GW would be the share of grid-connected solar PV Rooftop. The share of resources in the total installed electricity capacity of 329 GW with renewable energy capacity of 57 GW in 2017 is shown in Fig. 3. In the renewable capacity, wind has the largest share of 57%, solar is 17%, biomass 14%, and small hydro 4.8%. In response to the new RE Plan, there is a great momentum to set large solar parks. Total capacity addition from these parks was approximately 30,500 MW in 2019. The status of solar parks across the country is indicated in Fig. 4. The projected growth of solar energy is enthusiastic,

**Fig. 4** Solar park locations across India (compiled by authors using data from MNRE and SECI 2018) [14]

as a major fall in solar energy tariffs from INR 12 per unit in 2010 to less than INR 2.5 in 2019 has been achieved.

The RE Plan target for achieving 450 GW capacities additions has been set for 2030. Adoption of 100% renewable energy by the year 2051 is seen as a strategy for India to mitigate $CO_2$ emissions and to bring down projected $CO_2$ emissions from 11.2 to 0.40 Gt from the energy sector [15]. A whole range of energy transformations as well as international cooperation and accelerated advancement of technologies would be needed in the growth sectors of the economy to achieve these targets.

## 4.2 *Initiatives in $CO_2$ Sequestration*

An important milestone in the promotion of international R&D in carbon capture has been the Carbon Sequestration Leadership Forum [16] launched by the Department of Energy, the USA, in 2003. Participated by both developed and developing economies and joined by 22 countries and the European Union, collaborative R&D toward improved cost-effective technologies for CCS has been targeted. India is founder member to this Ministerial-level international climate change initiative. Department of Science Technology (DST), Government of India, under its Inter-sectoral Science and Technology Programs jointly with the Ministry of Power catalyzed support to $CO_2$ sequestration research and several new initiatives began with stakeholders' participation. A large number of new research initiatives and demonstration projects took shape during 2004–2008. A concise overview of government-funded research with a three-pronged approach is discussed below.

### 4.2.1 Basic and Applied Research

In India, 'National program on $CO_2$ Sequestration Research' began in 2006 to intensify R&D support and provide thrust to academia-industry collaboration. It identified four thrust areas of research as (i) $CO_2$ sequestration through microalgae biofixation techniques; (ii) carbon capture process development; (iii) policy development studies; and (iv) network terrestrial agro-forestry sequestration modeling. Basic and applied research in carbon capture and sequestration began in several national R&D laboratories across the country. Some of the important initiatives included: The Central Fuel Research Institute, Dhanbad—assessed that the storage potential exists mostly in peninsular India; The Indian Institute of Petroleum, Dehradun—mounted research efforts in developing ionic solvents, novel amine-based adsorbents, and pressure swing absorption (PSA) technique for post-combustion $CO_2$ capture [17, 18], Central Salt and Marine Chemicals Research Institute (CSMCRI), National Chemical Laboratory (NCL), and National Environmental Engineering Research Institute (NEERI)—initiated research on new adsorbent materials development for $CO_2$ capture. National Geophysical Research Institute (NGRI) took lead in evaluating the Deccan basalt formation in India as a potential long-term $CO_2$ storage option [19].

To provide thrust to research, India organized the first International workshop on R&D challenges in carbon capture and storage technology in 2007. Scientific community from nineteen countries attended [20]. In this workshop young scientists were encouraged to participate to get a world view. An 'Indian $CO_2$ Sequestration and Applied Research Network' took shape in the DST to coordinate the $CO_2$ sequestration research and development activities within the country. It aimed at exchange of knowledge involving various stakeholders and academic institutions [21]. Besides the government, R&D support has been extended from the Council for Scientific and Industrial Research (CSIR) as well as the stakeholder PSUs.

### 4.2.2 Stakeholders' Participation

Stakeholders' leading role in $CO_2$ sequestration research also emerged. The Institute of Reservoir Studies, Oil and Natural Gas Corporation (ONGC) conceived setting up a pilot experimental enhanced oil recovery (EOR) project in Gujarat, using $CO_2$ captured from a gas processing plant at Hazira [22]. Emerging out of recommendations of the technology assessment study on clean coal technology conducted in 2006, the Department of Science and Technology directed Bharat Heavy Electrical Ltd. (BHEL) to create a 'Centre for Excellence in Coal Research.' The aim was to intensify research on clean coal technologies and $CO_2$ capture in the country. The BHEL articulated a roadmap for oxyfuel coal combustion research to facilitate clean energy and low-cost carbon capture technology development. In the power sector, the National Thermal Power Corporation (NTPC) as India's largest power company launched a research program on $CO_2$ sequestration under the newly created NTPC Energy Technology Research Alliance (NETRA) in 2008. Coal India Ltd. and Geological Survey of India envisioned to assess and develop ways and means to study the feasibility of $CO_2$ storage sites in unmineable coal fields to explore enhanced coal bed methane as a future energy resource.

The government of India planned to provide a greater thrust to technology development and proposed a ninth mission on Clean Coal Technology (CCT), under the ambit of National Action Plan on Climate Change (NAPCC) 2008, which had eight missions prioritizing both climate change mitigation and adaptation. At the same time, to implement a set of initiatives in CCT and CCS/$CO_2$ sequestration, BHEL as part of a three-year EU-funded project in collaboration with Tiruchi Regional Engineering College—Science and Technology Entrepreneurs Park (TREC-STEP) organized two EU-funded CCS training programs on 'Introduction to CCS and CCT' in December 2011 and August 2012. Two Skill Leverage Programs on CCT-CCS technologies were held in January 2012 and October 2012 [23]. The BHEL set new pace in collaborations with Indira Gandhi Centre for Atomic Research (IGCAR) and NTPC for development of advanced ultra-supercritical technology in a consortium led by the Department of Atomic Energy, Government of India [24].

The energy industry as a whole, both public/private, not only invested in research but also needed to work toward business models. Reliance Industries Ltd. took the lead in encouraging $CO_2$ sequestration basic and applied research in the country. The development of clean coal technologies, coal gasification to facilitate $CO_2$ capture, and research on sequestration through the microalgae approach has been pursued. The Carbon Clean Solutions Limited (CCSL) was founded in 2009 and has successfully demonstrated field applications of $CO_2$ capture from flue gas of a chemical plant in Tuticorin and utilized it for producing soda in a low-cost operation. The CDRMax is a commercially available carbon capture technology from industrial flue gases with $CO_2$ concentration ranging from 3 to 25 vol.%. The technology enables >90% $CO_2$ capture rates from the flue gases or off-gases and delivers industrial quality $CO_2$ for re-use or sequestration [25]. The ONGC and the Indian Oil Corporation Ltd (IOCL) have recently launched a demonstration project for the development of a business model by capturing $CO_2$ from Koyali refinery in Gujarat and injecting it in the Gandhar oil field for increasing domestic oil production through $CO_2$-EOR [26]. The Institute of Reservoir Studies (IRS) under ONGC actively pursued R&D and carried out a feasibility study.

The Climate Change Research Institute, a non-governmental organization, has delved into trainings and public education for about a decade. The first workshop was convened in 2009 for the youth in industry and academia. Since then, the Institute has organized a series of short-term capacity development workshops held every three years with the support from the government of India and industry. These workshops have catalyzed new research groups and gained international peer recognition. MANIT Bhopal became the first Institute to start a course curriculum on $CO_2$ sequestration for the postgraduate students [27]. Academia-industry exchange of the current research in the field of anthropogenic $CO_2$ sequestration and utilization provided synergy toward low-carbon growth strategy [28]. The organizations engaged in $CO_2$ sequestration research in India are depicted in Fig. 5.

### 4.2.3 International Collaboration

The first international collaboration between DST, India, and the Research Council of Norway (RCN), Norway, in $CO_2$ sequestration research was launched under the Agreement of Cooperation in Science and Technology way back in 2007. Indian-Norwegian joint research projects in climate research included an innovative work area of metal–organic frameworks at the Indian Institute of Petroleum, Dehradun. The CSIR-NGRI, Hyderabad, joined hands with the Battelle Pacific North-West National Laboratory, USA, to study geological storage in basalts. Bureau of Energy Efficiency and Central Electricity Authority, India collaborated with Germany on the estimation of emissions from thermal power plants and taking remedial measures. Indian Institute of Technology Mumbai (IITB) became part of the Consortium for Clean Coal Utilization with Washington University at St Louis (WUStL), and several projects relating to carbon dioxide sequestration [30] were taken up. Indian Institute of Management, Ahmedabad, has conducted studies in assessing the storage potential in the Indian peninsula, source—sink matching, and modeling of $CO_2$ emissions [31, 32].

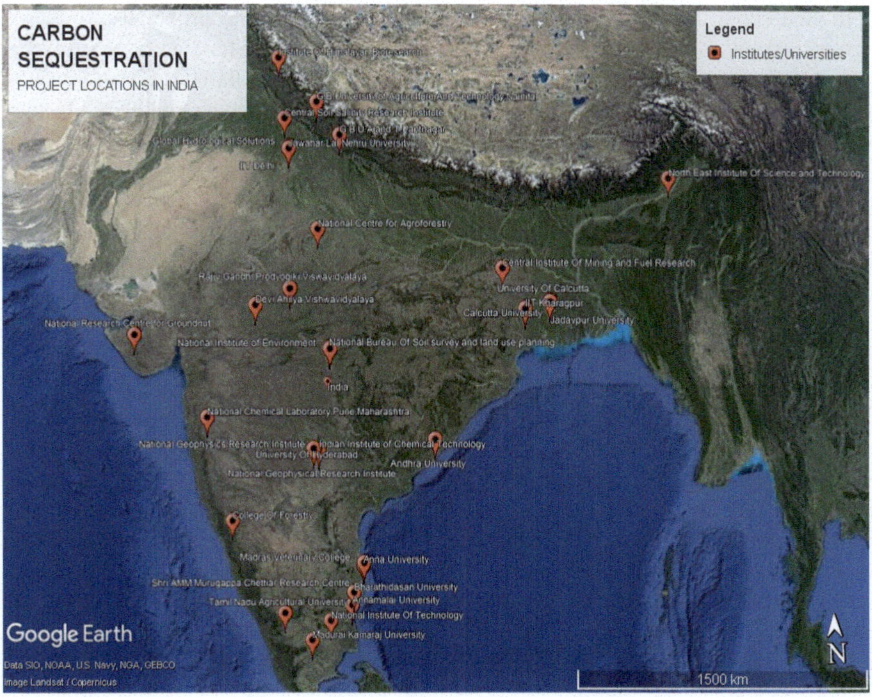


**Fig. 5** Carbon sequestration projects locations/institutions across India (compiled by authors from the data from various sources and TERI [29]

### 4.2.4 Global Trends

Globally, in 2018, the international research portfolio had culminated in 33 large-scale demonstration projects, with research publications increasing manifold. The USA having the largest number of RD&D projects is moving ahead in clean coal technologies as well as the application of carbon capture research with the launch of a program on Coal FIRST: Flexible, Innovative, Resilient, Small, and Transformative Coal. The aim is to develop 50–350 MW plants that will include $CO_2$ sequestration, be nimble, flexible, and modular and operate at nearly 45% efficiency [33]. Canada and Norway are in the forefront in $CO_2$ sequestration research and demonstration, with Norway showcasing the first CCS project at Sleipner (designed to inject 1 million ton of $CO_2$ per year into the ocean) operating since 1996, and Canada having begun the Weyburn EOR project soon after in 1998, is having the second largest number of $CO_2$ sequestration projects after the USA. Also, Australia has showcased one of the world's largest CCS/$CO_2$ sequestration plants: the Gorgon project, to separate carbon dioxide from natural gas and bury as much as 4 million metric tons of $CO_2$ annually.

In clean coal technologies, the CoolGen project in Japan is aiming to produce a high-efficiency low-emissions (HELE) plant using integrated gasification combined cycle (IGCC) combined with integrated gasification fuel cells (IGFC) and carbon capture. With the plant efficiency of 40% (HHV) and capturing 90% $CO_2$ in a commercial scale IGCC, the aim is to make it a zero-emission plant. A Roadmap Study for Carbon Capture and Storage Demonstration and Deployment in China conducted by Asian Development Bank has identified unique low-cost opportunities for $CO_2$ sequestration demonstrations during 2016–2020 and recognized that many crucial barriers remain to be overcome, such as excessive energy penalty, high capital costs, perceived and real technical risks, weak $CO_2$ off-take agreement regime, etc. [34]. It suggested a dual-track approach (i) taking large-scale $CO_2$ sequestration demonstrations in utilizing captured $CO_2$ and (ii) intensifying research aimed to overcome cost and energy penalty hurdles.

The high cost of $CO_2$ sequestration requires both policy support and higher investment in R&D to progress development and deployment of technology. It was noted that the $CO_2$ sequestration technology growth in key industry sectors has been slow compared with the use in fuel transformation, but there may be growing momentum in the recent years [35]. Between 2010 and 2016, the world spending on renewable energy was $2.3 trillion, while on $CO_2$ sequestration only $10 billion [36]. Incremental growth in $CO_2$ mitigation potential of renewable power and $CO_2$ sequestration technology is expected with the increasing investment [37]. To promote carbon utilization research and investment internationally, a $20 million NRG COSIA Carbon XPRIZE was announced to develop breakthrough technologies that will demonstrate on-scale conversion of $CO_2$ emissions (200 kg per day for three consecutive days) from power plants/or and industrial facilities into valuable products like building materials, alternative fuels and other items of use [38].

## 5 Scenario Analysis for $CO_2$ Emissions (2017–2030)

The premises are, India is moving toward fossil-free power production, dependence on coal as a fuel for meeting the basic needs of energy is likely to continue until 2040. The coal-based power generation capacity of 194 GW in 2017 would go up to 330–441 GW by 2040 [39]. Very few studies are reported in the literature to assess the impact of $CO_2$ sequestration in mitigating emissions from the energy industry. A $CO_2$ emission trend analysis from increasing the renewable energy share in the Indian power sector was reported in 2012 using data of operating thermal power plants.

### 5.1 $CO_2$ Emission Trend Analysis

Mittal et al. [40] gave an early $CO_2$ emission scenario based on certain assumptions adopting a bottom-up approach for region-wise distribution of power plants and their corresponding efficiencies. Using Central Electricity Authority data analysis of four zones, i.e. east, west, north, and south of the power plants in operation and future installation was reported. Different plants use different grades of coal, and $CO_2$ emission scenarios are generated based on the carbon content as obtained from the elemental analysis of the coal and the excess air used for future growth trends. The assumptions in business-as-usual (PC-BAU) and best-case-scenario (PC-BCS) are as follows.

(i) PC-BAU scenario is based on coal consumption in 2020 for thermal power generation projected by the Planning Commission of India under the business-as-usual (BAU) scenario.
(ii) PC-BCS scenario is based on future coal consumption projection for thermal power generation in India in 2020 under the best-case scenario (BCS), which includes increasing use of renewable energy sources and efficient coal technologies.

The growth profiles of $CO_2$ emissions from thermal power generation during 2001–2021 using a bottom-up approach for the two scenarios as above are projected in Fig. 6. With the PC-BAU scenario, electric power generation from coal- and lignite-based power plants will continue to grow and was expected to rise as 950 TWh and $CO_2$ emissions rise to 850,000 Gg (850 Mt) by 2020–21. While, with the PC-BCS scenario, the application of alternate technologies including renewable energy and improvement in energy efficiency will lead to the lower generation from coal to approximately 850 TWh, it was predicted that $CO_2$ emissions will rise to only 700 Mt by 2020–21. About 17% reduction in $CO_2$ emissions was predicted with extensive use of alternate technology.

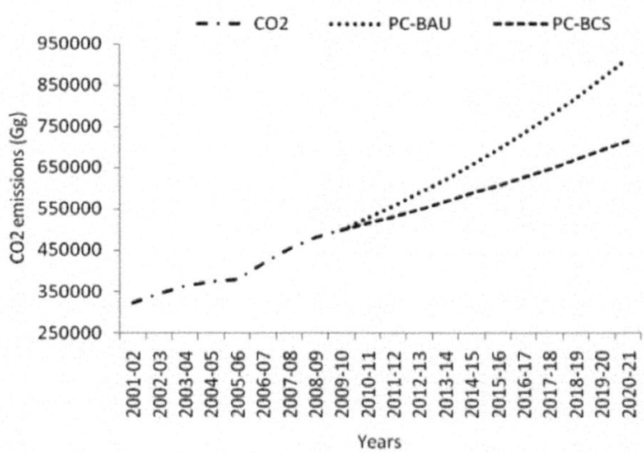


**Fig. 6** Scenarios of $CO_2$ emissions from coal-based thermal power plants (*Source* Mittal et al. [40], https://www.epa.gov)

### 5.2 *NitiAayog Study*

Using the MESSAGEix platform of IIASA with the IESS 2047 calculator of Niti-Aayog, a study has indicated that cumulative $CO_2$ emissions from the power sector will reduce by 375Mt $CO_2$ under the RE target by 2030 as compared to 2017 [41]. The total emissions however decline slowly, as it gets dampened due to the cumulative increase taking place in the absolute emissions. It was predicted that emissions intensity would improve significantly during this period falling by 41% in 2032, relative to the 2017 level. Deeper decarbonization methods should be adopted to slow down the growth in total $CO_2$ emissions.

Establishing an advanced $CO_2$ sequestration strategy as a climate engineering tool is seen as an important step for climate change mitigation in India and achieving a transition toward sustainable development with the use of fossil fuels. The $CO_2$ emission calculator has a computed $CO_2$ emission scenario for the penetration of $CO_2$ sequestration (from low to high) until 2047 [42] as follows.

Scenario 1: In the first case, CCS/$CO_2$ sequestration usage in power plants is slow and almost zero in 2025.

Scenario 2: The second case assumes that CCS/$CO_2$ sequestration starts penetration early and achieves 1 GW in 2022. The capacity with $CO_2$ sequestration increases to 35 GW in 2047.

Scenario 3 and Scenario 4: The third and fourth scenarios are assuming the case of faster adoption of CCS/$CO_2$ sequestration and reach 3 GW and 5 GW, respectively, in 2022. In both scenarios 3 and 4, the $CO_2$ sequestration large-scale deployment picks up after 2030 and becomes 80 GW and 90 GW in 2047, respectively.

The $CO_2$ sequestration-based capacity addition and reduction in $CO_2$ concentrations in all scenarios are plotted in Fig. 7. With $CO_2$ sequestration capacity addition of 90 GW in scenario 4, an emission reduction by 409Mt has been predicted in 2047. It is seen that the rate of penetration of $CO_2$ sequestration technology has a maximum contribution to the increasing reduction in $CO_2$ emissions. Innovations in carbon capture technology leading to a reduction in its cost could further add to the reduction.

## 6 Assessment of $CO_2$ Sequestration Potential Based on $CO_2$ Emission Factor

Central Electricity Authority (CEA) baseline Combined Margin $CO_2$ Emission Factor (CMEF) has been derived as 0.91t $CO_2$/MWh [44]. The $CO_2$ emitted per unit of electricity is the weighted average emission factor for thermal power generation in the country. The average is calculated by dividing the absolute $CO_2$ emissions of all power plants by the total net generation including hydro and nuclear stations (which qualify as low-cost/must-run sources) in the denominator. In the current study,

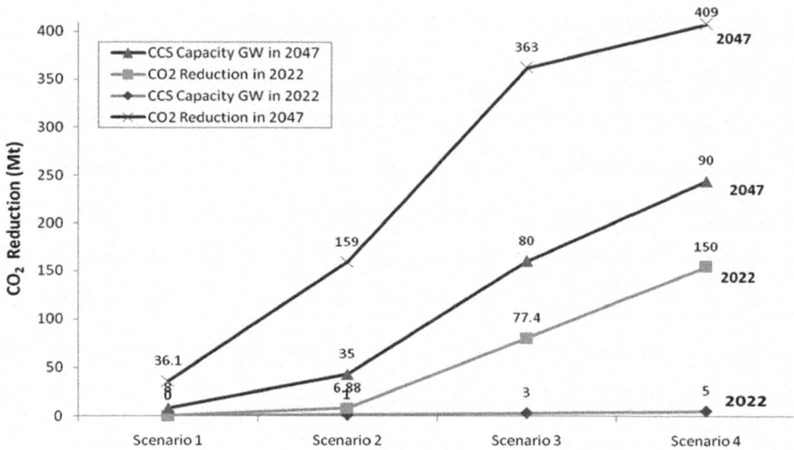


**Fig. 7** Projections for $CO_2$ emission calculator by use of $CO_2$ sequestration in four scenarios, compiled by authors from Ref. [43]

we assess the potential of $CO_2$ sequestration using CEA Baseline data. Coal Vision perspective scenarios; which depend on economic growth, energy efficiency, and renewable energy as alternate to coal; are implicitly taken as basic assumptions in three scenarios for $CO_2$ emissions up to 2030.

Scenario 1: High Coal Use.

Coal dependence continues to grow, and the use of coal grows from 856 Mt in 2017–18 to 1908 Mt in 2030.

Scenario 2: Business-as-Usual (BAU).

Both coal and RE use are growing in parallel, and this case use of coal grows from 856 Mt in 2017–18 to 1457 Mt in 2030.

Scenario 3: High Renewable Energy.

Taking RE aggressive targets of reaching 450 GW in 2030, the coal use grows from 856 Mt in 2017–18 to 1303 Mt in 2030.

A computation of $CO_2$ emissions during 2017–2030 is made for these scenarios. The emission from the high coal scenario shows a steep increase in emissions starting from the year 2020 (Fig. 8). We assume that $CO_2$ sequestration technology penetration starts from 2022 onward in all the scenarios. The emissions with the application of sequestration technology are shown in Fig. 9. Decrease in total emissions in the three scenarios is seen to be 20%, 14%, and 9%, respectively.

As expected maximum reduction is achieved in the high coal scenario, while the high renewable energy scenario is affected least by $CO_2$ sequestration penetration. In a high RE scenario, a reduction of 9% is possible by the application of $CO_2$ sequestration. To achieve these benefits in a coal-based generation, the maturity of the

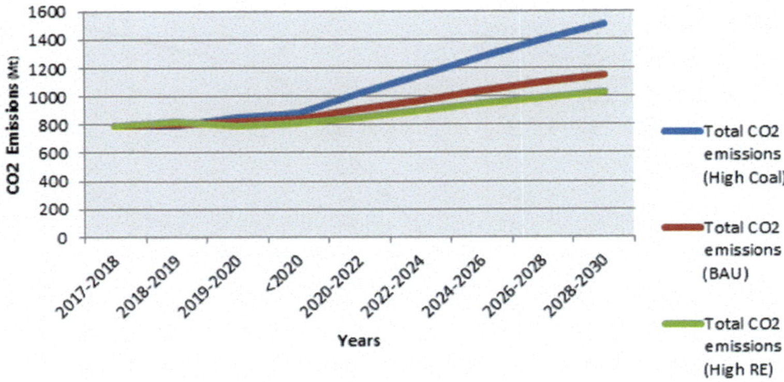


**Fig. 8** $CO_2$ emissions projections in three scenarios (2017–2030)

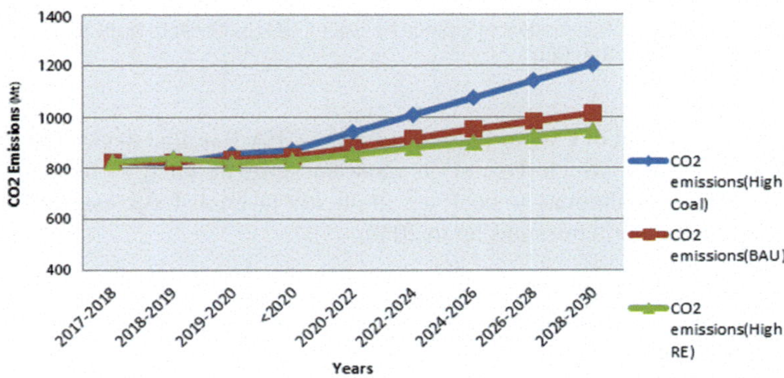


**Fig. 9** $CO_2$ emissions with the adoption of the $\mathbf{CO_2}$ sequestration technology

carbon capture technology and economic implications of $CO_2$ sequestration would have important roles. The changing market price of anthropogenic $CO_2$ (captured and stored away from the atmosphere) would provide further incentive for the adoption of $CO_2$ sequestration technologies. Main enabler for the application of $CO_2$ sequestration in India could be the pricing of carbon that reaches US$ 130 per ton $CO_2$ by 2050 [12]. Strategies for deep decarbonization pathways vary depending on the national development perspectives and policies. Government regulations with sector-specific decarbonization pathways would therefore be crucial for coal-based energy systems to incentivize and facilitate technology development as well as the adoption of $CO_2$ sequestration.

## 7 Conclusions

Carbon management and energy security are key challenges before the developing economies—striving to meet basic energy needs for their population on one hand and climate change mitigation with high financial implications on the other. India is on a path of accelerated economic growth and has set ambitious renewable energy targets. Coal is currently a dominant source of energy contributing 79% to power sector and almost 100% to industry. The government of India adopted a policy to intensify $CO_2$ capture and sequestration research. High cost of $CO_2$ capture, lack of potential geological sites, energy penalty and technical hurdles are some of the challenges. A proof-of-concept CO2-EOR field project was recently launched between the oil companies for validation of the research data.

We describe $CO_2$ emission projection studies from coal-based power plant efficiencies in India and scenario analysis to determine the impact of CCS penetration on $CO_2$ emissions. Based on the $CO_2$ emission factor defined by CEA, the $CO_2$ reduction in scenarios of high coal, business-as-usual, and high renewable energy with and without $CO_2$ sequestration is assessed. With the application of $CO_2$ sequestration technology, the reduction in $CO_2$ emissions in 2030 is seen to be 20%, 14%, and 9%, respectively, in the three scenarios. For the high RE scenario, a reduction of 9% in the overall $CO_2$ emission is projected by 2030. Adoption of renewable energy as sources of energy not only for generation, in the power and industry sectors (to make them sustainable), but also in $CO_2$ utilization for conversion processes (to make energy balance an attractive option), would continue to be challenging for both policymakers and industry.

Maturity of the carbon capture technology and the financial implications of both renewable resources and $CO_2$ sequestration on energy sector would determine future course of action. Cost repercussion studies for the adoption of $CO_2$ sequestration over and above RE technologies should be carried out to determine their relative contribution to the reduction of emissions. At the same time research investment, to achieve further innovations in green chemistry for $CO_2$ capture and utilization and to enhance academia-industry interaction, need to be mounted.

## References

1. Global Temperature. NASA (2016) GISS. https://climate.nasa.gov/vitalsigns/global-temperature/
2. Nell G (2017) 2016 was the hottest year yet. Scientists Declare. NPR. https://www.npr.org/sections/thetwo-way/2017/01/18/510405739/2016
3. Adams S, Aich V, Albrecht T, Baarsch F, Boit A, Trujillo N, Hare B (2014) Turn down the heat: confronting the new climate normal. The World Bank
4. Ma Q (1998) Greenhouse gases: refining the role of carbon dioxide. NASA Goddard Institute for Space Studies; Mahesh A, Shobha Jasmin KS (2013) Role of renewable energy investment in India: an alternative to $CO_2$ mitigation. Renew Sustain Energy Rev 26:414–424. https://www.giss.Nasa.gov/research/intro/ma

5. Hannah R, Max R (2017) $CO_2$ and greenhouse gas emissions. Published online at OurWorldInData.org. https://ourworldindata.org/co2-and-other-greenhouse-gas-emissions
6. NitiAayog (2017) Draft national energy policy (as of 27.06.17). Accessed at https://niti.gov.in/
7. Energy Statistics (2018) Ministry of Statistics and Programme Implementation Government of India. https://mospi.nic.in/sites/default/files/publication_reports/
8. https://energy.economictimes.indiatimes.com/news/coal/indias-thermal-coal-output-seen-growing-4-3-annually-till-2028-report/69437697
9. Coal Vision 2030, Stakeholders Consultation, Coal India Ltd. (2017). CVI-2030. https://www.coalindia.in/DesktopModules/DocumentList/documents/Coal_Vision_2030_document_for_Coal_Sector_Stakeholders_Consultation_27012018.pdf
10. Sharma
11. WRI (World Resource Institute) (2014) Climate analysis indicators tool: WRI's climate data explorer (World Resources Institute, Washington, DC)
12. Shukla P (2015) Pathways to deep decarbonization in India, SDSN—IDDRI. https://www.researchgate.net/publication/316986750_Pathways_to_deep_decarbonization_in_India
13. UNEP (2018) The emissions gap report 2018. United Nations Environment Programme, Nairobi, vol 97, pp 1625–1633
14. Ministry of New & Renewable Energy, Solar Parks approved under the Solar Park scheme of MNRE as of 30-06-2018
15. Shrivastava L (2014) TERI Discussion paper 'What would India need for moving to a 100% renewable energy scenario by 2050?'
16. https://www.cslforum.org/cslf/
17. Goel M (2009) Recent approaches in $CO_2$ fixation research in India and Future Perspective towards zero-emission coal-based power generation. Curr Sci 97:1625–1633
18. Garg M (2011) Carbon capture and storage/use (Research Priorities for CCS).CSIR—Indian Institute of Petroleum, Dehradun. Presented at Carbon Capture and Storage Workshop organized by the Ministry of Power, Government of India in New Delhi on September 27, 2011
19. Goel M, Bhandari A, Charan S (2008) $CO_2$ sequestration: recent Indian research, Glimpses of geoscience research in India. In: Singhvi AK, Bhattacharya A, Guha S (eds) The Indian Report to IUGS 2004–2008. INSA Platinum Jubilee Publication. https://www.iypeinsa.org/updates-09/art-11.pdf
20. Kumar B, Goel M, Charan N (2007) International workshop on R&D challenges in carbon capture and storage technology for sustainable energy future. J Geol Soc India 70(1):174–175
21. Peter V, Andreas V, Holger L, Daniel V, Samuel H (2011) Future CCS implementation in India: a systematic and long term analysis. Energy Proc 4:2708–2715
22. Chakraborty A (2008) $CO_2$ capture and storage: ONGC's perspective and plan. ONGC
23. https://carboncap-cleantech.com/capacity.html
24. India to build ultra-supercritical thermal power plant. The Economic Times, February 27, 2012 (2020). PTI. https://articles.economictimes.indiatimes.com/2012-02-27/news/31104416_1_power-plant-carbon-dioxide-ntpc
25. https://carboncleansolutions.com/
26. Mishra G (2019) Planning India's first $CO_2$-EOR project as carbon capture utilization and storage: a step towards sustainable growth, SPE-194629-MS was-the-hottest-year-yet-scientists-declare
27. Goel M, Johri V (2015) Carbon capture, storage, and utilization—an appraisal of current situation in coal-based economies, in carbon capture, storage and utilization. In: Goel M, Sudhakar M, Shahi RV (eds). TERI Press, New Delhi ISBN 9788179935682
28. Goel M, Sudhaker M (2017) Carbon utilization—applications for the energy industry. Springer Singapore, Green Energy and Technology Series, ISBN 978-981-10-3351-3
29. T E R I (2013) India CCS scoping study: final report New Delhi: The Energy and Resources Institute. 42pp [Project Report No. 2011BE02]
30. Consortium for Clean Coal Utilization (CCCU) Washington University in St. Louis. Ongoing Research Projects. https://cccu.wustl.edu/research-proj.php

31. Holloway S, Garg A, Kapshe M, Deshpande A, Pracha A, Khan S (2008) A regional assessment of the potential for $CO_2$ storage in the Indian subcontinent. IEA GHGR&D Programme
32. Garg A, Shukla P (2009) Coal and energy security for India: role of carbon dioxide ($CO_2$) capture and storage (CCS). Energy 34(8):1032–1041
33. IEA Clean Coal Newsletter (IN) (2020). https://www.iea-coal.org/
34. Asian Development Bank (2015 Nov) Roadmap for carbon capture and storage demonstration and deployment in the People's Republic of China. ISBN 978-92-9257-042-2 (Print), 978-92-9257-043-9 (e-ISBN) Publication Stock No. RPT157521-2. https://www.adb.org/sites/default/files/publication/175347/roadmap-ccs-prc.pdf
35. Transforming Industry through CCUS (2019). A Report by International energy agency. https://www.iea.org/reports/transforming-industry-through-ccus
36. Rathi A (2017) Humanity's fight against climate change is failing. One technology can change that, Quartz. https://qz.com/1144298/humanitys-fight-against-climate-change-is-failing-one-technology-can-change-that/
37. Mahesh A, Jasmin K (2013) Role of renewable energy investment in India: an alternative to $CO_2$ mitigation. Renew Sustain Energy Rev. https://doi.org/10.1016/J.RSER.2013.05.069
38. A New Perspective. $CO_2$. https://carbon-xprize.cosia.ca/
39. Energy Demand Projections, Department of Atomic Energy. https://www.dae.gov.in/node/128
40. Mittal M, Singh R, Sharma C (2012) Estimates of $CO_2$ from thermal power plants in India, 20th Emission inventory Conference, August 13–16, 2012. Tampa, Florida, USA
41. Thambi S, Oliver F, Anindya B (2017) India's energy and emissions outlook: results from India energy model. NitiAayog Working Paper. https://niti.gov.in/sites/default/files/2019-07/India%E2%80%99s-Energy-and-Emissions-Outlook.pdf Accessed on 14 Feb 2020
42. ISGF: India Energy Scenarios 2047 (2015) Documents prepared by ISGF for Planning Commission. Available at https://indiasmartgrid.org/reports/India%20Energy%20Security%20Scenarios%202047.pdf
43. IESS, User Guide for India's 2047 Energy, Carbon Capture and Storage Calculator. Available at https://www.google.com/search?q=IESS%2C+User+Guide+for+India%E2%80%99s+2047+Energy+Calculator
44. $CO_2$ Baseline Database for the Indian Power Sector, User Guide, Version 14.0, December 2018 (2015). Central Electricity Authority

# Challenges for $CO_2$ Management Through Renewable Energy, Conventional Power and System Reliability in Indian Power Sector

**V. S. Verma**

**Abstract** The subject of integration of renewable sources of energy, namely wind and solar (mainly) into the grid, has assumed significance in view of the fact that the renewable sources of power are now being adopted on a large scale to bring down the generation of huge quantities of $CO_2$ from fossil-fired power plants. Govt. of India has set grand targets of renewable energy source capacity additions. More proportion of renewable energy in the grid creates the problem of grid security and reliability issues. To avoid any load shedding into system due to variable nature of the RE sources, the higher ramping up and ramping down rates per minutes are required to be incorporated in the existing and new coal-/gas-fired conventional power plants. Further, the standards of minimum loading of the thermal power plants without oil support need to be brought down from the present value of 55% to further lower values to save on the import of huge quantities of diesel and crude oil and consequent drains of foreign exchange. The new technologies to ignite the coal directly are now commercially available internationally. These would need to be seriously considered. All these aspects have been discussed in the paper in detail so that there could be smooth change over from the present scenario of low renewable to high renewable and consequently prepare a solid ground for accepting without problem areas a higher share of renewable energy into our system.

## Abbreviations

| | |
|---|---|
| CCS | Carbon capture and storage |
| CEA | Central Electricity Authority, India |
| CERC | Central Electricity Regulatory Commission India |
| POSOCO | Power System Operation Corporation |
| ROE | Return on equity |

V. S. Verma (✉)
Central Electricity Regulatory Commission, B-01, Swati Apts., 12 IP Extn., New Delhi 110092, India
e-mail: vermavs2@gmail.com

 
M. Goel et al. (eds.), *Climate Change and Green Chemistry of $CO_2$ Sequestration*, Green Energy and Technology, https://doi.org/10.1007/978-981-16-0029-6_4

# 1 Introduction

$CO_2$ management in the power sector has become an important issue with increasing awareness of greenhouse effect due to consequences of power generation from burning of fossil fuels like coal, gas, oil, etc. globally. More than 80% of world power generation comes from fossil fuels and rest comes from hydro, nuclear and small portion from renewable energy sources. This has focused the attention of the international community to bring down $\mathbf{CO_2}$ production from the coal-fired thermal power generating stations substantially by adopting technologies such as carbon capture and storage (CCS). India has explored other options like adoption of supercritical and ultra-supercritical technologies on new coal-fired power plants, retiring the old smaller size coal-fired power stations consisting of mainly non-reheat units and which have already lived their useful life of 25–30 years (about 15,000 MW capacity consisting of 30, 60, 62.5, 82.5, 100, 110, 120, 140 and some of the 200 MW units), renovation and modernization of rest of the 200 and 500 MW units to bring back their efficiencies to at least near to design values, and induction of renewable sources of energy like wind and solar, etc.

These options were analyzed for our conditions in the year 2003–04. The Bureau of Energy Efficiency and CEA started bringing out the data on the $CO_2$ emission from the power sector in India in collaboration with GTZ Germany. This data/information were available for public consumption for planning their activities relating to the abatement of $CO_2$ emissions. India, however, did not agree to install the CCS technology in the power sector due to reasons of economics and efficiency aspects as well as disaster management. On the other hand, it was agreed to adopt the alternate measures to reduce $CO_2$ emissions from the power sector. Research institutions in India were directed to develop technologies to convert $CO_2$ into useful building materials or other chemicals.

## *1.1 Renewable Energy Targets*

Introduction of renewable source of energy, namely wind, solar, biogas, small hydro, waste to energy, etc., will replace conventional power generation during the day time and save fuel and consequently reduce $CO_2$ generation. Under mission mode, a target of 20,000 MW of renewable energy capacity addition by the year 2020 was set. In 2014, it was revised to 175,000 MW consisting of 100,000 MW of solar energy by 2022. The target has been set to achieve 450,000 MW by the year 2030. Large proportion of such capacity is already commissioned in the country, as discussed in the subsequent paras. As a consequence of the above policy guideline, more and more solar and wind power stations have started pouring in.

### 1.2 Challenges Ahead

It is pertinent to mention that intermittent solar and wind power capacity does not help meeting our peak demand, for which we will have to depend on the conventional power; be it coal-fired or nuclear power since the peak requirement generally occurs at a time when solar insolation is either not there or very minimal. Introduction of renewable source of power is therefore challenging for the grid managers and poses the risks of grid stability and reliable power supply issues. This will require the conventional power sources to be capable of appropriately ramp down or ramp up to suit the requirements. The above issues need to be addressed appropriately by all concerned.

## 2 The Present Electricity Capacities in the Country

The installed generation capacities in the country as on April 30, 2020, are given in Table 1. The total installed capacity has reached 370,347 MW. Out of this 87,268 MW is from renewable sources, consisting of 34,811 MW solar, 37,745 MW wind and rest consisting of small hydro, biomass, waste to energy, etc. Thermal power including coal, lignite, gas and diesel totals to about 230,600 MW, hydro—45,700 MW and nuclear—6780 MW. India has ventured into substantial power capacity addition in the form of renewable source of power. It is pertinent to mention that the major share of renewable capacity addition is in the form of solar and wind power, and both of these are of intermittent and varying nature. The grid managers find it difficult to maintain the absolute reliability and integrity of the grid. At a time when the solar or the wind power is not available, balancing power needs to be made available to stabilize the grid. Accordingly, dependence upon the conventional fossil-fired power stations or nuclear power stations or hydropower during peak power requirement continues. Since hydro and nuclear generation capacity being very small in comparison with the total demand, we will have to mainly depend on thermal power coal-fired.

## 3 Characteristics of Coal-Fired Thermal Power Stations vis-a-vis the Grid Requirements (Ramping-Up and Ramping-Down Rates)

**3.1** It is interesting to note that a thermal power station unlike the hydropower station has limitations in regard to operation below a certain load beyond which secondary oil support is needed for firing of coal. However, the Regulatory Commissions so far have not agreed to reimburse such costs below 55%-unit load. Accordingly, the

**Table 1** All India installed capacity (in mw) of power stations (as on 30.04.2020) (utilities)

| Region | Ownership/sector | Mode-wise breakup | | | | | | | | Grand total |
|---|---|---|---|---|---|---|---|---|---|---|
| | | Thermal | | | | | Nuclear | Hydro | RES * (MNRE) | |
| | | Coal | Lignite | Gas | Diesel | Total | | | | |
| Northern Region | State | 16,659.00 | 250.00 | 2879.20 | 0.00 | 19,788.20 | 0.00 | 5777.25 | 701.01 | 26,266.46 |
| | Private | 22,425.83 | 1080.00 | 558.00 | 0.00 | 24,063.83 | 0.00 | 2817.00 | 15,790.10 | 42,670.93 |
| | Central | 14,354.96 | 250.00 | 2344.06 | 0.00 | 16,949.02 | 1620.00 | 11,491.52 | 379.00 | 30,439.54 |
| | Sub-total | 53,439.79 | 1580.00 | 5781.26 | 0.00 | 60,801.05 | 1620.00 | 20,085.77 | 16,870.11 | 99,376.93 |
| Western Region | State | 21,740.00 | 1040.00 | 2849.82 | 0.00 | 25,629.82 | 0.00 | 5446.50 | 555.54 | 31,631.86 |
| | Private | 32,847.17 | 500.00 | 4676.00 | 0.00 | 38,023.17 | 0.00 | 481.00 | 24,821.29 | 63,325.46 |
| | Central | 18,347.95 | 0.00 | 3280.67 | 0.00 | 21,628.62 | 1840.00 | 1695.00 | 666.30 | 25,829.92 |
| | Sub-total | 72,935.12 | 1540.00 | 10,806.49 | 0.00 | 85,281.61 | 1840.00 | 7622.50 | 26,043.13 | 120,787.24 |
| Southern Region | State | 19,512.50 | 0.00 | 791.98 | 159.96 | 20,464.44 | 0.00 | 11,774.83 | 586.88 | 32,826.15 |
| | Private | 12,747.00 | 250.00 | 5340.24 | 273.70 | 18,610.95 | 0.00 | 0.00 | 41,344 74 | 59,955.69 |
| | Central | 11,835.02 | 3240.00 | 359.58 | 0.00 | 15,434.60 | 3320.00 | 0.00 | 541.90 | 19,296.50 |
| | Sub-total | 44,094.52 | 3490.00 | 6491.80 | 433.66 | 54,509.99 | 3320.00 | 11,774.83 | 42,473.52 | 112,078.34 |
| Eastern Region | State | 7450.00 | 0.00 | 100.00 | 0.00 | 7550.00 | 0.00 | 3537.92 | 275.11 | 11,363.03 |
| | Private | 6153.00 | 0.00 | 0.00 | 0.00 | 6153.00 | 0.00 | 96.00 | 1214.05 | 7463.05 |
| | Central | 13,682.05 | 0.00 | 0.00 | 0.00 | 13,682.05 | 0.00 | 1005.20 | 10.00 | 14,697.25 |
| | Sub-total | 27,285.05 | 0.00 | 100.00 | 0.00 | 27,385.05 | 0.00 | 4639.12 | 1499.16 | 33,523.32 |
| North Eastern Region | State | 0.00 | 0.00 | 497.71 | 36.00 | 533.71 | 0.00 | 422.00 | 233.25 | 1188.95 |
| | Private | 0.00 | 0.00 | 24.50 | 0.00 | 24.50 | 0.00 | 0.00 | 101.39 | 125.89 |
| | Central | 770.02 | 0.00 | 1253.60 | 0.00 | 2023.62 | 0.00 | 1155.00 | 30.00 | 3208.62 |

(continued)

**Table 1** (continued)

| Region | Ownership/sector | Mode-wise breakup | | | | | | | | Grand total |
|---|---|---|---|---|---|---|---|---|---|---|
| | | Thermal | | | | | Nuclear | Hydro | RES * (MNRE) | |
| | | Coal | Lignite | Gas | Diesel | Total | | | | |
| | Sub-total | 770.02 | 0.00 | 1775.81 | 36.00 | 2581.83 | 0.00 | 1577.00 | 364.64 | 4523.46 |
| Islands | State | 0.00 | 0.00 | 0.00 | 40.05 | 40.05 | 0.00 | 0.00 | 5.25 | 45.30 |
| | Private | 0.00 | 0.00 | 0.00 | 0.00 | 0.00 | 0.00 | 0.00 | 7.84 | 7.84 |
| | Central | 0.00 | 0.00 | 0.00 | 0.00 | 0.00 | 0.00 | 0.00 | 5.10 | 5.10 |
| | Sub-total | 0.00 | 0.00 | 0.00 | 40.05 | 40.05 | 0.00 | 0.00 | 18.19 | 58.24 |
| All India | State | 65,361.50 | 1290.00 | 7118.71 | 236.01 | 74,006.21 | 0.00 | 26,958.50 | 2357.03 | 103,321.74 |
| | Private | 74,173.00 | 1830.00 | 10,598.74 | 273.70 | 86,875.45 | 0.00 | 3394.00 | 83,279.41 | 173,548.85 |
| | Central | 58,990.00 | 3490.00 | 7237.91 | 0.00 | 69,717.91 | 6780.00 | 15,346.72 | 1632.30 | 93,476.93 |
| | Total | 198,524.50 | 6610.00 | 24,955.36 | 509.71 | 230,599.57 | 6780.00 | 45,699.22 | 87,268.74 | 370,347.52 |

Figures at decimal may not tally due to rounding off
Abbreviation: SHP = small hydro project ($\leq$525 MW); BP = biomass power; U&I = urban & industrial waste power; RES = renewable energy sources
*Note* 1. RES include SHP, BP, U&I, solar and wind energy. Installed capacity in respect of RES (MNRE) as on 30.04.2020 (as per latest information available with MNRE)

power stations are generally not interested in backing down below 55% generation. The plants, rather, prefer to shut down instead. This situation is very scary and difficult and must be avoided. Secondly, with reference to the integration of solar and wind power into the grid specially keeping in view the integrity and security of grid, it will be important to focus on the requirements of ramping-up and ramp-down rate required to deal with the varying nature of the renewable sources of power.

Coal-fired thermal power stations take about six hours of time to start in a hot restart condition. It takes about two more hours to reach full load condition. In case the unit has to back down as per the requirement of the grid operation to support the renewable energy induction, the unit can be safely brought down to 55% generation as permitted by central regulations without really calling for any secondary oil support.

**3.2** The contribution of renewable sources of generation is bound to increase in time to come; backing down the coal-fired generation below 55% level will be rather imminent and a necessity. The study by Palchak et al. [2] suggested changes in operational practice to reduce the cost of operation of the power system and reduce RE curtailment. Coordinating scheduling and dispatch over a broader area is the largest driver to reduce costs, and a saving of INR 6300 crore (USD 980 million) annually can be expected. It suggested that lowering minimum operating levels of coal plants (from 70 to 40%) is the biggest driver to reduce RE curtailment—from 3.5 down to 0.76%.

With this in view, Central Electricity Authority (CEA) made recommendations [1] and favoured reduction of generation from coal-fired thermal power stations to a level of 35% without any oil support, which due to resistance from various generators as well as looking into various contingencies of operation were not agreed. These have led to the recommendations that this is high time to adopt alternative technologies which could enable the direct firing of coal without oil support. One such technology known since a very long time is called plasma ignition combustion system (PICS) and other one is called minimum oil ignition system (MOIS).

Retrofitting of these technologies into the existing boilers results into huge oil savings during start-ups as well as during the operation of units at part loads without oil support. Ramping up and ramping down of generating units is possible without any requirement of oil support. In large sized units of 660 MW in the private sector, these have been tested in India. Their large-scale application needs serious consideration. There is also a need to investigate further from the point of view of indigenous manufacturing, availability of spare parts, etc. Presently, these have been supplied by China as a major supplier of the same. There are other suppliers of the same in USA, UK, etc.

## 4 Need for Managing the Balancing Power Almost Instantly

From the forgoing, it is vividly clear that in case the reliability and security of grid management are to be ensured, one has to seriously think of the urgent requirement of the balancing power which would need to be introduced immediately as soon as the solar and wind power intensity is reduced due to any reason, or it completely goes out of grid due to inclement weather conditions. In such a situation, the ramping-up and ramping-down rates of conventional coal-fired power stations acquire greater significance since nuclear stations are generally not comfortable for varying load conditions, hydrocapacity is limited, and reliable battery systems are very expensive.

The balancing power thus could be supplied only through hydropower, gas-fired power stations or ramping up or ramping down of already running coal-fired power stations as the case may be. The reverse situation would arise in case the renewable power becomes available quickly again. As an example, Delhi has a peak demand of 7000 MW and let us imagine that 40% of this comes from renewable sources of power. Weather turns inclement especially in the peak summer season and renewable power supply cannot be met. In that case, almost 2800 MW power is to be immediately arranged/replenished from the conventional coal-fired stations so as to maintain the reliability of power supply and to avoid any load shedding condition.

The options available to the grid operators are as follows to immediately resorting to ramping up of the coal-fired stations, pumping the hydrostations as per their availability, open cycle or combined cycle gas-fired stations, reliable batteries systems or any other such sources depending on the availability. All this has to be preconceived and earmarked for the situation. In case of non-compliance due to any reason, load shedding will be imminent. Reverse will occur in case the must run renewable power comes back in the grid. If the ramping down to the required level does not take place, it will cause the frequency-related grid disturbance, etc.

## 5 Ramping-Up and Ramping-Down Capabilities of Our Existing Thermal Stations

The POSOCO carried out a study under a pilot project on all our coal-fired thermal stations in July 2019 [3] and concluded that the ramping-up and ramping-down rates feasible in our coal-fired power stations vary between 0.8 and 1.2% per minute, whereas the recommended rates by CEA are around 3%. Thus, it indicates that when the share of renewable energy increases, we will have difficulty in handling the grid stability and reliability issues with our existing thermal stations.

The new thermal stations which are yet to be ordered will have to incorporate this requirement as a part of the specifications. Study needs to be carried by the CEA, NTPC and one State GENCO along with the manufacturers as to what modification would be needed, to be carried out in the existing stations to achieve the higher ramping-up and ramping-down rates to cater to the continuously increasing share of the renewable sources of generation.

## 6 Regulatory and Other Legal Interventions Requirements

(a) The CEA would need to incorporate in their technical standards the requirements of ramping-up/-down rates to be complied by all new power stations to be set up in future and make modifications in the existing power stations to achieve the required rates as per the outcome of the studies to be carried out by the committee to be set up by the Govt. for the purpose.
(b) Appropriate penalties for not complying with the CEA Technical Standards need to be included in the appropriate regulations.
(c) The CERC has already announced incentive scheme to offer higher ROE rates for confirming and offering the higher ramping-up and ramping-down rates. These need to be maintained and might need further review.
(d) The solution of providing the storage batteries needs to be considered in the long run, though expensive at this stage, the prices might come down in future.
(e) As nuclear plants capacity is limited, this needs to be exempted from ramping-up and ramping-down requirements, also from technical considerations. The nuclear stations one more comfortable with constant base load operations.

## 7 Conclusions and Recommendations

To overcome challenges for the smoother grid operation with high renewable energy input, the following options would be the future line of action for integration of grid with renewable energy. The grid security and reliability without avoidable load shedding could be more favorable than.

(a) Set up a study group consisting of manufacturers, plant designers and operation groups, to study the existing power plants and equipments and come out with recommendations and feasible solutions to make the existing 500 MW and above rating units to give at least $\pm 3\%$ ramping rates per minute along with cost implications to facilitate the regulators to reimburse such costs in reasonable manner.
(b) The CEA should issue amendments to their technical standards and incorporate ramping rates of $\pm 5\%$ per minute for all new power stations to give more stable operation.
(c) Appropriate guidelines are required to be issued by CEA/Ministry of Power/CERC for judicious operations of hydropower stations in tandem with solar and wind power stations to supply the balancing power.
(d) The conventional mode of operation of hydropower stations should change to suit the new emerging scenario. There is a need to encourage development of pumped storage hydroschemes for new hydrostations wherever there is a scope of adopting such schemes.
(e) CERC to consider enough incentives for development of the pump storage schemes of hydrostations.

(f) To facilitate decreasing generation on coal-fired stations to accommodate the higher share of renewable energy sources, without secondary oil support, below 55%, introduction of Plasma Direct Coal Firing Technology and The Minimum Oil Ignition Technologies needs to be seriously considered to be incorporated in the existing and new coal-fired station to be set up. CEA would need to consider such recommendations in their technical standards. Such technologies are already successfully operating around the world on more than eighty hundred thousand MW capacity power stations. These have also found application in many of our 660 MW supercritical units and working satisfactorily in India.

(g) It needs to be evaluated as to what are the cost implications of operating the coal-fired stations on reduced load while ramping up and down especially with reference to the impact on thermal reduced efficiencies in such operations. The optimal operation of the total power system would need to be considered, and realistic costs have to be worked out to be more transparent.

(h) Effects on the $CO_2$ reductions from coal-fired stations would also need to be evaluated.

## References

1. Central Electricity Authority (2019) Flexible operation of thermal power plant for integration of renewable generation. https://www.cea.nic.in/reports/others/thermal/trm/flexible_operation.pdf
2. Palchak D, Cochran J, Ehlen A, McBennett B, Milligan M, Chernyakhovskiy I, Deshmukh R, Abhyankar N, Soonee SK, Narasimhan SR, Joshi M, Sreedharan P (2017) GREENING THE GRID: pathways to integrate 175 gigawatts of renewable energy into India's electric grid, vol. I—national study. https://www.nrel.gov/docs/fy17osti/68530.pdf
3. POSOCO (2019) Fast response ancillary services (FRAS) implementation in Indian grid. https://posoco.in/wp-content/uploads/2019/08/POSOCO_FRAS_Feedback.pdf

# CFD Analysis of Chemical Looping Combustion with Special Emphasis on Indian Coal: A Review

**Pulkit Kumar and Ajit Kumar Parwani**

**Abstract** Soaring of average temperature and change in climate pattern is posing challenge to find out a reliable and economical solution of reducing carbon dioxide percentage from the atmosphere. Out of various methods of capturing carbon dioxide, chemical looping combustion (CLC) is proving to be the most efficient and economical method. Although many researchers did experimental investigation of CLC, prior to experimentation, it is necessary to perform numerical simulation to determine the optimum working parameters. The literature on CFD simulation of CLC is very limited. This paper reviews the recent work done by researchers on CFD simulation of CLC with different qualities of coal including Indian coal.

**Keywords** CLC · Computational fluid dynamics (CFD) · Multiphase modelling

## Abbreviations

| | |
|---|---|
| AR | Air reactor |
| CD-CLC | Coal direct CLC |
| CFD | Computational fluid dynamics |
| CLC | Chemical looping combustion |
| CLOU | Chemical looping with oxygen uncoupling |
| DEM | Discrete element method |
| FR | Fuel reactor |
| OC | Oxygen carrier |
| PDE | Partial differential equation |
| TG-DSC-MS | Thermal analyser-differential scanning calorimeter-mass spectrometer |
| TGA | Thermal-gravimetric analyser |

P. Kumar · A. K. Parwani (✉)
Institute of Infrastructure Technology Research and Management, Ahmedabad, Gujarat, India
e-mail: ajitkumar.parwani@iitram.ac.in



M. Goel et al. (eds.), *Climate Change and Green Chemistry of $CO_2$ Sequestration*, Green Energy and Technology, https://doi.org/10.1007/978-981-16-0029-6_5

# 1 Introduction

In 1896, Arrhenius has discovered a connection between earth surface temperature and the concentration of carbon dioxide ($CO_2$) [4]. Because of consistent outflow of $CO_2$ from the thermal power plants and different sources which are anthropogenic, concentration of $CO_2$ in the climate has ascended from an estimation of 280 ppm (543 mg/m$^3$) amid the time when there was no industry to around 416 ppm (806.75 mg/m$^3$) nowadays [9]. Since we do not have a reliable alternative to fossil fuels, this increase in the concentration of $CO_2$ percentage is inevitable, and it may lead to more adverse climatic and environmental issues. So, this serious concern has led the researchers to find a solution which will help in separating $CO_2$ from the exhaust of different sources and then storing it for isolation or utilizing it for different purposes. This developed technique is called as carbon capture and sequestration (CCS). Thus, CCS is a technique in which $CO_2$ is isolated from the emission containing other gases and then storing it in a suitable geographical region, in such a way to ensure long-term isolation.

Various approaches were proposed to capture $CO_2$ from different sources which can be broadly classified as (1) pre-combustion, where $CO_2$ is separated before combustion (2) oxy-combustion, where $CO_2$ is separated during combustion and (3) post-combustion, where $CO_2$ is captured after the combustion by using sorbent-based absorption method. These technologies are successful in reducing the $CO_2$, but there is significant energy penalty attached with them which decreases the overall efficiency of power generation. These penalties can be projected up to 15% of the total energy use [15, 16]. Hence, there should be a technology which can capture the carbon efficiently and should not be costlier. Chemical looping combustion (CLC) has turned out as a good alternate in recent years which is highly efficient and economic as the cost of capturing carbon is very low as compared to other methods.

CLC is an intrinsic $CO_2$ capture method. Generally, the CLC system has two fluidized bed reactors which are interconnected (1) fuel reactor (FR) and (2) air reactor (AR). Figure 1 shows that FR operates as a bubbling bed and AR operates in a fast fluidization regime. A metal oxide which is also called as oxygen carrier (OC) is utilized to supply the oxygen required for the combustion of fuels and circulates between the FR and AR units. Into the FR, the fuels are supplied where it is combusted after reacting with oxygen which is obtained by reducing the OC. The product of this combustion reaction is $CO_2$ and steam. In the next step, OC gets oxidized into the AR, and it recirculates into FR where it again participates in the reaction. In this way, this whole loop of the process continues [17].

The reactions inside the FR and the AR are given by Eqs. (1) and (2), respectively [8].

$$\text{nMexOy} + \text{Fuels} \rightarrow \text{nMexOy-1} + CO_2 + H_2O \tag{1}$$

$$\text{MexOy-1} + 0.5O_2 \rightarrow \text{MexOy} \tag{2}$$

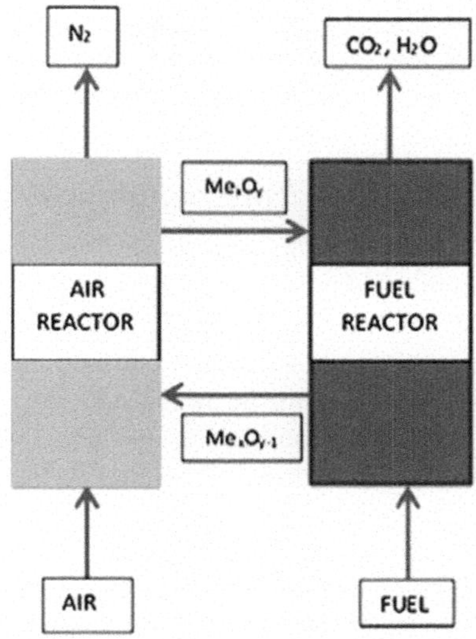


**Fig. 1** Conceptual diagram of CLC [38, 39]

where *m* and *n* represent the number of moles of reactants and products. Since the ignition of fuels in the FR takes place without air, only oxygen being utilized for the combustion which is given by the OC, the vent stream from the FR is not polluted or weakened by different gases like nitrogen. This gives a high-virtue $CO_2$ stream accessible for catch at the outlet of FR without the requirement for costlier process of air separation. Energy required for the solid recirculation is the only expense related to CLC method of carbon capture. On the basis of gasification and reaction schemes, CLC units can be suitably classified into mainly three types: (1) syngas CLC; (2) in situ CLC; and (3) chemical looping with oxygen uncoupling (CLOU).

## 1.1 Syngas CLC

In syngas CLC process, coal is not directly fed into the CLC unit. Initially, coal is gasified and converted into syngas, and, then it is released in the FR unit of CLC. In this method, an air separation unit is required which is energy exhaustive, and hence, it is the biggest disadvantage.

## 1.2 In Situ CLC

Pulverized coal is acquainted specifically with the FR where the gasification of coal and further reactions with OC will happen all the while in a similar reactor. Since coal is infused directly, separation unit for air is not required yet issues may happen because of ash deposition and agglomeration. It is also called as coal direct chemical looping process (CD-CLC) as pulverized coal is fed directly into reactor.

### *1.3 Chemical Looping with Oxygen Uncoupling (CLOU)*

In this technology of CLC, solid fuels are burnt with oxygen without the requirement of any separation unit for air. It involves three stages. The first stage starts with conversion of suitable metal into metal oxide (also called as OC) after it reacts with oxygen from the air in the AR.

In the second stage, in a FR, the OC discharges oxygen in the gas phase, and in the third stage, again in a FR, the gas-phase oxygen reacts with solid fuels. This process demands a metal which reacts with the only oxygen from the air in the AR to form metal oxide and reduces easily to provide oxygen required for combustion in FR. The CLC is an extremely encouraging technology and holds the key to the long run of inexpensive $CO_2$ capture technologies. Nevertheless, the applicability of CLC will rely upon accessibility of profoundly effective and economically practical OC.

## 2 Oxygen Carriers (OC) for CLC

The OC is generally by and large made out of a metal oxide (active phase) and inert phase. The benefit of inert phase is that it provides more surface area for the reaction by porous support. Porous support also helps in increasing mechanical strength of the particles. In general, CLC process, the oxygen transporter ought to have the capacity to change over the fuel gas totally into $CO_2$ and $H_2O$ in the FR.

Furthermore, OC for CLC should have the following properties:

- Highly reactive during oxidation by air and reduction by fuels gas,
- Resistance towards carbon formation,
- Mechanically strong to bear the stresses generated due to particle circulation,
- Do not support agglomeration effect,
- Low production cost.

The OCs which are frequently examined in the literature are iron (Fe), copper (Cu), cobalt (Co), manganese (Mn) and nickel (Ni). Since development of OC is an important factor in CLC, therefore lots of research are going on, and several researches have been already conducted studies to see the nature of its reaction with different gaseous and solid fuels.

Mattison et al. [24], have experimented with iron oxide ($Fe_2O_3$) as OC and methane as Fuels in CLC. They have used a fixed-bed quartz reactor at 950 °C temperature. Analysis of four gases, namely $CH_4$, $CO_2$, CO and $O_2$, was carried out during the combustion. The results obtained from their experimental investigation confirmed the suitability of $Fe_2O_3$ in terms of reaction rates for oxidation and reduction.

Leion et al. [21], have investigated feasibility of employing a variety of solid fuels in CLC. The OC chosen for this investigation was natural mineral ilmenite and a mixture of 60% $Fe_2O_3$ and 40% $MgAl_2O_4$. Oxidation and reduction of sample were performed alternatively in a fluidized bed reactor system of solid fuels. Temperature and steam content in fluidizing gas of the reactor were examined. Also, the conversion rate of different fuels was investigated. They concluded that the OC containing mixture of 60% $Fe_2O_3$ and 40% $MgAl_2O_4$ and ilmenite shows similar results with different fuels.

Rubel et al. [31], have investigated the different OC for their potential use in the CLC of solid fuels. Thermal analyser-differential scanning calorimeter mass-spectrometer (TG-DSC-MS) was used for analysis. Various factors like rate of reaction, changes in reaction rates with frequent oxidation–reductions, exothermic heats during oxidation and the effect of changing reduction gas composition were considered in the study. They have concluded that that the iron oxide powder and iron-based catalyst are suitable for CLC of solid fuels because its reduction rate is fast. Their reduction rates further increased with increase in reducing gas concentrations in the FR.

Cuadrat et al. [8], have done the feasibility analysis of coal with different ranks in a continuous CLC process. For this, they have considered ilmenite as OC. From their experiment, they have evaluated that carbon capture efficiency was higher in case of lignite followed by bituminous and anthracite. The performance of CLC was affected by the volatile content and reactivity of char.

Hassan and Shamim [13], did multistage analysis of power plant by employing CLC method for $CO_2$ capture using different OC and fuels. For the analysis purpose, they used ASPEN Plus software. They have chosen natural and synthetic gas as a fuel and checked the feasibility of nickel, copper and iron as OC. Their results have concluded that iron and nickel show maximum thermal efficiency when used as OCs. But iron requires higher mass flow rate than nickel and copper.

He et al. [14], used bimetallic oxygen carrier in experimental study of chemical looping gasification of solid fuels. They suggested to use a secondary oxygen carrying metal oxide which is copper oxide (CuO) with the iron-based oxygen carrier. Experiments were carried out on a thermal-gravimetric analyser (TGA) which shows that when a small amount of CuO (5% by weight) is mixed with other oxygen carrier, then it shows more effective coal char conversion.

From the reviews of the available literatures mentioned above, it can be concluded that researchers mainly carried out laboratory-scale experiments. But setting up and executing a laboratory examination can be a costly and difficult process. Whereas CFD simulation of reactors for the analysis of energy balance and chemical kinetics is an efficient method, it can also analyse the complexities in the fluidization behaviour of the bed particles. Although lots of review have been done by researcher on chemical looping process, review on CFD simulations of CLC is limited. Therefore, the main focus of this paper is to present a review of the recent development made by the researcher at reactor-level simulations of CLC. This is followed by some review of results of CFD simulation on FR for in situ gasification CLC with Indian coal.

## 3 CFD Simulations of CLC

The governing equation of the flow of fluids (liquid or gas) for the conservation of mass, momentum and energy is represented by partial differential equation (PDE). In computational fluid dynamics (CFD), PDE systems are replaced by algebraic equations which can be read and resolved by computers. We can do the qualitative analysis of fluid flows in CFD by the following steps

- Firstly, mathematical modelling is done in terms of PDE.
- In the next step, discretize that PDE by using some method and solve that equation by a definite solver.
- Lastly, software tools like solver and post-processor are used to generate the result.

By the help of CFD, researchers and engineers can perform many numerical experiments before doing or implementing it on large scale. Some of the discretization methods used to do the analysis are:

- Finite volume method: It is a very well-known approach utilized in CFD codes, as it has benefit of memory usage and convergence speed, particularly for complicated problems and turbulent flows. During this methodology, the governing PDE is recast in an exceedingly conservative type and then solved over discrete control volumes.
- Finite element method: It is used in analysing the loads of physical structures and their components. However, its application is not limited to physical structures, but it is also applicable to fluids. FEM needs lots of computer memory, and therefore, it is slow, but it gives stable solution than the FVM.
- Finite difference method: It is easy to programme and presently utilized in few specialized codes that handle advanced pure mathematics. This method uses the concept of fixed boundaries or overlapping grids to generate the highly accurate solution efficiently.

CFD studies are carried out to perform better analysis of method and system of various subjects. Although lots of software are available commercially as well as open source, to conduct the CFD study efficiently and effectively, ANSYS is the most widely preferred.

Conclusively, it is clear from the above discussion that CFD is an effective technique to analyse the CLC system performance and to characterize chemical kinetics and the fluidization in the system. The set-up of single reactor is not practical with solid fuels like coal since it is very difficult to remove the coal from the system earlier before switching to the AR configuration to avoid combustion of coal in air; also, the design and operation of the hot gas switching system remain problematic. As the solid fuels like coal are likelihood to remain dominant fossil fuels in the near future

[36], the design of a CLC system operating on coal should be the dual fluidized bed set-up. Now, since two or more than two phases will be involved in this simulation, multiphase modelling consisting of granular solid and gas is required.

Multiphase modelling can be done by various levels of accuracy with varying computational expenses. Generally, two approaches are used for doing multiphase modelling which is Eulerian or Lagrangian approach.

### *3.1 Eulerian Approach*

Reliable solution can be achieved by using the Eulerian or multi-fluid approach in which the demand of computational power is avoided for particle-based models.

In this model, the solid phase is approximated as a continuum fluid, and variables of particle such as velocity, temperature and mass are calculated by averaging over a control volume that is comparatively bigger than the particle size [6]. Only bulk behaviour of the solid phase can be accounted using this approach. For modelling the interactions between gas and solid phases, solid phase viscosity and solid phase pressure are needed. These viscosity and pressure equations can be obtained using the concept of the kinetic theory of granular flow.

The kinetic theory of granular flow is an extension of the classical kinetic gas theory that includes the inelastic particle–particle interactions [28, 29]. The investigation of fluid properties using multi-fluid models is very simple compared to the particle-based method and has been implemented successfully in an all-around application of fluidization such as bubbling fluidized beds [11], particle mixing [7], down-flow reactors [37] and spouted beds [12]. This approach was firstly used by Jung and Gamwo [19], for modelling the CLC process. However, the disadvantage of the Eulerian multi-fluid model is that it cannot determine the exact dynamics of particle in the system precisely.

### *3.2 Lagrangian Approach*

Force balance calculation of the particle is used to resolve the trajectory of each individual particle in the Lagrangian particle-based model or discrete element method (DEM). In this approach, particle and the fluid interaction is analysed separately for each particle. Tracking of the particle is coupled with CFD solution for the fluid phase. Properties such as temperature, composition, position and velocity can be modelled by the DEM with high accuracy, but it is limited to specifics of the particle collision parameters which is employed in the model. As the number of particles increases drastically due to collisions of particle, the DEM solution is constrained to around $10^5$ particles given current computational capabilities. Therefore, only few researches are available on particle-based models for CLC in the literature due to high computational cost.

## 4 Review of Results from CFD Simulation

The computational approach is mainly based on CFD-based numerical simulation, thermodynamic computation, ASPEN Plus simulation, etc. These simulations are intended for modelling of process and to study the behaviour of variation of design and operating parameters.

A review of CFD modelling technique for combustion and gasification in fluidized beds was performed by Singh et al. [33]. They also discussed the fundamental equations used for developing CFD model. Hydrodynamics of gas and solid particles were studied by Shuai et al. [32], by developing a 2D CFD model for chemical looping combustion using interconnected fluidized beds. Eulerian continuum two-fluid model was considered for both gas and solid phases. The mixing and segregation behaviour of binary mixtures of particles in bubbling fluidized bed were investigated by Peng et al. [30], by carrying out a numerical study on 10 kWth chemical looping combustor. DEM approach was considered in their investigation to track the motion of the particle, while gas–particle interaction was investigated by a two-way coupling method. Abad et al. [1], developed a mathematical model for the fuel reactor to assess the key parameters such as reactor temperature, solids circulation rate and solid inventory on the efficiency of carbon dioxide capture. They validated their simulated results against a 100 kWth chemical looping combustion unit. Their result showed carbon dioxide capture efficiency as 98.5% when operating temperature of fuel reactor was 1000 °C.

Mahalatkar et al. [22, 23], have carried out CFD studies of the chemical looping FR utilizing gaseous fuels and NiO and $Fe_2O_3$ as oxygen carriers, with substantial attention on the heterogeneous chemical reactions. Gas and solid phases were described by a continuum two-fluid model. The CFD simulations were conducted to see the applicability of the CFD model to determine variations in concentrations of gas at outlet by varying the number of parameters like metal oxide concentration, superficial velocity, temperature of reactor, etc. The obtained simulation results were compared with the experimental results of Mattison et al. [24], and concluded that CFD simulations provided a reasonable match with this data.

However, cold-flow simulations using the coupled CFD/DEM model have been considered to provide consistently matching the dynamics of particle with several laboratory-scale fluidized bed experiments using comparatively large particles [3, 34, 40]. Parker [27], has also used the DEM approach for developing model of circulating reactor system with reacting flow for CD-CLC. But due to high complication in the work, it took 81 days for simulation of 50 s.

In the literature review, it was found that integrating the chemical reactions into a CFD/DEM model has received very less attention. Almost all of the modelling works which are performed on CD-CLC featuring its chemical reactions are based on either Eulerian simulation [2, 5, 22, 23] or currently on process simulations from the perspective of energy balance using software like Aspen Plus [20, 35, 38, 39].

## 4.1 CFD Simulation of CLC Using Indian Coal

A lot of CFD studies are going on to simulate the steady and unsteady state flow behaviour of system, reactions taking place inside the reactor and to analyse the performance of the system under various operating conditions. Numerous studies have been carried out on CFD analysis on CLC using ANSYS in order to get a full hand idea prior to experimentation. But one thing that was common in almost all the CFD simulation (in case of solid fuels) was quality of coal.

Most of the research work on CFD analysis of CLC has been done on high-grade coal. But in a country like India where coal is expected to play a major role in addressing some of India's energy challenges, high-grade coal is not available. Ash content in Indian coal is very high, and it varies from 35 to 45%, while in other countries, it remains around 15%. High content of silica, alumina and quartz is found in the ash. When ash will get fused in the syngas cooling system, it may erode the system because of its highly abrasive nature.

Fusion temperatures of Indian coals were studied by Dutta [10]. In his research, he found that the fusion temperature of coal varies from 1050 to 1320 °C.

He could not propose a definite relationship between fusion point and composition of coal, but he mentioned that ratio of acidic and base component in the coal is related to the fusion point.

Relation between them confirms that if the ratio is higher, the fusion point will also be higher and vice versa. However, the sulphur content of the Indian coal is very low which is approximately about 0.5%. So, their removal is not important, and also, it is economically not justifiable. High ash Indian and Turk coals were investigated by Jayaraman et al. [18], to understand the pyrolysis and kinetics of gasification. They have used a coupled mass spectrometer and thermal analyser system for the identification of the volatiles evolved and the thermal characterization of the coal samples during the heating of the samples. TGA results obtained enhanced the knowledge on high ash coal thermal characterization and gasification.

Mishra et al. [26], compiled the studies conducted in the literature on pyrolysis and gasification of coal (high ash Indian coal). He made conclusion on the gasification of coal which states the steps involved during gasification of coal. Pyrolysis, gasification and combustion were the three steps suggested by him which are required for complete gasification process. He also concluded that the key processing parameters to control the gasification of char would be pressure, temperature and type of gasifier.

However, many research articles have been published on CLC including numerical (or CFD) simulations of CLC systems; only few articles available in the literature are based on Indian coal. Menon and Patnaikuni [25], have carried out CFD simulation of FR for in situ gasification CLC of Indian coal. The FR considered for the simulation was cylindrical in shape with conical bottom. For entrance of gasifying agent and exit of product gases, single inlet and outlet were provided. Solid and gaseous phases were modelled using multiphase Eulerian approach. Calculations were carried out by considering one gas phase and two solid phases. Turbulent flow behaviour of the components was characterized by the k-ε model.

Gunn correlation was used to define the heat transfer between the phases available in fluent database. Since the heat transfer between the solid phases is small, it is neglected. The models used in the system were taken from ANSYS theory guide 18 and Mahalatkar et al. [22, 23]. In this work, three OCs namely, $Fe_2O_3$, CuO and a combination of $Fe_2O_3$ and CuO, were studied. From the simulation result, it was observed that when effect of ash is excluded, CuO shows better performance than mixed carrier and $Fe_2O_3$, while mixed carrier consisting of CuO and $Fe_2O_3$ showed better performance when ash effect was considered. These studies also concluded that in case of stable performance with change in temperature, $Fe_2O_3$ is more preferable than mixed carrier and CuO. $H_2O$ and $N_2$ are chosen as an oxidizing agent instead of $H_2O$ and $CO_2$ to check the conversion rates of coal. These results of conversion rates were compared for the three OCs. Finally, their studies also reveal that mixed carrier of $Fe_2O_3$ and CuO is more preferable in the case when production of pure hydrogen is also one of the objectives along with CLC.

## 5 Conclusions and Outlook

In this study, the CFD simulation of CLC reactor using multiphase modelling approach has been reviewed. Potential of ANSYS Fluent software in CFD simulation of CLC is discussed in detail using two different approaches. Also, the importance of OC in CLC has been discussed. Work done by various researchers using different approaches of multiphase modelling is presented along with their result. The following are the conclusions drawn from their work.

(a) Most of the simulations performed are based on two-dimensional model of the reactors, while it is very important to consider three-dimensional model to achieve actual results.
(b) Simulation of complete CLC system consisting of air reactor, fuel reactor and cyclone separator will give us the actual idea of the process. Very few literatures are available where complete loop simulation of the CLC has been considered.
(c) While performing simulation using Indian coal or low-grade coal, it is important to consider ash agglomeration effect because high amount of ash is produced during combustion. The presence of high amount of ash in coal has adverse effect on combustion and $CO_2$ capture efficiency. Also, the fuel flow rate increases proportionately to maintain required process conditions for CLC.
(d) While selecting the OC for simulation, its availability and fast oxidation as well as reduction nature should be kept in mind.
(e) The problem of high ash content in the Indian coal can be mitigated by selecting suitable OC and blending low-grade coal with high-grade coal. This can be done by optimizing the process using advanced computational tools and software like ANSYS prior to experimentation.

## References


1. Abad A, Adánez J, de Deigo LF, Gayán P, Labiano FG, Lyngfelt A (2013) Fuels reactor model validation: assessment of the key parameters affecting the chemical looping combustion of coal. Int J Greenhouse Gas Control 19:541–551
2. Ahmed B, Lu H (2014) Modelling of chemical looping combustion of methane using a Ni-based oxygen carrier. Energy Fuels 28(5):3420–3429
3. Alobaid F, Ströhle J, Epple B (2013) Extended CFD/DEM model for the simulation of circulating fluidized bed. Adv Powder Technol 24(1):403–415
4. Arrhenius S (1896) On the influence of carbonic acid in the air upon the temperature of the ground. Philos Mag 41(251):237–277
5. Banerjee S, Agarwal RK (2016) An Eulerian approach to computational fluid dynamics simulation of a chemical-looping combustion reactor with chemical reactions. J Energy Resour Technol 138(4):042201
6. Banerjee S, Agarwal RK (2018) Computational fluid dynamics modeling and simulations of fluidized beds for chemical looping combustion. In: Breault RW (ed) Handbook of chemical looping technology, pp 303–332
7. Cooper S, Coronella CJ (2005) CFD simulation of particle mixing in a binary fluidized bed. Powder Technol 151:27–36
8. Cuadrat A, Abad A, García LF, Gayán P, de Diego LF, Adánez J (2012) Relevance of the coal rank on the performance of the in situ gasification chemical-looping combustion. Chem Eng J 195–196:91–102
9. Dlugokencky E, Tans P (2016) ESRL global monitoring division–global greenhouse gas reference network. www.esrl.noaa.gov/gmd/ccgg/trends/. Accessed 18 May 2020
10. Dutta RKR (1940) A critical study of some Indian coal ashes. Geological Survey of India (report)
11. Enwald H, Almstedt AE (1999) Fluid dynamics of a pressurized fluidized bed: comparison between numerical solutions from two-fluid models and experimental results. Chem Eng Sci 54:329–342
12. Gryczka O, Heinrich S, Deen NG, van Sint Annaland M, Kuipers JAM, Jacob M, Mörl L (2009) Characterization and CFD-modeling of the hydrodynamics of a prismatic spouted bed apparatus. Chem Eng Sci 64(14):3352–3375
13. Hassan B, Shamim T (2013) Effect of oxygen carriers on performance of power plants with chemical looping combustion. Procedia Eng 56:407–412
14. He F, Galinsky N, Li F (2013) Chemical looping gasification of solid fuels using bimetallic oxygen carrier particles: feasibility assessment and process simulations. Int J Hydrogen Energy 38:7839–7854
15. Hong J, Chaudhry G, Brisson JG, Field R, Gazzino M, Ghoniem AF (2009) Analysis of oxy-fuels combustion power cycle utilizing a pressurized coal combustor. Energy 34(9):1332–1340
16. Hong J, Chaudhry G, Brisson JG, Field R, Gazzino M, Ghoniem AF (2009) Performance of the pressurized oxy fuels combustion power cycle with increasing operating pressure. In: 34th international technical conference on clean coal and fuels systems, Clearwater, 31 May–4 June 2009
17. Ishida M, Jin H (1996) A novel chemical-looping combustor without $NO_x$ formation. Ind Eng Chem Res 35(7):2469–2472
18. Jayaraman K, Bonfari E, Marlo N, Gokalp T (2013) High ash Indian and Turkish coal pyrolysis and gasification studies in various ambiences, Cesme Izmir, pp 8–13
19. Jung J, Gamwo I (2008) Multiphase CFD-based models for chemical looping combustion process: fuels reactor modeling. Powder Technol 183:401–409
20. Kramp M, Thon A, Hartge E (2012) Chemical looping combustion of solid fuels—modeling and validation. In: International conference on chemical looping, Darmstadt
21. Leion H, Mattison T, Lyngfelt A (2008) Solid fuels in chemical looping combustion. Int J Greenhouse Gas Control 2:180–193

22. Mahalatkar K, Kuhlman J, Huckaby ED, O'Brien D (2011) CFD simulation of a chemical looping fuels reactor utilizing solid fuels. Chem Eng Sci 66:3617–3627
23. Mahalatkar K, Kuhlman J, Huckaby ED, O'Brien T (2011) Computational fluid dynamic simulations of chemical looping fuels reactors utilizing gaseous fuels. Chem Eng Sci 66(3):469–479
24. Mattison T, Lyngfelt A, Cho P (2001) The use of iron oxide as an oxygen carrier in chemical looping combustion of methane with inherent separation of $CO_2$. Fuels 80:1953–1962
25. Menon KG, Patnaikuni VS (2017) CFD simulation of fuels reactor for chemical looping combustion of Indian coal. Fuels 203:90–101
26. Mishra A, Gautam S, Sharma T (2014) Gasification of non-coking coals. In: International conference of advance research and innovation (ICAR), pp 1–6
27. Parker J (2014) CFD model for the simulation of chemical looping combustion. Powder Technol 265(47):47–53
28. Patil DJ, Van Sint Annaland M, Kuipers JAM (2005) Critical comparison of hydro-dynamic models for gas–solid fluidized beds—part I: bubbling gas–solid fluidized beds operated with a jet. Chem Eng Sci 60(1):57–72
29. Patil DJ, Van Sint Annaland M, Kuipers JAM (2005) Critical comparison of hydro-dynamic models for gas–solid fluidized beds—part II: freely bubbling gas–solid fluidized beds. Chem Eng Sci 60(1):73–84
30. Peng Z, Doroodchi E, Alghamdi Y, Moghtaderi B (2013) Mixing and segregation of solid mixtures in bubbling fluidized beds under conditions pertinent to the fuels reactor of a chemical looping system. Powder Technol 235:823–837
31. Rubel A, Liu K, Neathery J, Taulbee D (2009) Oxygen carriers for chemical looping combustion of solid fuels. Fuels 88:876–884
32. Shuai W, Guodong L, Huilin L, Juhui C, Yurong H, Jiaxing W (2011) Fluid dynamic simulation in a chemical looping combustion with two interconnected fluidized beds. Fuels Process Technol 92:385–393
33. Singh RI, Brink A, Hupa M (2013) CFD modeling to study fluidized bed combustion and gasification. Appl Therm Eng 52:585–614
34. Sutkar VS, Deen NG, Mohan B, Salikov V, Antonyuk S, Heinrich S, Kuipers JAM (2013) Numerical investigations of a pseudo-2D spout fluidized bed with draft plates using a scaled discrete particle model. Chem Eng Sci 104:790–807
35. Thon A, Kramp M, Hartge E (2012) Operational experience with a coupled fluidized bed system for chemical looping combustion of solid fuels. In: 2nd international conference on chemical looping, Darmstadt
36. U.S. Energy Information Administration (2010) International energy outlook 2010. Technical report DOE/EIA-0484. U.S. Department of Energy, Washington
37. Vaishali S, Roy S, Mills PL (2008) Hydrodynamic simulation of gas-solids downflow reactors. Chem Eng Sci 63:5107–5119
38. Zhang X, Banerjee S, Agarwal RK (2015) Process simulation and maximization of energy output in chemical-looping combustion using Aspen Plus. Int J Energy Environ 6(2):201–226
39. Zhang X, Banerjee S, Agarwal RK (2015) Validation of chemical-looping with oxygen uncoupling (CLOU) using Cu-based oxygen carrier and comparative study of Cu, Mn, and Co based oxygen carriers using Aspen Plus. Int J Energy Environ 6(3):247–254
40. Zhang Z, Zhou L, Agarwal RK (2014) Transient simulations of spouted fluidized bed for coal-direct chemical looping combustion. Energy Fuels 28(2):1548–1560

# Feasibility Study of $CO_2$ Control in Automobile Sources

**P. Senthilkumar**

**Abstract** In the recent years, due to the rapid escalation in the vehicle population and deforestation, the $CO_2$ level in the atmospheric has been increased drastically. Since $CO_2$ is formed due to complete combustion of fuel, which is desirable for getting more millage, the abnormally increased $CO_2$ level causes climate change problems. To mitigate the $CO_2$ level in the atmosphere, different approaches like absorption, adsorption and membrane separation can be adopted in the automobile sector. This chapter provides the information about the $CO_2$ emissions worldwide and nationwide. The experimental results show that the phase change fluids like amine solutions have better absorption characteristics up to 90%, as compared with activated carbon and zeolites. The regeneration of amines after absorption is, however, a big challenge, especially when it is used in vehicles. Possible technologies to alleviate the $CO_2$ emission from the automotive sector are also discussed.

**Keywords** Carbon dioxide · Automotive · Vehicle · Absorption · Amines

## 1 Introduction

Increase in mobility and vehicle population reflects the increase in the usage of fossil fuel consumption in the world. The statistics show that currently in India, the vehicle population increased by four times in comparison with the year 2000. Fossil fuel consumption in India is also increased drastically related to vehicle population. Most of the automobile vehicles are powered with the diesel, gasoline, liquefied petroleum gas (LPG) and compressed natural gas (CNG). These carbon-based fossil fuels have more impact on the environment and human health, if the usage of these fuels is increased, and then there will be drastic effects on climate change. At present, many countries are initiating strategies to control $CO_2$ emission from vehicles.

P. Senthilkumar (✉)
Department of Automobile Engineering, Madras Institute of Technology Campus, Anna University, Chromepet, Chennai, Tamil Nadu 600044, India
e-mail: drpsk@mitindia.edu

 
M. Goel et al. (eds.), *Climate Change and Green Chemistry of $CO_2$ Sequestration*, Green Energy and Technology, https://doi.org/10.1007/978-981-16-0029-6_6

## 2 World Scenario of $CO_2$ Emissions

$CO_2$ is the most dangerous gaseous emission which absorbs and emits thermal radiation. $CO_2$ is a greenhouse gas, along with its methane and $NO_x$ emissions are also a product of fossil fuel combustion. Initially, the UK was the first large-scale $CO_2$ emitter which is followed by the USA and China. By 2015, China had 15% of total cumulative $CO_2$ emissions and India had 3%. Total cumulative emissions are not emissions in particular year but the cumulative sum of country's historical emissions causes climatic change problem. The recent data from global carbon project suggested that there is a 2.7% increase in emissions by 2018.

At present, China is the largest $CO_2$ emitting country followed by the USA, India, Russia, Indonesia, Brazil, Japan, Canada and Mexico. Emissions from all over the world increased from 2 billion tonnes of $CO_2$ in 1990 to over 36 billion tonnes 15 years later. The $CO_2$ emission from the transport sector worldwide contributes to about 23% of the overall share (Fig. 1). In low- and middle-income economies trucks and buses consume more fuel and emit more $CO_2$. To control $CO_2$ emissions, some countries like China, Singapore and Hong Kong have decided to shift the transport to low-carbon emission for example instead of a cars or buses, the metro trains can be used [1]. The average concentration of carbon dioxide in the atmosphere has increased to 412 ppm in 2018 surpassing 410 ppm just a few years after first passing the 400 ppm mark. Before the industrial age began, the $CO_2$ concentration was only around 280 ppm. The average annual increase has risen from less than 1 ppm a year in the 1950s, when measurements began, to well over 2 ppm today. The $CO_2$ emission projected growth for different regions is shown in Fig. 2 and from different countries in Table 1.

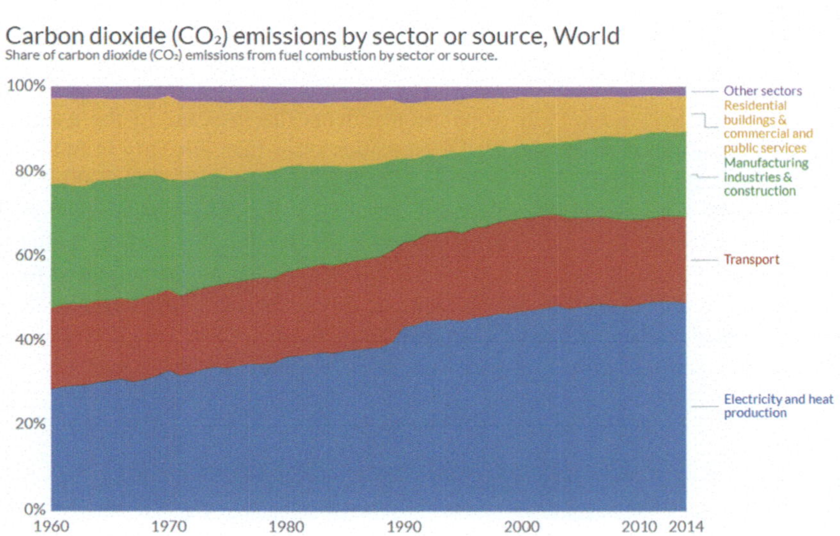


**Fig. 1** Carbon dioxide emissions by sector or source worldwide [1]

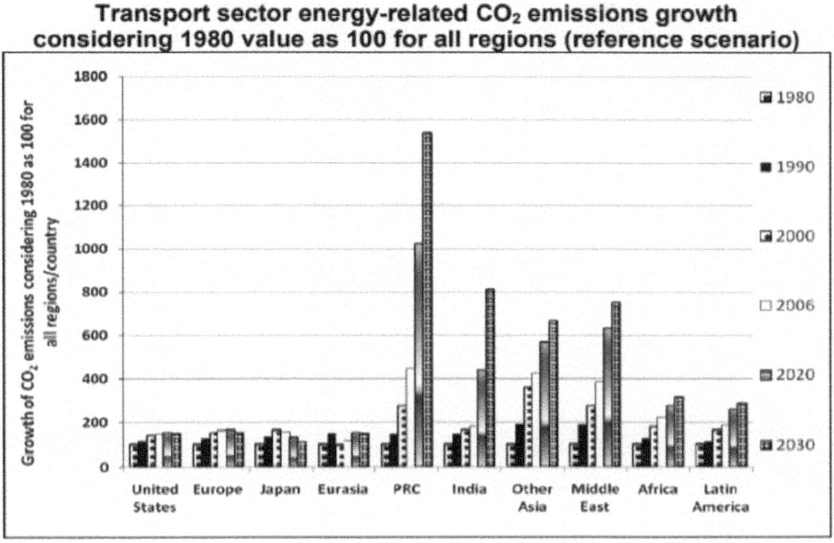


**Fig. 2** $CO_2$ emissions projected growth with area [2]

**Table 1** $CO_2$ emissions from different countries [1]

| Country | Total $CO_2$ emissions (million metric tonnes) |
|---|---|
| China | 10,709.48 |
| United States of America | 5011.69 |
| India | 2533.63 |
| Japan | 1239.60 |
| UK | 367.86 |

## 3 Indian Scenario with $CO_2$ Emissions

In India, the number of vehicles and their emissions in different states varies accordingly. Data from transport research wing from Ministry of Road Transport Highways presents that the total number of vehicles sold in the year 2012 is approximately 16 crores, which includes two wheelers, three wheelers, cars, jeeps, taxis buses and goods carrier vehicles. The total revenue from road vehicles is close to Rs. 133,840 crores in the year 2012–13 from sales tax, accessories and import and excise duty [3]. The road transport sector has contributed about 94.5 and 53.3% of total transport emissions of $CO_2$ and CO. Maharashtra is the most $CO_2$ emitting state in India (11.8%), and it is followed by Tamil Nadu (10.8%), Gujarat (9.6%), Uttar Pradesh (7.1%), Rajasthan (6.22%) and Karnataka (6.19%). In these six states, 51.8% $CO_2$ is from road transport. Vehicular emissions in major metropolitan cities are shown in Fig. 3. Table 2 shows $CO_2$ emissions in different categories in India [3].

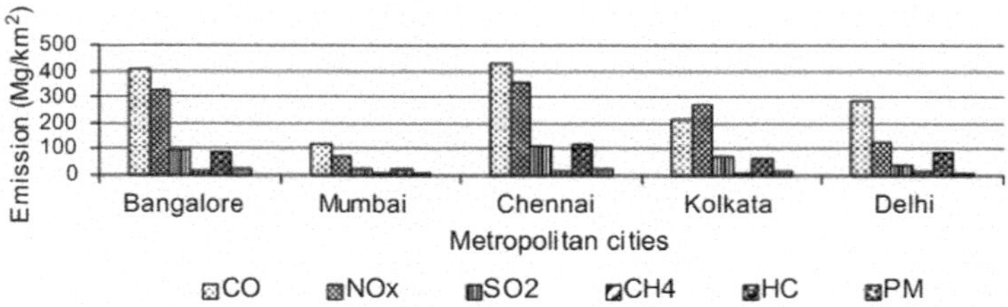


**Fig. 3** Vehicular emissions in major metropolitan cities. *Source* Ramachandra [3]

**Table 2** Emissions from different types of vehicles in India [3]

| Categories | $CO_2$ emissions (in tonnes) |
|---|---|
| Bus | 28,748.16 |
| Omni buses | 8508.42 |
| Two wheelers | 8701.08 |
| Light motor vehicles (passengers) | 4378.10 |
| Car and jeeps+ | 23,901.22 |
| Taxi | 2367.08 |
| Trucks and lorries | 70,288.92 |
| Light motor vehicles (goods) | 44,654.48 |
| Trailers and tractors | 46,563.85 |
| Others | 5705.22 |

## 4 Effect of Fossil Fuel Combustion and Engine Exhaust Emission

The basic parameter that fuel should have is that it should produce a moderate amount of heat when it is combusted. Most of the fossil fuels are carbon-based fuels and during proper combustion produce large amounts of $CO_2$, and improper combustion produces carbon monoxide which is toxic gas and more hazardous to the environment and human health. Combustion of coal and diesel also produces $NO_x$ and sulphur dioxide causing more health problems like cancer, mental health problem, asthma, infectious disease, psychological trauma, etc. The increase in the amount of $CO_2$ causes the earth to warm up and the oceans to become more acidic. From 1990 to 2016, greenhouse gas emissions have increased by 40%. By using natural gas, air pollution can be decreased to some extent, but during transportation of methane, if leakage occurs, it is 34 times more dangerous than $CO_2$. Global coal consumption is likely to decline between 2015 and 2021 [4].

## 4.1 $CO_2$ Share from Carbon-Based Fuels

Burning 1 liter of gasoline in an automobile produces approximately 2.3 kg of $CO_2$, normally when 2000 liter gasoline is burned every year and releases about 4600 kg of $CO_2$. During combustion, hydrogen from fuel and oxygen from the air give water, and similarly, carbon from fuel and oxygen from the air give carbon dioxide and the generated heat gives mechanical energy, therefore oxygen from the air makes exhaust gas heavier.

$$C_XH_Y + O_2 + \text{Spark} \rightarrow H_2O + CO_2 + \text{Heat}$$

For denser hydrocarbon, fuels like diesel which has more carbon produce more carbon dioxide for the given volume of fuel. Gasoline is a complex mixture with several types of hydrocarbon molecules. It varies with the crude source, age of the product and storage condition. If hydrocarbon is not fully oxidized, it produces carbon monoxide [5]. Tables 3, 4 and 5 consist of $CO_2$ per unit for carbon based fuels, $CO_2$ tail pipe emissions and carbon dioxide emission coefficients.

**Table 3** $CO_2$ per unit for carbon-based fuels

| Fuel type | kg of $CO_2$ per unit (mmBtu) |
|---|---|
| Natural gas | 53.06 |
| Diesel fuel | 73.84 |
| Petrol | 75.36 |
| Coal | 113.67 |

*Source* Emission factors for greenhouse gas inventories [6]

**Table 4** $CO_2$ tailpipe emissions

| Fuel type | $CO_2$ tailpipe emission (kg/l) |
|---|---|
| Gasoline | 2.29 |
| E10 (10% ethanol and 90% gasoline) | 2.21 |
| E85 (85% ethanol and 15% gasoline) | 1.61 |
| Diesel | 2.65 |
| B5 (5% biodiesel and 95% diesel) | 2.65 |
| B20 (20% biodiesel and 60% diesel) | 2.62 |

*Source* Autosmart; fuel consumption and $CO_2$ (https://www.nrcan.gc.ca/sites/www.nrcan.gc.ca/files/oee/pdf/transportation/fuel-efficient-technologies/autosmart_factsheet_6_e.pdf)

**Table 5** Carbon dioxide emission coefficients [7]

| Carbon dioxide factors | Pounds $CO_2$ (per unit volume or mass) | Kilograms $CO_2$ (volume or mass) |
|---|---|---|
| Propane | 12.70 gallon | 5.76 gallon |
| Butane | 14.80 gallon | 6.71 gallon |
| Butane/propane mix | 13.70 gallon | 6.21 gallon |
| Home heating and diesel fuel (distillate) | 22.40 gallon | 10.16 gallon |
| Kerosene | 21.50 gallon | 9.75 gallon |
| Coal (all types) | 4631.50 short tonne | 2100.82 short tonne |
| Natural gas | 117.10 thousand cubic feet | 53.12 thousand cubic feet |
| Gasoline | 19.60 gallon | 8.89 gallon |

## 4.2 *$CO_2$ Emission in Vehicles*

The average gasoline vehicle has a fuel economy of about 22 miles/gallon and drives around 11,500 miles per year, and when every gallon of gasoline burns, it creates about 8887 g of $CO_2$. A normal passenger vehicle emits about 4.6 metric tonnes of carbon dioxide per year. $CO_2$ emitted by tailpipe for driving one mile is 404 g. Mostly, the vehicles create tailpipe gas emissions, by converting the fuel to vehicle power generated but it also creates greenhouse gas. The electric vehicle has no tailpipe emissions but the production and distribution of electricity used to fuel the vehicle will create indirect emissions. Figure 4 shows emissions for different automobile manufacturers with changing fuel economy. In the year 2017, Honda has led the industry with the lowest $CO_2$ emissions and the highest fuel economy (https://www.nrcan.gc.ca/sites/www.nrcan.gc.ca/files/oee/pdf/transportation/fuel-efficient-technologies/autosmart_factsheet_6_e.pdf).

## 4.3 *Strategies Taken by Government of India to Reduce $CO_2$ in Automobile Sector*

India is facing many challenges for low-carbon energy transformation and taking several initiatives to reduce carbon dioxide emissions. Indian railways are the largest electricity consumer, and they are trying to reduce carbon emissions as well as to improve energy by reducing fossil fuels and increasing renewable energy capacity. Bharat V emission standards, skipped and Bharat VI emission standard will be implemented directly by 2020 to reduce emissions. According to India's Biennial Update Report (BUR) to reduce industrial carbon intensity by 30% as in 2005, it can achieve this by 2.5 billion tonnes of carbon sequestration through tree plantation.

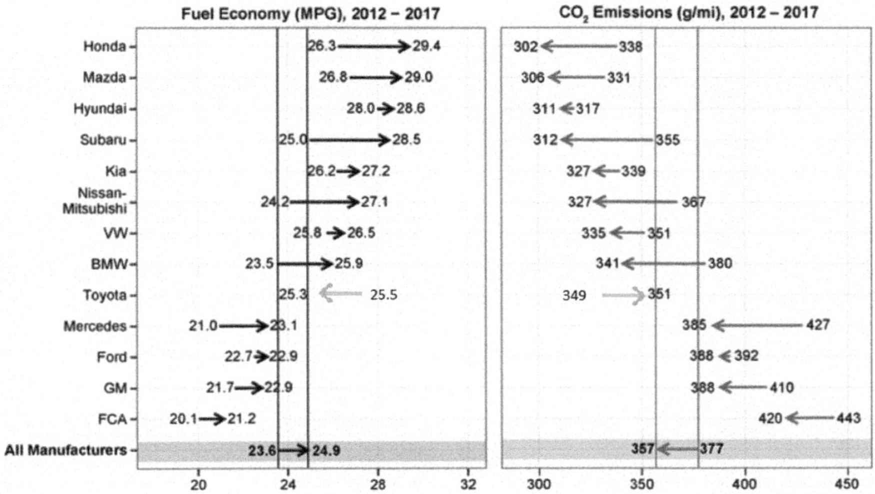


**Fig. 4** Changes in estimated real-world fuel economy and $CO_2$ for large manufacturers (https://www.nrcan.gc.ca/sites/www.nrcan.gc.ca/files/oee/pdf/transportation/fuel-efficient-technologies/autosmart_factsheet_6_e.pdf)

## 5 Feasible Technologies Controlling $CO_2$ Emissions

To prevent the climatic change due to $CO_2$ emissions, the approaches like the reduction of energy generated by fossil fuels, generating energy by using non-fossil fuel energy sources like nuclear energy, solar energy, wind energy and using carbon capture and storage/utilization technology are taken. Usage of low-carbon fuels can also help to minimize $CO_2$ emissions. Low-carbon fuels or clean fuels are fuels which are produced in a sustainable manner and result in lower-carbon pollution in comparison with fossil fuels. Advanced biofuels can be classified under low-carbon fuels. The International Council on Clean transportation has set low-carbon fuel standards (LCFS) which are performance-based regulations aimed to reduce the life cycle carbon intensity of the transport fuel supply [8]. By implementing minimum carbon reductions on fuel suppliers which increase over time, they force some combination of the adoption of low-carbon alternative fuels and reductions in the carbon intensity of conventional fossil fuel supplied.

Today many companies are taking efforts to minimize $CO_2$ emissions. India's climate change report 2017 based on responses from 51 companies suggests that in these 51 sectors about 275.92 million tonnes $CO_2$ is contributed of greenhouse gas emissions. Six companies joined the World Bank's carbon pricing leadership coalition which includes Infosys, YES bank, Mahindra & Mahindra, Dalmia cement, Arvind and Hindustan Construction Company [9]. India has prepared a national action plan for climate change and to reduce emission intensity of GDP by 20–25%, over 2005 levels, by 2020 [10].

At present, fossil fuels like coal, oil and natural gas supply around 85% of world's energy needs, and these are the major source of $CO_2$. International energy agency predicts that fossil fuels will be widely used around the world, so the demand for energy will be increasing and also $CO_2$ emissions will be more. To reduce atmospheric concentration of $CO_2$, it needs many solutions to reduce emissions, including energy efficiency and also to use less carbon-intensive fuels and also to produce renewable energy from wind, earth, sun and tides. The CCS or $CO_2$ sequestration plays a major role in reducing greenhouse gas emissions, and this is the current technology to reduce emissions in automobiles. It can reduce 90% emissions from fossil fuel burning power stations which can be gas or coal fired. Around 2000 CCS facilities would be needed around the world by 2040 to meet global climate targets [11].

In CCS technology, $CO_2$ is separated from the source and used either in other processes or it is stored in underground and ocean storage. The important technologies for $CO_2$ capture are pre-combustion, post-combustion and oxy fuel combustion.

In pre-combustion process, the carbon dioxide is removed before combustion takes place, in the upstream treatment of fossil fuels before their combustion for the recovery of heat (via steam), or the production of electric power or hydrogen (Fig. 5). This is also called as decarbonization of fuels. In such cases, the fossil fuel is partially oxidized, for instance in a gasifier. The resulting syngas (CO and $H_2O$) is converted into $CO_2$ and more $H_2$. The resulting $CO_2$ can be captured from a relatively pure exhaust stream. The $H_2$ can be used as a fuel. The technology for pre-combustion is widely applied in fertilizer, chemical, gaseous fuel ($H_2$, $CH_4$) and power production.

In oxy combustion technique (Fig. 6), the fuel is burned in the presence of oxygen which eliminates the nitrogen from the flue gases and combustion temperature of about 2500 °C. The net flue gas has a concentration of 80–90% $CO_2$. The oxy-fuel combustion $CO_2$ capture system has an efficiency of 100%. This process involves the usage of no extra solvents, it simply consists of compression and cooling steps [6].

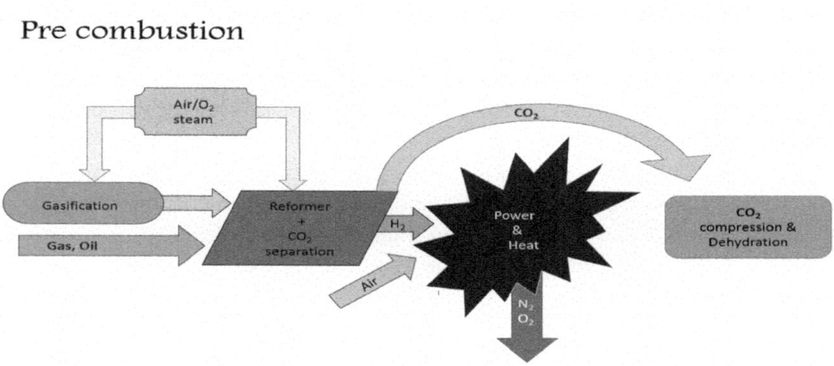


**Fig. 5** Carbon capture through pre-combustion process

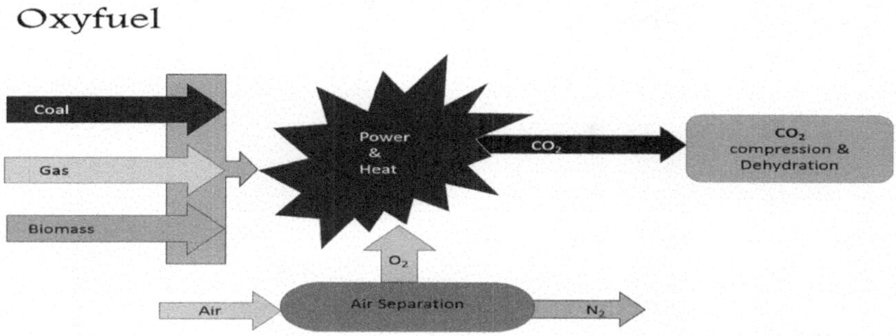


**Fig. 6** Carbon capture through oxy-fuel combustion process

Post-combustion capture (Fig. 7) involves the removal of $CO_2$ from the flue gas produced after combustion. Existing power plants use air, which is almost 78% of nitrogen, for combustion and generate a flue gas which is at atmospheric pressure and typically have a $CO_2$ concentration of less than 15%. The thermodynamic driving force for $CO_2$ capture from flue gas is low, post-combustion carbon capture has the greatest near-term potential for reduction of greenhouse emissions because it can be easily retrofitted to existing units that generate two-third of the $CO_2$ emissions in the power sector. $CO_2$ capture in industrial processes using fossil fuel is shown in Fig. 8. Post-combustion carbon capture can be further classified as shown in the flow chart (Fig. 9).

In the adsorption process, $CO_2$ first selectively sticks to the surface of the material without forming a chemical bond while other gases pass through. This happens under either increased pressure or decreased temperature. In the second phase, the $CO_2$ is

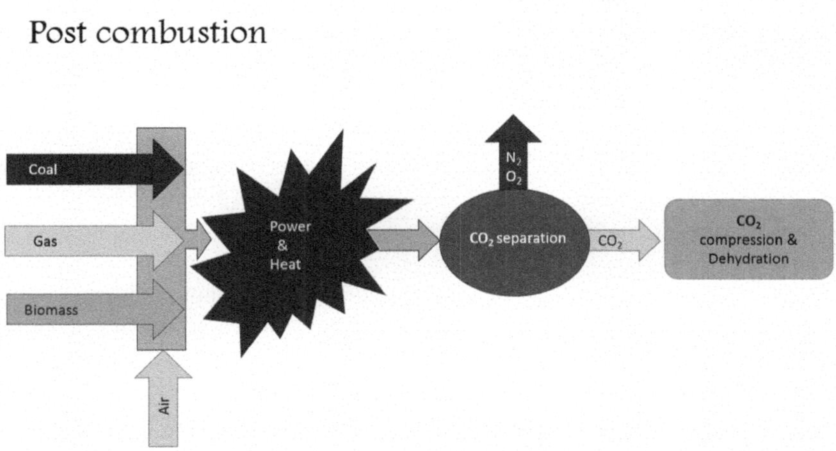


**Fig. 7** Working of carbon capture through post-combustion process

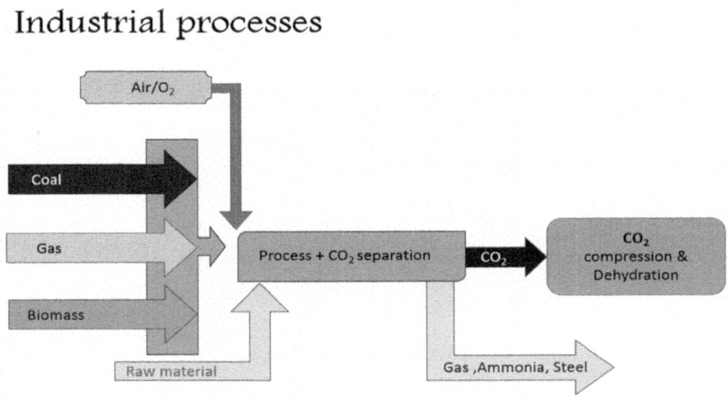


**Fig. 8** Carbon capture in conventional industrial process

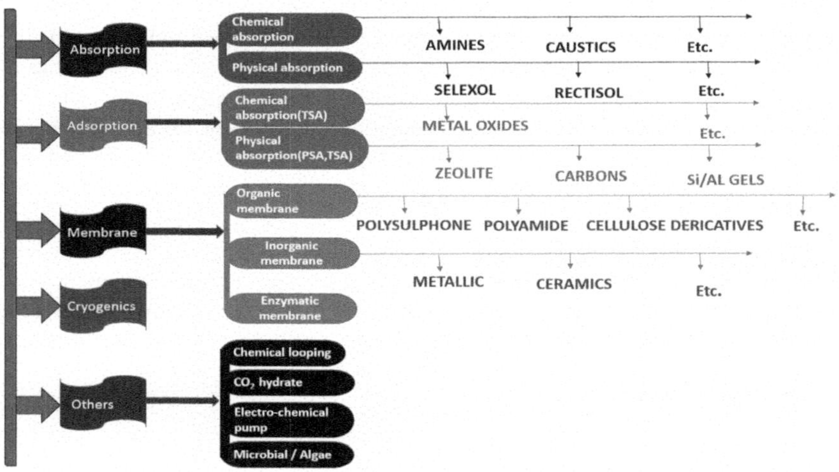


**Fig. 9** Various post-combustion CCS technologies

separated by reduction of pressure or increasing the temperature, allowing the $CO_2$ to be drawn off.

Zeolites are used in the petrochemical and petroleum refining industries as ion exchangers and adsorbents (Fig. 10). The characteristics of zeolites that are of interest when partitioning an analysed gas from a mixture of gases, such as the separation of $CO_2$ from air, include uniform molecular pore size, polarity, reversible and selective adsorption (different cationic forms of zeolite can lead to significant differences in the adsorption of a given gas) and adsorption capacity. Desorption of a gaseous adsorbate from zeolites can be effectively controlled by the application of adequate temperature or pressure otherwise a hysteretic effect may occur. Finally, zeolites

**Fig. 10** Zeolite pellets and $CO_2$ trap

possess a high adsorption capacity at ambient temperature and pressure, even at low adsorbate concentrations.

Many research works carried out on zeolites suggest that the porous nature of the material locks the $CO_2$ molecules inside them [12]. The properly designed trap has to be fabricated based on the $CO_2$ flow rate and tested on a single-cylinder diesel engine. The Trap filled with zeolites shows the positive results on $CO_2$ absorption. By the optimized traps, it can absorb approximately 60% of $CO_2$ emissions from the single-cylinder diesel engine (Fig. 11).

$CO_2$ absorption using chemical reaction is a common process in the chemical industry, which has been applied in the treatment of industrial gas streams containing pollutants like $H_2S$, $NO_x$ and $CO_2$. In absorption, exhaust gases are first passed through a liquid medium into which the carbon dioxide selectively dissolves. A second step is required to remove the carbon dioxide from the solution. This is generally done by heating the solution to remove the carbon dioxide. This method is

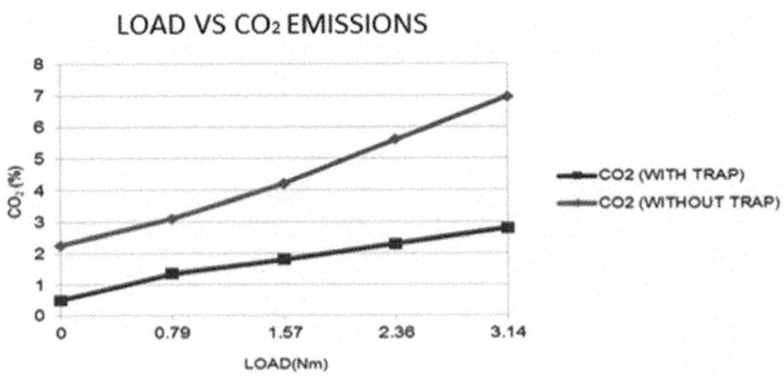


**Fig. 11** $CO_2$ reduction by adsorption with zeolite trap [13]

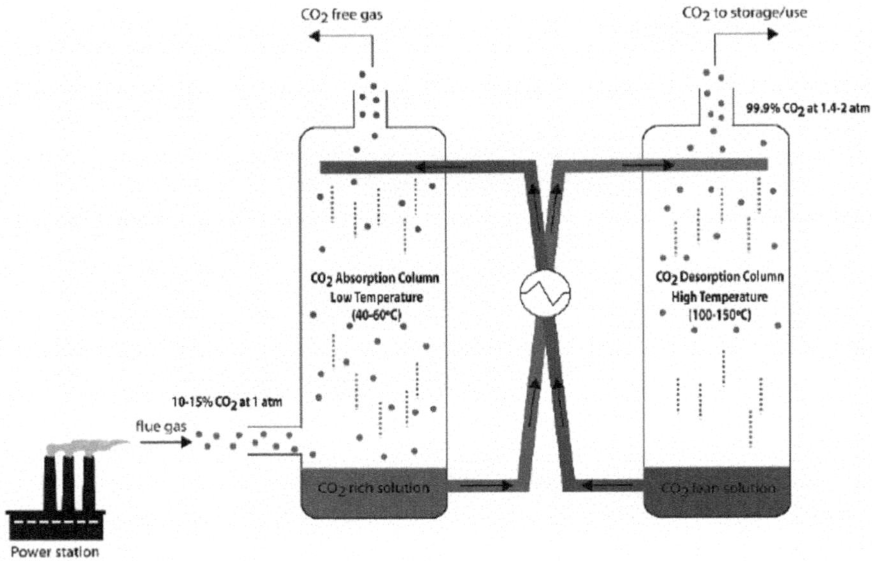


**Fig. 12** $CO_2$ reduction using absorption using amines

commonly used for carbon capture on a small scale and is being adapted for use in large-scale coal-burning electrical power operations (Fig. 12).

Metal amines are commonly used to remove $CO_2$ from natural gas through the absorption process; the majority of the captured $CO_2$ will result in the formation of bicarbonates in the liquid amine capture system. Phase change solvents, a new class of solvents, have emerged and been developed into one of the most promising technologies for $CO_2$ capture. Such phase change technology removes $CO_2$ from power plant flue gases using a solvent which when it reacts with $CO_2$ rapidly forms two distinct phases: a $CO_2$-rich phase and a $CO_2$-lean phase. Only the $CO_2$-rich phase will then undergo regeneration. By regenerating only the $CO_2$-rich phase, significantly less energy may be needed for the whole process.

From Fig. 13, it is seen that the metal amines like L aspartic acid have very good absorbent properties because at low and high load condition 90% of $CO_2$ emission were reduced due to the higher molecular weight. At low load conditions, the 2.45 percentage per volume of $CO_2$ emission reduced, and at high load conditions, 5.88 percentage per volume of $CO_2$ emissions were reduced. The L serine + L aspartic acid is a very sensitive absorbents, it absorbs 90% of $CO_2$ emissions from the exhaust at low load conditions due to the amino acids are basic material, and it absorbs the acidic gases. At low load condition, 1.2% per volume and at high load condition, 3% per volume of $CO_2$ emissions were reduced by the L serine + L aspartic acid combination.

In the membrane separation process (Fig. 14), $CO_2$ is separated from other exhaust gases using a semipermeable membrane which allows $CO_2$ to pass through more

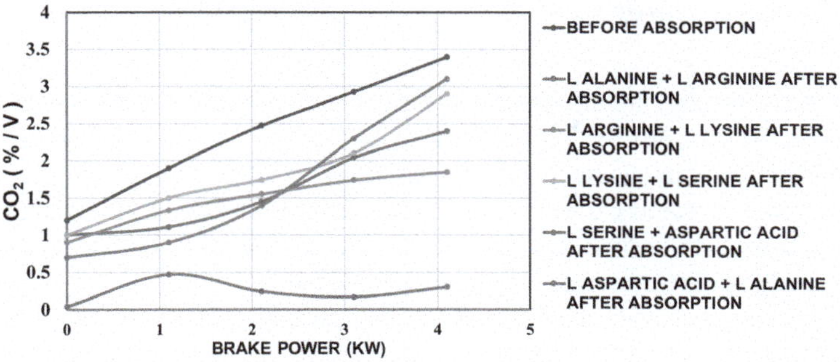


**Fig. 13** $CO_2$ reduction using absorption using ammines [13]

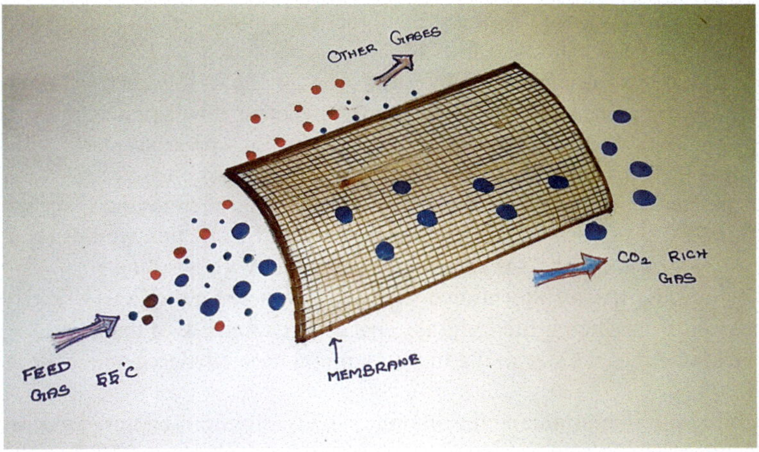


**Fig. 14** Membrane separation process for $CO_2$ reduction [14]

easily than other gases in the exhaust stream. The separated $CO_2$ is then captured for later storage. This process requires high pressure to drive the separation. Membranes, which generally consist of thin polymeric films, owe their selectivity to the relative rates at which chemical species permeate. Differences in permeation rates are generally due to the relative sizes of the permeating molecules, their solubilities and/or diffusion coefficients in the membrane material. Because permeation rates vary inversely with membrane thickness, membranes are made to be as thin as possible without compromising mechanical strength.

### 5.1 *Concerns About Carbon Capture in Vehicles*

Use of CCS technology in automobile vehicles can reduce global carbon emissions because automobiles are the second-largest source for increase in global carbon emissions. Some of the various absorbents used in CCS technology are zeolite, membranes and activated carbons control $CO_2$ emissions [15]. Molecular sieve 5A of 1.5 mm is used to reduce the emission rate from automobile exhaust, and this should be compared with normal engine exhaust [16].

Combustion of one gallon of petroleum is equivalent 8.9 kg of $CO_2$. If all the emitted $CO_2$ is captured after combustion in automobiles, not only will it contribute to an increase in the vehicle's weight, but the volume required for $CO_2$ storage will also exceed the volume taken by the fuel tank. This issue can be solved in a variety of ways, including requiring more frequent offloading of collected $CO_2$, allocation of more storage volume for $CO_2$ or reduction of the amount of $CO_2$ captured during the combustion of one tank of petroleum.

It might assume that storage volume allocation for $CO_2$ is the same as for gasoline storage for illustration. However, it is uncertain whether this is practical, or if some alternative method might be developed to reclaim the space vacated by the fuel tank. Still, if it is assumed that the stored $CO_2$ can be compressed and stored in a compressed state, the resulting volume would be about three times the space required for gasoline. If the same volume is used for fuel and $CO_2$, this would require the vehicle operator to offload $CO_2$ three times per refuel with gasoline.

The second alternative, allocation of additional storage for the $CO_2$, would require the consumer to sacrifice vehicle storage space. To capture all of the $CO_2$, a volume of at least three times the size of the fuel tank might be needed to manage the storage demand.

The third option, minimizing the amount of $CO_2$ emissions captured, reduces the value of the environmental contribution of the carbon capture process. For example, if only 33% of carbon emissions are captured, the resulting accumulated $CO_2$ might occupy the same volume as a tank of gasoline. While customers seem to generally value low emission vehicles, this may be based on a broad heuristic sense that it is generally good to curb consumption of an item known to be in limited supply.

## 6 Carbon Storage

Carbon storage occurs both naturally and as a result of anthropogenic activities. It naturally refers to the storage of carbon that has the immediate potential to be converted into carbon dioxide gas. In consideration to the growing concerns about climate change due to increased carbon dioxide concentrations in the atmosphere, considerable interest has been given to the likelihood of increasing the rate of carbon sequestration through changes in land usage, forestry and geo-engineering techniques such as carbon capture and storage.

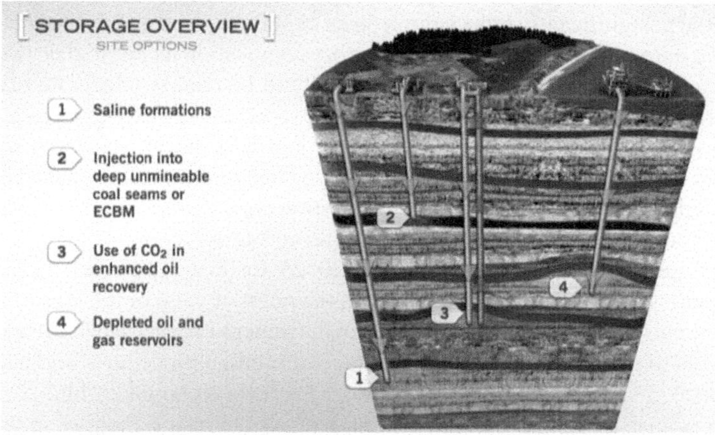


**Fig. 15** Geological storage of $CO_2$ [17]

Rock has enough millimetres sized voids or pores which have the capacity to store $CO_2$. The pores in the rocks are feasibly connected so they have the capability to accept $CO_2$ when it is injected and it spreads. There will be barrier or caprock at the top which helps to store $CO_2$ permanently. Some carbon storage methods are deep saline formation to store $CO_2$ permanently. Coal bed methane, in this $CO_2$, is injected into the coal bed to exchange $CO_2$ with methane; currently, this method is in research and it is not under operation. Another method is enhanced oil recovery (EOR); $CO_2$ is injected to increase oil production. Some depleted oil or gas fields cannot be used for oil or gas production but they can be used for storage (www.gns.cri.nz) (Fig. 15).

## 7 Conclusions

The potential impact from climate change brought about by increasing atmospheric concentrations of greenhouse gases is a global problem that requires urgent global action. $CO_2$ sequestration will need to play an important role within the portfolio of approaches required to achieve a material reduction in $CO_2$ emissions for two reasons as follows:

a. The continued importance of fossil fuels to future energy supply;
b. The scale of $CO_2$ emissions from industries where there are limited other abatement options.

The integration of capture, transport, storage and utilization in full-scale projects is needed to gain the knowledge and experience required for a more widespread deployment of $CO_2$ sequestration technologies. R&D is also needed to improve knowledge of emerging concepts and enabling technologies for $CO_2$ capture that have the potential to significantly reduce the costs of capture for new and existing facilities. Among the capture mechanisms reviewed, post-combustion technologies were given focus because these mechanisms are most likely to be readily adaptable to operate with existing internal combustion engines. Three separation processes were described as follows: absorption, membrane separation and adsorption. Research has been carried out with adsorption materials and phase change Absorption materials know the potential of $CO_2$ capture in internal combustion engines and also the regeneration.

Even though, there is no stringent current norms for carbon dioxide in automobiles, future norms are proposed by 2020–25 for heavy vehicles and tractors. Considerations about operational costs from the point of view of the consumer have to be taken into account, including storage management of captured $CO_2$, additional energy costs to support separation and storage, discharge procedures and additional maintenance costs. All the above considerations have to be taken into account to make $CO_2$ sequestration a viable after-treatment system in automobiles in the forthcoming years. Motivation to use hybrid vehicles and electrical vehicles is required with the utmost priority.

## References


1. Ritchie H, Roser M (2017) $CO_2$ and greenhouse gas emissions: our world in data. Available from: https://ourworldindata.org/co2-and-other-greenhouse-gas-emissions
2. Schipper L, Leather J, Fabian H (2009) Transport and carbon dioxide emissions: forecasts, options analysis, and evaluation. Available from: https://www.indiaenvironmentportal.org.in/files/Transport-CO2-Emissions.pdf
3. Ramachandra TV (2009) Emissions from India's transport sector: statewise synthesis. Atmos Environ 43(34):5510–5517. Available from: https://www.sciencedirect.com/science/article/abs/pii/S1352231009005871?via%3Dihub
4. Sydbom A, Blomberg A, Parnia S, Stenfors N, Sandström T, Dahlen SE et al (2001) Health effects of diesel exhaust emissions. Eur Respir J 17(4):733–746. Available from: https://erj.ersjournals.com/content/17/4/733.article-info
5. Andres RJ, Boden TA, Bréon FM, Ciais P, Davis S, Erickson D, Gregg JS, Jacobson A, Marland G, Miller J, Oda T et al (2012) A synthesis of carbon dioxide emissions from fossil-fuel combustion. Biogeosciences 9(5):1845–1871. Available from: https://backend.orbit.dtu.dk/ws/portalfiles/portal/12504956/Biogeosciences.pdf
6. Environmental Protection Agency (2014) Emission factors for greenhouse gas inventories: stationary combustion emission factors. Available from: https://www.epa.gov/sites/production/files/2015-07/documents/emission-factors_2014.pdf
7. U.S Energy Information Administration (2016) Carbon dioxide emission coefficents. Available from: https://www.eia.gov/environment/emissions/co2_vol_mass.php
8. International Council on Clean Transportation (2019) Low-carbon fuels. Available from: https://www.theicct.org/issues/low-carbon-fuels
9. Goswami U (2017) India on track to achieve set targets under Paris agreement. Economic Times. Available from: https://economictimes.indiatimes.com/india-on-track-to-achieve-set-targets-under-paris-agreement/articleshow/61598846.cms?from=mdr
10. India Environment Portal (2014) KNOWLEDGE FOR CHANGE. The final report of the expert group on low carbon strategies for inclusive growth. Available from: https://www.indiaenvironmentportal.org.in/content/392618/the-final-report-of-the-expert-group-on-low-carbon-strategies-for-inclusive-growth/

11. Energyworld.com (2018) How carbon utilization will be a key step in India's climate strategy. Available from: https://energy.economictimes.indiatimes.com/energy-speak/how-carbon-utilization-will-be-a-key-step-in-india-s-climate-strategy/3024
12. Susan M (2007) Isotopic (13C and 14C) tracers in ecosystem respiration. Available from: https://theses.gla.ac.uk/74022/1/10753931.pdf
13. Senthil Kumar P (2009) Recent advancement in $CO_2$ capture technologies. Available from: https://ccri.in/pdf/acbccu2018/Senthilkumar.pdf
14. Sullivan JM, Sivak M (2012) Carbon capture in vehicles: a review of general support, available mechanisms, and consumer acceptance issues. UMTRI. Available from: https://deepblue.lib.umich.edu/bitstream/handle/2027.42/90951/102855.pdf
15. Jenoris Muthiya S, Amarnath V, Senthil Kumar P, Mohan Kumar S et al (2014) Carbon capture and storage from automobile exhaust to reduce $CO_2$ emission. Int J Innov Res Sci Eng Technol 3(2). Available from: https://www.ijirset.com/upload/2014/special/tapsa/7_TAPSAAUTO001.pdf
16. Balan KN, Valarmathi TN, Reddy MSH, Reddy GA, Srinivas JKS et al (2017) Analysis of $CO_2$, CO and HC emission reduction in automobiles. IOP Conf Ser Mater Sci Eng 197(1):012012. IOP Publishing. https://iopscience.iop.org/article/10.1088/1757-899X/197/1/012012
17. Carbon capture and storage. Available at https://www.gns.cri.nz/Home/Our-Science/Energy-Futures/Carbon-Capture-and-Storage

# Patent Landscape Analysis in Carbon Dioxide Capture Technologies

**Prashrita Kaushal, Sanjeev Majumdar, and Hanumanthu Purushotham**

**Abstract** $CO_2$ sequestration is the process of capturing, transporting and storing $CO_2$ from major emission sources. Alarming levels of $CO_2$ in atmosphere are responsible for creating huge climate change and risks like global warming. To reduce $CO_2$ and other greenhouse levels, world should shift to renewable resources. The Paris Agreement sets goal of reducing increase in temperature of earth which could be achieved by capturing and storage of $CO_2$ in parallel of reducing use of fossil fuels for industries. The $CO_2$ capturing methods are pre-combustion, post-combustion, etc., and storage sites are geological, ocean storage, industrial uses, etc. This chapter presents patent landscape analysis of $CO_2$ capture technologies. Analysis includes patents filed in the field of carbon capture technologies worldwide and their geographical distribution, top countries, top assignees and top inventors in the world and India. It further discusses categorization of patents based on technology domain, commercialized products worldwide and Indian scenario.

**Keywords** $CO_2$ sequestration · Carbon capture · Patent · Assignees · Inventors · Patent technology · CCS facilities

## Abbreviations

| | |
|---|---|
| CCS | Carbon dioxide capture and storage |
| $CO_2$ | Carbon dioxide |
| US | United States of America |
| WIPO | World Intellectual Property Organization |

P. Kaushal · S. Majumdar · H. Purushotham (✉)
National Research Development Corporation, New Delhi, India
e-mail: h.purushotham@gmail.com

 
M. Goel et al. (eds.), *Climate Change and Green Chemistry of $CO_2$ Sequestration*,
Green Energy and Technology, https://doi.org/10.1007/978-981-16-0029-6_7

## 1 Introduction

CCS or $CO_2$ sequestration is the process of capturing, transporting and storing $CO_2$ from major emission sources. According to IPCC(Intergovernmental Panel on Climate Change) special report on $CO_2$ capture and storage, 2014—"Carbon dioxide capture and storage (CCS) is a process consisting of the separation of $CO_2$ from industrial and energy-related sources, transport to a storage location and long-term isolation from the atmosphere" [1]. The alarming levels of greenhouse gases mainly $CO_2$ are creating huge climate change risks like global warming. $CO_2$ levels crossed 400 ppm in 2012 [2] and increasing since and world's primary source of energy is still fossil fuels which is predicted to remain major energy source for at least middle of next century. So, the solution is to replace non-renewable energy sources with renewable energy sources. To achieve the Paris Agreement goal of keeping a global temperature rise, this century is well below 2 °C above pre-industrial levels, and to pursue efforts to limit the temperature increase even further to 1.5 °C, there is a need of capturing and storage of $CO_2$ in parallel to reducing use of fossil fuels for industries. Major $CO_2$ sources are power sector, and other fossil fuel-related industries (refineries, oil and gas industries, metallurgy-related industries) [2]. Carbon capture technologies are required to produce pure $CO_2$ stream at high pressure for transportation. Capturing methods are—*pre-combustion* (capturing carbon before combustion of fuels), *post-combustion* (capturing carbon after combustion of fuels), *oxy-fuel combustion* (capturing carbon during combustion of fuels) and *industrial separation* (natural gas processing or ammonia production). For transportation of captured $CO_2$, pipeline or shipping is used. Storage is done in various ways like.

- *Geological storage*—enhanced oil recovery (EOR), gas or oil fields, saline formations, enhanced coal bed methane recovery (ECBM).
- *Ocean storage*—direct injection (dissolution type), direct injection (lake type)
- *Mineral carbonation*—natural silicate minerals and waste materials
- *Industrial uses of $CO_2$.*

CCS will help in reducing $CO_2$ levels of atmosphere and help in maintaining sustainable climate.

## 2 Patent Analysis

This patent landscape analysis is done for understanding technology trends for $CO_2$ capture technologies for last 15 years (January 2003–August 2018). The focus is on understanding what are most explored technologies for $CO_2$ captures, top countries, top inventors, top assignees and top research institutes of the world carrying out research activities in this field.

The database used for patent analysis is orbit patent database. The search is carried out using various combinations of keywords or search strings. Search strings used are "carbon capture technologies", "capturing carbon", "$CO_2$ capturing" and "carbon dioxide capturing". These search strings resulted in 8080 patents for the period of January 2003–August 2018. Analysis of these patents is the basis for preparing patent landscaping report on $CO_2$ capture technologies.

## 3 Growth in Patents

The growth in published patents with respect to publication year is illustrated in Fig. 1. As it can be seen, there is an astute increase in published patents from 2011 to 2012, and this level was maintained till 2016. After 2016, there was little decline in published patents in 2017 and 2018.

## 4 Geographical Distribution

According to the WIPO, a patent family is "a collection of published patent documents relating to the same invention, or to several inventions sharing a common aspect, that are published at different times in the same country or published in different countries or regions"[3]. The number of patent families protected country-wise throughout the world is shown in Fig. 2. USA has maximum number of patent families (92 patent families) which is followed by China (84 patent families), European Patent Office (48 patent families), Canada (37 patent families), Japan (33 patent families), India (29 patent families), Korea (28 patent families), Denmark (25 patent families), Great Britain (24 patent families), Australia (16 patent families) and Norway (12 patent families) and other countries, respectively. It can also be seen that worldwide India is at sixth position with 29 patent families being protected

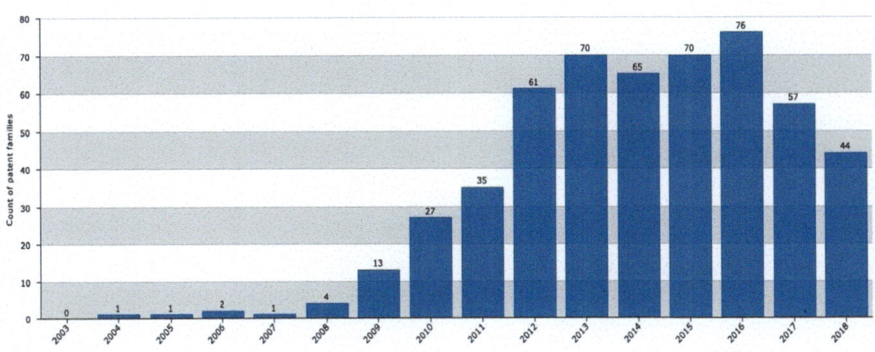


**Fig. 1** Annual growth of patents related to carbon capture technologies with respect to publication year

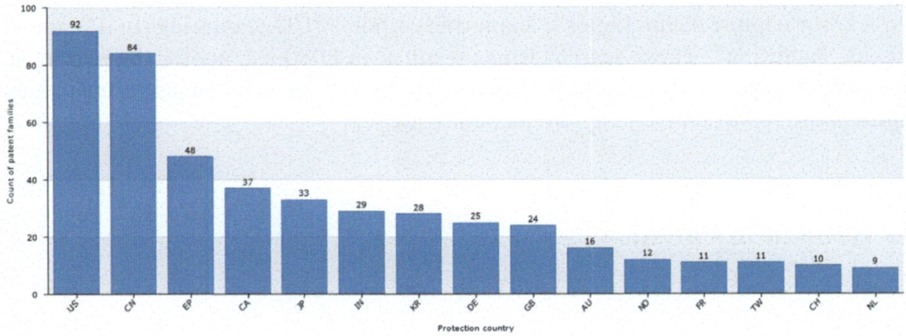


**Fig. 2** Patent families worldwide distribution with respect to protection countries

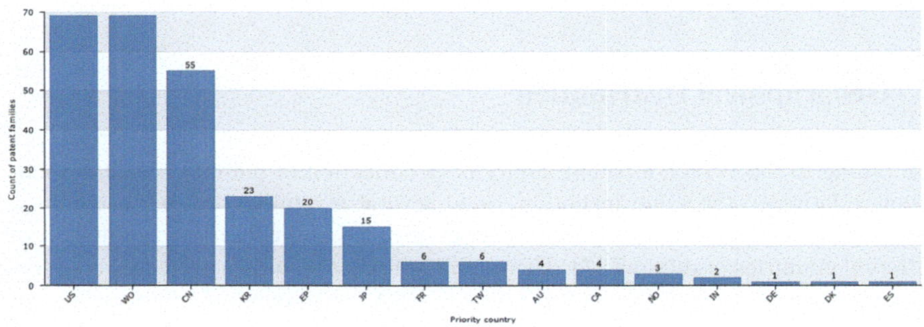


**Fig. 3** Patent families worldwide distribution with respect to priority countries

here. Figure 3 depicts patent families with their respective priority countries. Priority countries are countries where patents are filed first and then they are filed in foreign countries. Maximum number of patent families (69 patent families) are having their priority in USA and WIPO (Patent Cooperation Treaty PCT applications which are filed first in WIPO and subsequently filed in different countries). China is at third position with 55 patent families having priority here, following Korea with 23 patent families, Europe with 20 patent families, Japan with 15 patent families, France with 6 patent families, Taiwan with 6 patent families, Australia with 4 patent families and Canada with 4 patent families.

## 5 Assignee Analysis

The top players in the field of carbon capture technologies around the world are depicted in Fig. 4. Top three assignees of the world are General Electric Technology with 14 patent families followed by Alstom Technology with 10 patent families and

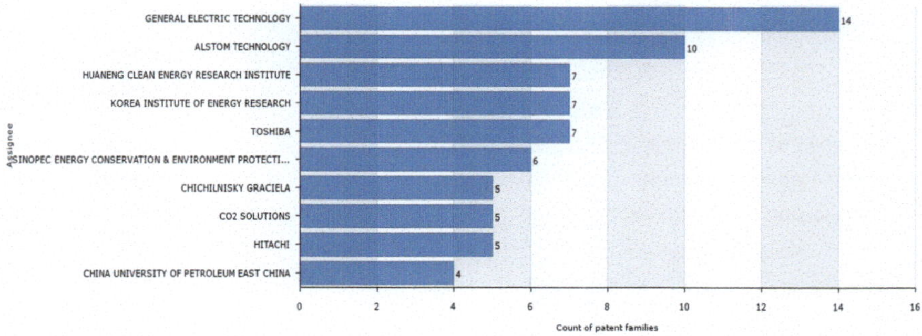


**Fig. 4** Top assignees of the world in the field of CCS

Huaneng Clean Energy Research Institute, Korea Institute of Energy Research and Toshiba each with 7 patent families.

Figures 5 and 6 illustrate top research universities assignees and top institutes assignees in the world. Huaneng clean energy research institute is at top with 37 patent families. Korea Institute of Energy Research (34 patent families), China Petroleum and Chemical (15 patents families), Research Institute of Nanjing Chemical Industry (14 patent families) and Research Institute of Industrial Science and Technology (10 patent families) are in list of top assignees. Tianjin University(13 patent families), Rice University (9 patent families), University of Texas (8 patent families), Zhejiang University (8 patent families) followed by Columbia University, Dalian University of Technology, Southeast University Nanjing and Tsinghua University each with 6 patent families are among top 10 universities as assignees in the world. It is evident from Fig. 7 that corporate assignees have highest percentage (69.63%), while univer-

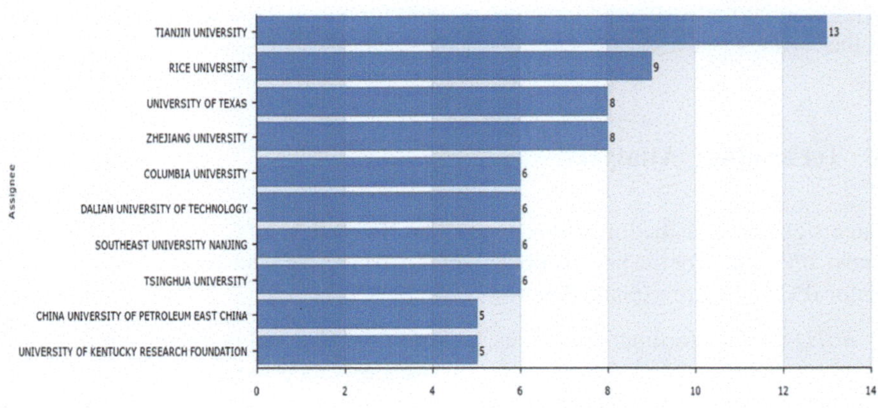


**Fig. 5** Top university assignees of the world in the field of CCS

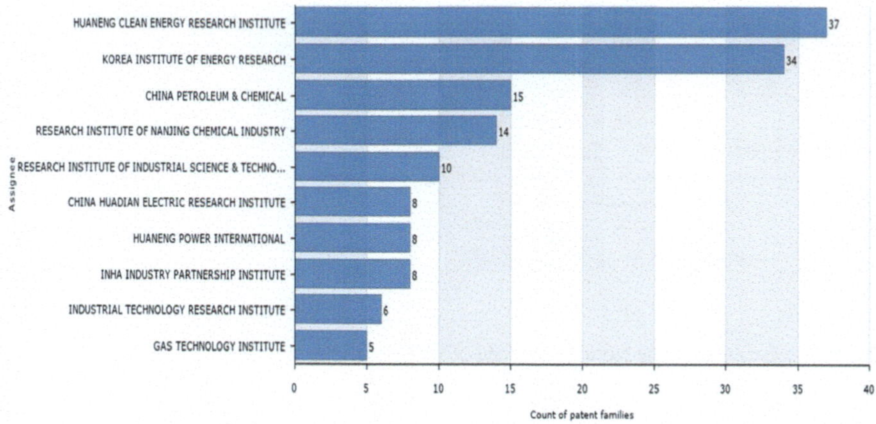


**Fig. 6** Top institute assignees of the world in the field of CCS

**Fig. 7** Categories of assignees worldwide

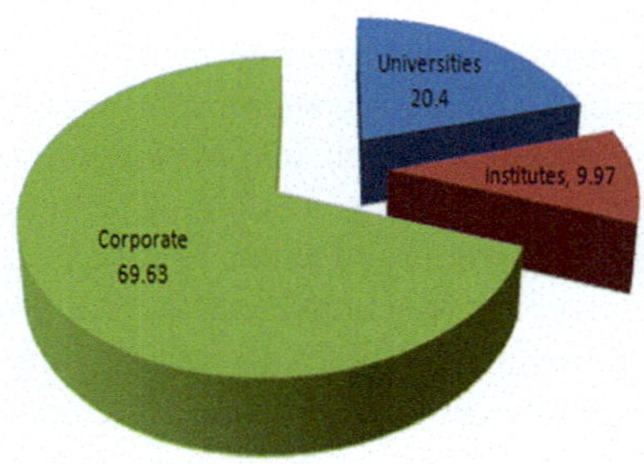


sities account for 20.4% and institutes account for only 9.97% of the total assignees in the world.

## 6 Technology Analysis

Patents can be classified in various ways like IPC codes, CPC codes, etc. Figure 8 shows IPC codes for the patents obtained for carrying out landscaping analysis. The major IPC codes are explained below:

*B01D-053*: Performing operations; transporting, physical or chemical processes or apparatus in general, separation of gases or vapours; recovering vapours of volatile solvents from gases; chemical or biological purification of waste gases, e.g., engine exhaust gases, smoke, fumes, flue gases or aerosols.

*B01D-053/62 (18.27%)*: By chemical or biological purification of waste gases done by removing components of carbon oxide.

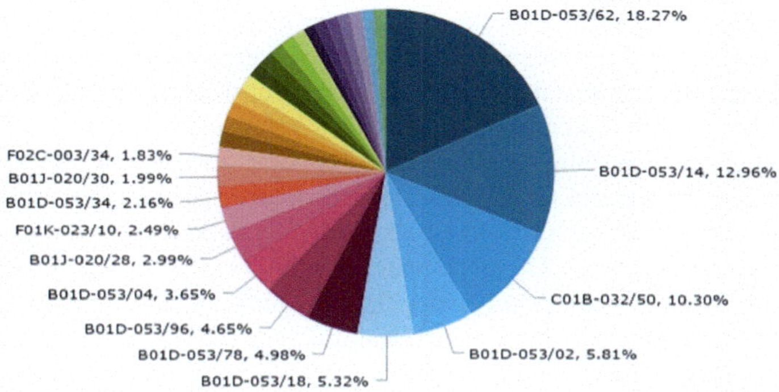


**Fig. 8** Classification of patents based on IPC codes

B01D-053/14(12.96%): By absorption.
*B01D-053/02 (5.81%)*: By adsorption, e.g., preparative gas chromatography.
*B01D-053/18(5.32%)*: Absorbing unit; liquid distributors.
*C01B-032/50 (10.30%)*: Chemistry; metallurgy, chemistry, inorganic chemistry, non-metallic elements; compounds thereof, carbon; compounds thereof, carbon dioxide [3].

Patent families can also be classified based on technology domains. Figure 9 presents technology domain distribution. As it can be seen, 170 patent families come

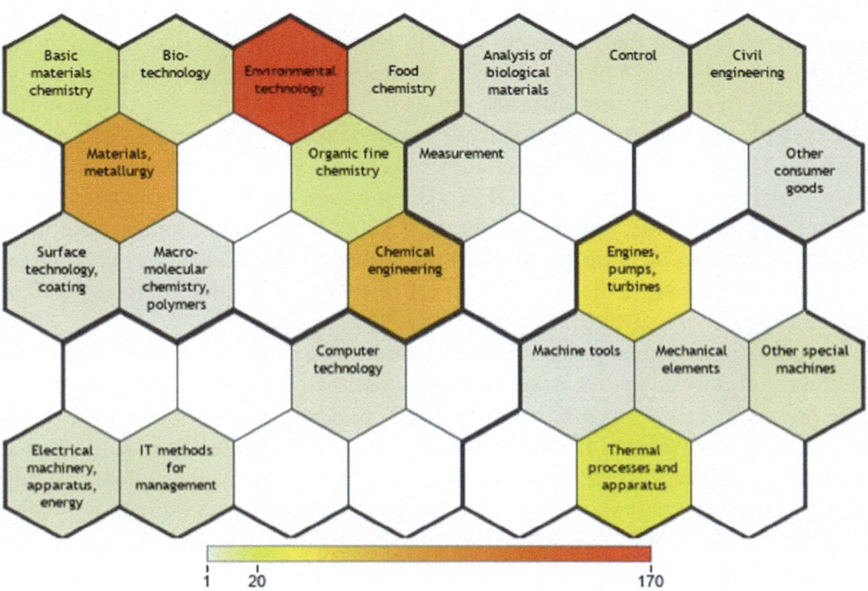


**Fig. 9** Technology beehive map showing patents related to various technology domains

under environment and technology domain. Among these few examples are the regeneration system comprising of regeneration tower, a reboiler and a gas–liquid separator is used for coal-fired power station boiler, and it fully utilizes waste heat of a regenerating solution and provides easy release of $CO_2$ at relatively low pressure. Thus, the regeneration energy consumption of unit $CO_2$ capture is reduced (CN106975324). Some other are solar-based $CO_2$ capture and recovery system. This utility model uses optical energy sequestration of $CO_2$ to effectively capture $CO_2$ and uses photocatalytic to convert capture $CO_2$ to methanol (CN207591581U). Integrated particle fluidized $CO_2$ capture method and device for a power plant have the advantages of being scientific, reasonable, high in applicability, capable of realizing large-scale application, stable and reliable in operation, high-efficiency, energy-saving, low in operation and maintenance cost, etc. (CN107138024).

The next major technology domain is materials, and metallurgy is having 75 patent families. Some of the patents are cement production of $CO_2$ capture device. The advantages that the $CO_2$ trapping device, a particular environmental setting for the production of cement for the filtration, purification, drying or the like, so that the final $CO_2$ purity of the product is high, subject to the collection of $CO_2$, into the air discharge thereof being reduced (CN108217653). A $CO_2$ capture system includes a first capture system including a first absorber, and a second capture system causes $CO_2$ contained in a combustion exhaust gas to be absorbed in a second absorbing liquid (US20170165603). This invention relates to a method for producing sodium carbonate by integration of a $CO_2$ capture process with an ammonia-soda process (EP2757071).

Chemical engineering domain contains 65 patent families. $CO_2$ can be captured by method of preparing an organic–inorganic hybrid compound (KR101638109). There are devices for thermal control of the process of liquification for capturing atmospheric $CO_2$ (ES2549364). Process is developed for formation of a carbon dioxide-selective absorption material and/or sulphur-selective absorption material (EP2214814). Another major category is engines, pumps and turbines (39 patent families). This $CO_2$ capture regeneration system is integrated with a machine set steam–water system where thermal degradation loss of the solution is captured and the operating cost of the system is reduced (CN104791031B). Another example is gas turbine system including a first gas turbine generator, a heat recovery steam generator and a steam turbine generator heat rejection system for capturing carbon dioxide from flue gas (US20170268425).

The domain of thermal processes and apparatus consists of 27 patent families. In this comes oil-derived hydrocarbon converter comprising a catalytic cracking vessel in the presence of catalyst particles in fluidized phase and a regenerator, for regenerating said catalyst particles by burning off the coke deposited on them, said catalyst circulating between said cracking vessel and said regenerator, said regenerator being a reactor integrated into a combustion installation for steam generation comprising $CO_2$ capture (EP1969092). A method of controlling a power system includes monitoring a load on an electrical grid, monitoring a power generation from one or more power producers connected to the electrical grid, wherein at least one of the one or more power producers is a power plant including a $CO_2$ capture system, and utilizing the $CO_2$ capture system of the one or more power producers as an operating reserve in response to an increase in the load on the electrical grid or a reduction in the power generation from one or more power producers connected to the electrical grid (WO201359236).

Environmental technology domain includes patents dealing mostly with improving energy efficiency by using renewable energy or by utilizing waste gases. Metal, metallurgy and chemical engineering domain deal with patents in metallurgy-related industries. Chemical engineering consists of patents capturing carbon dioxide through organic or inorganic compounds or by development of improved absorbents. Engines, turbines and pump domain have patents related to improving efficiency of gas turbines, steam turbines or application of pumps and engines for developing carbon capture plants.

## 7 Commercialized Products

Operational CCS facilities are of different industrial scales and having different end targets like only testing or fully commercialized. According to Global CCS Institute, large-scale integrated CCS facilities are defined as facilities involving the capture, transport and storage of $CO_2$ at a scale of:

- at least 800,000 tonnes of $CO_2$ annually for a coal-based power plant, or
- at least 400,000 tonnes of $CO_2$ annually for other emissions-intensive industrial facilities (including natural gas-based power generation).

At present among total 38 facilities, eighteen (18) are operating, five (5) are in construction, four (4) are in advanced development stage and eleven (11) are in early development stage. The analysis of operational CCS facilities is as follows:

The industry types are ethanol production (1), fertilizer production (2), hydrogen production (2), iron and steel production (1), natural gas processing (9), power generation (2) and synthetic natural gas (1). Countries with number of facilities are USA (9), Canada (3), Norway (2), China (1), Brazil (1), UAE (1) and Saudi Arabia (1) [4].

Various methods of capture type used in these facilities are 16 facilities use industrial separation and 2 use post-combustion method. Also, different methods of storage are implemented in these facilities—14 facilities use enhanced oil recovery (EOR), 2 facilities use dedicated geological storage (offshore deep saline formations) and 2 facilities use dedicated geological storage (onshore deep saline formations).

Apart from large CCS facilities, there are also some other projects. One of them is pilot and demonstration, i.e., lesser-scale CCS projects of various types around the world which provided valuable information to assist in the design and development of large-scale $CO_2$ capture plants, to advance the understanding of the behaviour of $CO_2$ in the subsurface and to contribute to public outreach activities. There are total 88 projects of such kind among which 39 projects focus on capture, 23 on storage and rest are full chain projects. Also 24 projects are operational, 49 are completed, 7 are under construction and 8 are in advanced development stage.

There are also few test centres and other initiatives especially in North America and Europe which helped in development of carbon capture. Last but not least is carbon dioxide utilization project which are getting more attention recently. There are 20 such projects worldwide which are at scale of 400 tonnes per day or more of $CO_2$ use. Some of the utilizations are in food and beverages, residue carbonation, algae cultivation, mineral carbonation and various others [4].

## 8 Indian Scenario

Presently, about 75% of the total primary energy demand in India is met by the fossil fuels. And as for electricity, coal is the largest source of electricity in India and will continue to be the major component in energy mix even by 2050. India ratified the Paris Agreement on climate change on 2 October 2016, marking the occasion of Gandhi Jayanti, which is also the International Day of Non-Violence. India, which is the world's fourth-largest carbon emitter accounting for 4.1% of the total global emission, is the 62nd nation to ratify the agreement. As part of the "Intended Nationally Determined Contribution (INDC)", India has committed to reduce the emission intensity by 33–35% by 2030. The major focus of the mitigation strategy is on improving the energy efficiency and use of renewable sources of energy. The aim is to produce 40% of the total electricity from sources other than fossil fuels [5].

Absence of potential storage sites for carbon dioxide as in India we do not have fully exploited oil reserves. Also, there is reluctance in policies as CCS facilities are very new for the world and there are very few success stories. Thus, policy-makers are hesitant for promoting carbon dioxide storage underground. Cost is also a major issue in implementing capture facilities in power plants. Carbon clean solutions built a plant in Tuticorin in Southern India (in partnership with Tuticorin Alkali Chemicals and Fertilizers) that captures carbon dioxide from its coal-fired boiler and converts it into soda ash. It has filed six patents—five US patents and one international patent (in USA and Europe). Thus, carbon capture and utilization projects are more promising for Indian scenario [6].

The patent landscaping analysis on carbon capture technologies is presented here with respect to Indian scenario. Top assignees in the field of carbon capture technologies India are shown in Fig. 10. Council of Scientific & Industrial Research is at top with 12 patent families followed by Reliance Industries (8), Carbon Clean Sloutions Pvt (3), IIT Madras (3), Indian Oil (3), NTPC (3), Bharat Heavy Electricals (2) and others.

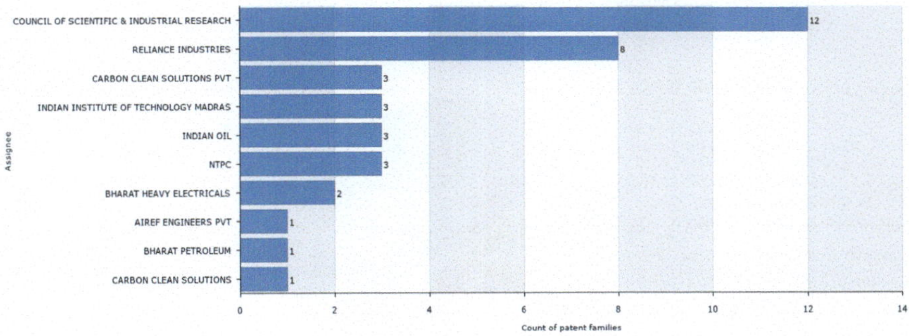


Fig. 10 Top assignees of India in the field of CCS

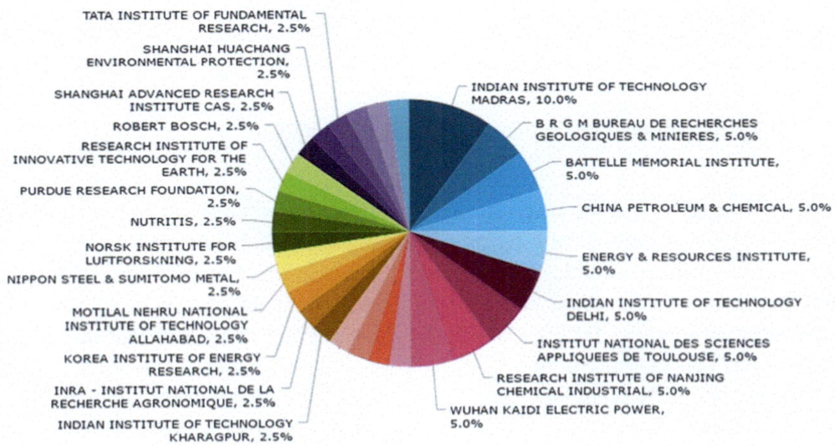


Fig. 11 Top global institute assignees in India in the field of CCS

Top research institutes and universities assignees in the field of carbon dioxide capture and storage (CCS) in India are shown in Figs. 11 and 12. Major institutes and universities are University of California, University of Regina, Jawahar Lal Nehru University, Princeton University, IIT Madras, IIT Delhi, IIT Kharagpur, Tata Institute of Fundamental Research, etc.

# 9 Conclusion

Major inferences drawn from patent landscaping analysis on carbon capture technologies are presented here. Most of the patents are filed in USA, China, Korea,

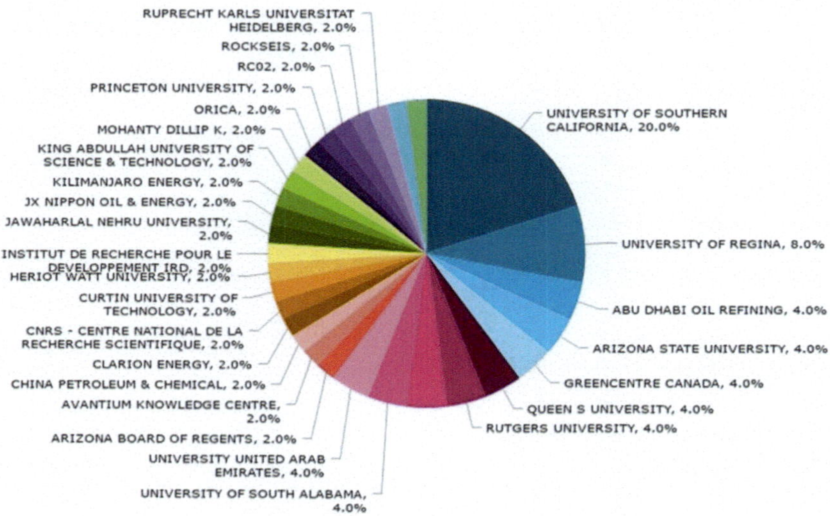


**Fig. 12** Top global university assignees in India in the field of CCS

Europe and Japan. It can be inferred that most explored technologies are technologies related to capturing carbon by process of absorption or adsorption. There was sharp increase (almost double) in published patents in this field during 2009–2012, which has stabilized in recent years. More than 50% of the assignees of the world are corporate assignees. Tables 1 and 2 [4] show the list of operational large-scale CCS facilities and in construction CCS facilities, respectively, all around the world. The most successfully commercialized technology for carbon capture and storage is EOR-based facilities, mainly in USA, Canada and Norway. China is the most recent country to start a large-scale CCS facility (2018), and many more facilities are set to start operating in near future in various parts of the world. As operating these facilities are still expensive and location dependent (EOR) so there is research going for non-EOR-based cost-effective facilities which can also be set up in developing countries. Also, research is directed towards carbon utilization like mineral carbonation (need of building materials) and making products like soda ash. Thus, CCS technology will greatly help in decreasing levels of carbon dioxide in atmosphere and thus creating sustainable environment on earth.

*Note*: *Figures are original and are drawn based on the patents data searched in the orbit patent database.*

**Table 1** Large-scale CCS facilities which are operational worldwide

| Facility name | Location | Operation date | Industry | Capture type | Capture capacity (Mtpa) | Primary storage type |
|---|---|---|---|---|---|---|
| Terrell natural gas processing plant (formerly val verde natural gas plants) | United States | 1972 | Natural gas processing | Industrial separation | 0.4–0.5 | Enhanced oil recovery |
| Enid fertilizer | United States | 1982 | Fertilizer production | Industrial separation | 0.7 | Enhanced oil recovery |
| Shute Creek gas processing plant | United States | 1986 | Natural gas processing | Industrial separation | 7.0 | Enhanced oil recovery |
| Sleipner $CO_2$ storage | Norway | 1996 | Natural gas processing | Industrial separation | 1.0 | Dedicated geological storage |
| Great plains synfuels plant and Weyburn-Midale | Canada | 2000 | Synthetic natural gas | Industrial separation | 3.0 | Enhanced oil recovery |
| SnØhvit $CO_2$ storage | Norway | 2008 | Natural gas processing | Industrial separation | 0.7 | Dedicated geological storage |
| Century plant | United States | 2010 | Natural gas processing | Industrial separation | 8.4 | Enhanced oil recovery |
| Air products steam methane reformer | United States | 2013 | Hydrogen production | Industrial separation | 1.0 | Enhanced oil recovery |
| CoffeyVille gasification plant | United States | 2013 | Fertilizer production | Industrial separation | 1.0 | Enhanced oil recovery |
| Lost cabin gas plant | United States | 2013 | Natural gas processing | Industrial separation | 0.9 | Enhanced oil recovery |

(continued)

**Table 1** (continued)

| Facility name | Location | Operation date | Industry | Capture type | Capture capacity (Mtpa) | Primary storage type |
|---|---|---|---|---|---|---|
| Petrobras santos basin pre-salt oil field CCS | Brazil | 2013 | Natural gas processing | Industrial separation | 1.0 | Enhanced oil recovery |
| Boundary dam carbon capture and storage | Canada | 2014 | Power generation | Post-combustion capture | 1.0 | Enhanced oil recovery |
| Quest | Canada | 2015 | Hydrogen production | Industrial separation | 1.0 | Dedicated geological storage |
| Uthmaniyah $CO_2$-EOR demonstration | Saudi Arabia | 2015 | Natural gas processing | Industrial separation | 0.8 | Enhanced oil recovery |
| Abu Dhabi CCS (phase 1 being emirates steel industries) | United Arab Emirates | 2016 | Iron and steel production | Industrial separation | 0.8 | Enhanced oil recovery |
| Illinois industrial carbon capture and storage | United States | 2017 | Ethanol production | Industrial separation | 1.0 | Dedicated geological storage |
| Petra nova carbon capture | United States | 2017 | Power generation | Post-combustion capture | 1.4 | Enhanced oil recovery |
| CNPC jilin oil field $CO_2$ EOR | China | 2018 | Natural gas processing | Industrial separation | 0.6 | Enhanced oil recovery |

*Source* www.globalccsinstitute.com [4]

**Table 2** Large-scale CCS facilities which in construction worldwide

| Facility name | Location | Operation date | Industry | Capture type | Capture capacity (Mtpa) | Primary storage type |
|---|---|---|---|---|---|---|
| Gorgon carbon dioxide injection | Australia | 2018 | Natural gas processing | Industrial separation | 3.4–4.0 | Dedicated geological storage |
| Alberta Carbon Trunk Line ("ACTL") with agrium $CO_2$ stream | Canada | 2019 | Fertilizer Production | Industrial separation | 0.3–0.6 | Enhanced oil recovery |
| Alberta Carbon Trunk Line ("ACTL") with North West redwater partnership's sturgeon refinery $CO_2$ stream | Canada | 2019 | Oil refining | Industrial separation | 1.2–1.4 | Enhanced oil recovery |
| Sinopec qilu petrochemical CCS | China | 2019 | Chemical production | Industrial separation | 0.4 | Enhanced oil recovery |
| Yanchang integrated carbon capture and storage demonstration | China | 2020 | Chemical production | Industrial separation | 0.4 | Enhanced oil recovery |

*Source* www.globalccsinstitute.com [4]

## References

1. IPCC fifth assessment report on Carbon Dioxide Capture and Storage, 2014
2. https://www.co2.earth/
3. www.wipo.int
4. www.globalccsinstitute.com
5. https://www.carbonbrief.org/the-carbon-brief-profile-india
6. https://carboncleansolutions.com

# Carbon Dioxide Capture and Green Chemistry of Its Conversion

# $CO_2$ Conversion into Chemicals and Fuel: India's Perspective

**Niharika, Yatendra S. Chaudhary, and Suddhasatwa Basu**

**Abstract** India has come up with many policies and programs to take action toward mitigation of carbon dioxide from the energy sector through rapid augmentation of renewable energy growth and energy efficiency improvement. However, there is also need to unify abatement and recycling measures like amplification of low-carbon technologies. Emissions from various industries can be captured and reused for the production of various value-added chemicals and fuels. This chapter briefly discusses all the chemical reactions which require carbon dioxide as a primary product, giving a brief detail of electrocatalytic and photocatalytic pathways in which the anthropogenic carbon dioxide can be activated and subsequently converted into value-added chemicals/fuels. Apart from that, certain issues related to scaling up the technology have been highlighted. The author concludes by putting light on various problems related to the technologies, which need efforts and research so as to achieve practically viable catalysts for $CO_2$ conversion to chemicals and fuel.

**Keywords** Carbon dioxide · Capture · Conversion · Electrocatalyst · Photocatalyst · $CO_2$ emission

## 1 Introduction

Commercial energy sector has been a major driving force in industrial and urban reformation of India since 1991 and has subsequently contributed to the climate change by increased level of greenhouse gases emission. Carbon dioxide ($CO_2$) emission has great concern among greenhouse gases caused by utilization of fossil fuels. In addition to the energy sector, industrial processes and product use (IPPU) sector, agriculture forestry and other land use (AFOUL) and waste also contribute significantly in $CO_2$ emission [1]. The second biennial update report (BURs) submitted to the United Nations Framework Convention on climate change in 2018 states that India has emitted 2607.49 Mt $CO_2$ (excluding land use, land-use change, and forestry

Niharika · Y. S. Chaudhary · S. Basu (✉)
Materials Chemistry Department, CSIR-Institute of Minerals and Materials Technology, Bhubaneswar 751 013, Odisha, India
e-mail: sbasu@immt.res.in



M. Goel et al. (eds.), *Climate Change and Green Chemistry of $CO_2$ Sequestration*, Green Energy and Technology, https://doi.org/10.1007/978-981-16-0029-6_8

(LULUCF)) and 2306.3Mt $CO_2$ (including LULUCF) in 2014 [2]. Out of this total emission, the energy sector alone accounts for 73% emission and others includes, IPPU for 8%, agriculture for 16% and waste sector for 3%, (Fig. 1). India is world's fourth largest emitter of $CO_2$ in the overall global emission of $CO_2$ (i.e., 32.31 Gt in 2016) [3]. There is further increase in the global $CO_2$ emission by 1.5%, when contribution of China, India, and European Union are accounted together (Fig. 2) [4]. In the U.N. Climate Change Conference held in Paris (COP 21, 2015), 195 countries and EU have come to an agreement to limit the temperature rise below 2°C between now and year 2100. In order to achieve this, two long-term goals being considered are: (i) a peaking of emissions as soon as possible and (ii) net greenhouse gas neutrality (a balance between anthropogenic emissions and their removal by sinks) [5]. Therefore, it is vital to reduce such anthropogenic emissions. On the other hand, with the rise in

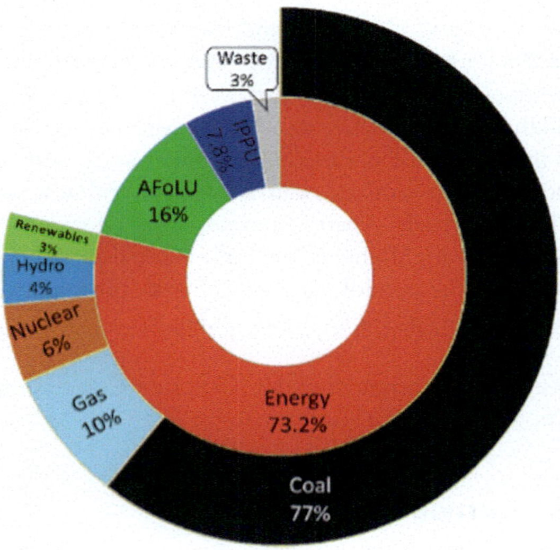


**Fig. 1** Distribution of $CO_2$ emission from different sources in India. Figure plotted using the data from INDIA second biennial update report to the United Nations Framework Convention on Climate Change, 2018. (Accessed from http://envfor.nic.in)

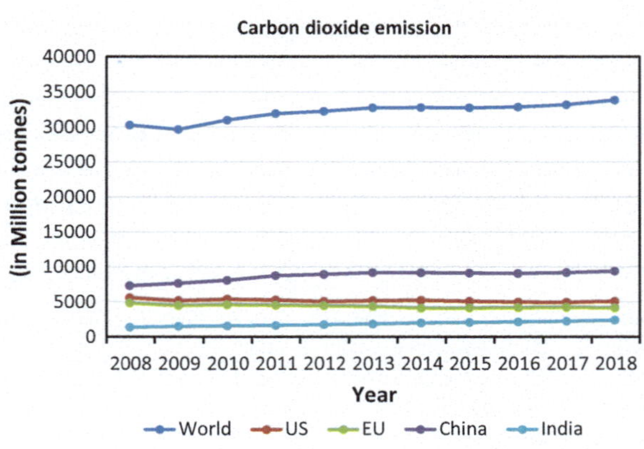


**Fig. 2** Carbon dioxide emission trend over the last 10 years. Figure plotted using the data from BP statistical review of world energy, 68th edition, 2019

the population and sustained techno-economic growth of the developing countries, particularly India, a significant increase in the energy demand is expected over the next decades, and thus, new and renewable energy are required to meet this ever increasing demand. The use of captured $CO_2$ as a feedstock to produce value-added chemicals and fuel can mitigate the problem of anthropogenic $CO_2$ emission and meet the energy demand challenge.

## 2 Initiatives Taken by India to Mitigate Anthropogenic Emission

In order to mitigate this problem, India has pledged to curtail the emission intensity by 33–35% of its gross domestic product (GDP) by 2030, as against of its current c.a. 21% reduction [6]. The Kaya identity expresses the role of the energy sector in $CO_2$ emission, and it relates to a country's emission to the gross domestic product (GDP), total primary energy supply (TPES), and population (Pop) [7] as given below.

$$CO_2 = \left(\frac{CO_2}{\text{TPES}}\right) \times \left(\frac{\text{TPES}}{\text{GDP}}\right) \times \left(\frac{\text{GDP}}{\text{Pop}}\right) \times \text{Pop} \quad (1)$$

The thermal plants being operated in India have a thermal efficiency of about 35% with an emission ratio of 0.90 kg/kWh of $CO_2$ emissions as stated by Central Electricity Authority (CEA) [8]. The reduction of anthropogenic emission by 33% intensity as pledged by India in COP-21 would translate to a decrease of $CO_2$ emissions to ~0.58 kg/kWh by 2030. Such decrease is conceivable only by unified abatement and recycling measures like amplification of low-carbon technologies (such as renewable and clean coal technologies and carbon capture and sequestration). While considering the energy penalty in Indian context, $CO_2$ sequestration is considered more prospective option than that of $CO_2$ storage, and still is of the order of 15%. It can be further reduced to 4–5%, if solar thermal device is used for producing steam for regenerating an amine solvent and stripping of $CO_2$.

In the energy sector, several policies and programs have been implemented to address climate change concerns, which include an ambitious plan to attain 40% cumulative electric power installed capacity from renewable energy resources by 2030, as per the Nationally Determined Contribution (NDC) submitted to Paris Agreement. As a result of India's sustained actions on climate change mitigation, some of the endeavors includes:

- The emission intensity of India's GDP has decreased by 21% over the period of 2005–2014.
- The current share of non-fossil fuel-based electricity generation installed capacity has increased to 35.5% as of June 2018.
- Solar installed capacity in India has increased by about 9 times from 2.63 Gigawatt (GW) to 23.28 GW between March 2014 and August 2018.

- The share of non-fossil sources has increased from 30.5% in March 2015 to 35.5% in June 2018 in installed capacity of electricity generation.

In addition, India aims to achieve 175 GW of renewable energy capacity by 2022 (including 100 GW of solar and 60 GW of wind energy capacity). By 2027, it is estimated to have installed 275 GW of renewable energy capacity which is more than a fourfold increase than that of current capacity [9] (Figs. 3 and 4).

In order to mitigate the $CO_2$, there has been wide interest in carbon capturing utilization and storage (CCUS) concept. As far as the carbon capture and storage (CCS) is concerned, it involves the selective removal of $CO_2$ from gas streams, followed by its compression into a supercritical condition, and subsequently its sequestration in geologic formations (including depleted oil and gas reservoirs or oceans) [10]. However, this process involves high cost that comes primarily from the capture and compression, accounting almost 75% of the total cost and thus has impacted its large scale deployment. The carbon capture and utilization (CCU) approach has gained attention for converting the captured $CO_2$, as feedstock to

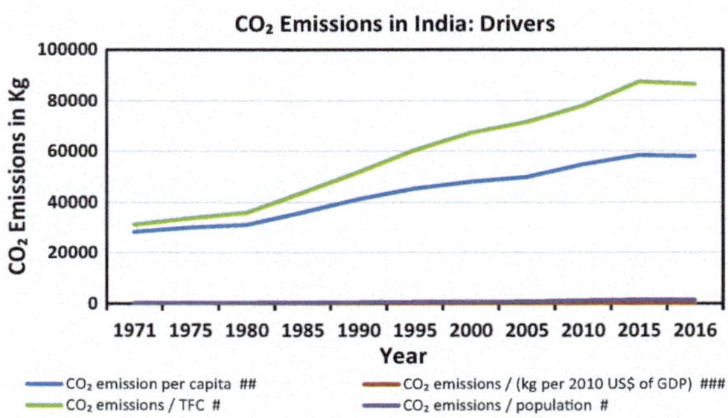


**Fig. 3** $CO_2$ emissions drivers in India. Figure plotted using the data from #OECD, IEA (2018). $CO_2$ emissions from fuel combustion highlights; ##https://knoema.com; ###https://data.worldbank.org (data available up to 2010)

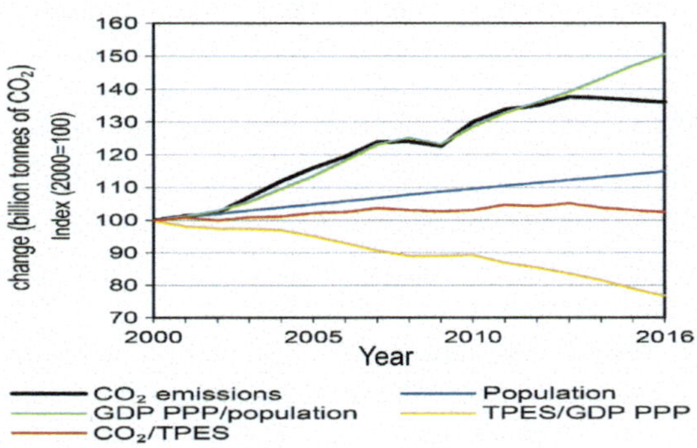


**Fig. 4** Global $CO_2$ emissions drivers. Reprinted with permission. *Source* IEA (2018). $CO_2$ emissions from fuel combustion highlights. All rights reserved

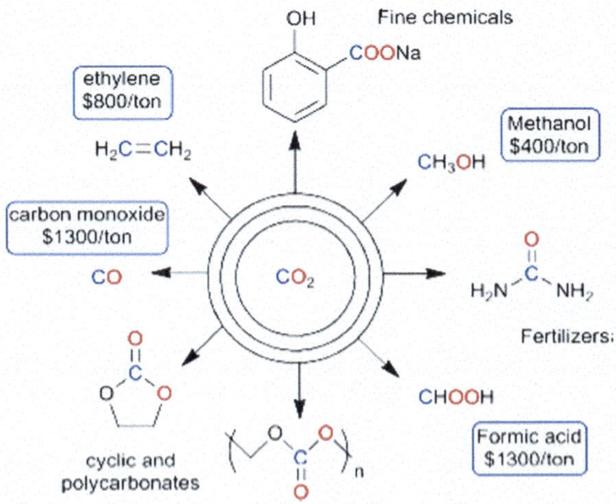


**Fig. 5** Carbon dioxide reduction to value-added chemicals. ([11]. Reproduced by permission of The Royal Society of Chemistry)

produce value-added chemicals and fuel instead of perpetually sequestering it, and can be treated as an alternate to the conventional petrochemical feedstock, Fig. 5.

## 3 $CO_2$ Conversion to Value-Added Chemicals and Fuel

In spite of several merits of CCU in comparison to that of CCS, conversion of $CO_2$ is challenging, mainly because $CO_2$ is linear and thermodynamically stable molecule. Its C = O bond possesses higher dissociation energy of ~750 kJ/mol$^{-1}$, eventually lowering the reduction efficiency of $CO_2$ and it often leads to formations of various products with different oxidation states of the carbon. The single-electron reduction of $CO_2$ is highly endergonic reaction (Eq. 2). However, the multi-electronic reduction comparatively requires less energy than single-electron reduction for the below reactions.

$$CO_2(g) \rightarrow CO(g) + 1/2\,O_2(g), \quad \Delta G^0 = 257\ kJ/mol,\ \Delta E^0 = 1.33\ V. \tag{2}$$

$$CO_2(g) + H_2O(l) \rightarrow HCOOH(l) + 1/2\,O_2(g), \Delta G^0 = 286\,kJ/mol, \quad \Delta E^0 = 1.48\,V. \Delta H^0 = 41.2\,kJ/mol \tag{3}$$

$$CO_2(g) + H_2O(l) \rightarrow HCHO(l) + O_2(g), \quad \Delta G^0 = 522 kJ/mol, \Delta E^0 = 1.35\,V. \Delta H^0 = 795.8\,kJ/mol \tag{4}$$

$$CO_2(g) + 2\,H_2O\,(l) \rightarrow CH_3OH\,(l) + 3/2\,O_2(g), \Delta G^0 = 703\,kJ/mol, \quad \Delta E^0 = 1.21V. \Delta H^0 = 727.1\,kJ/mol \tag{5}$$

$$CO_2(g) + 2\,H_2O(l) \rightarrow CH_4(g) + 2O_2(g), \quad \Delta G^0 = 818\,kJ/mol, \Delta E^0 = 1.06\,V. \Delta H^0 = 890.9 kJ/mol \tag{6}$$

The $CO_2$ capture is undertaken by different routes that include absorption, membrane separation, absorption, and chemical looping; however, this is beyond the scope of this chapter and thus not discussed in detail further. The hydrogenation of $CO_2$ is considered as one of the promising pathway that converts it into carbon monoxide, methane, methanol, and formic acid. Methane is an important constituent of natural gas. However, it is not considered as a valuable fuel, as it has low volumetric gas and high global warming potential. Another promising approach is the dry reforming of methane (DRM) to produce methanol, an important feedstock for the chemical industry and used in a variety of applications in paints, plastics, combustion engines, and organic solvents [12]. The reverse water–gas-shift (RWGS) reaction, converting $CO_2$ into CO, is also appealing, as CO can be used for the synthesis of methanol and a variety of other liquid fuels by the Fischer–Tropsch (FT) process [13]. The reactions that convert $CO_2$ to value-added chemicals can be classified under six reaction categories:

(i) Stoichiometric reactions: These are also known as redox, neutralization, and spontaneous reactions and do not require any external energy. These reactions use internal energy (i.e., chemical potentials) of the reactants to drive the reaction. Such conversion of $CO_2$ in stoichiometric reaction is not considered renewable [14–16].
(ii) Thermo-chemical reactions: These reactions are both non-spontaneous as well as spontaneous and require external energy in the form of heat to drive the reaction. The integration of these reactions with renewable energy resources may offer viable process [17–19].
(iii) Biochemical reactions: These include algae production that mainly utilizes sunlight, but the efficiency of this process is low (<1%) and are not commercially viable. Bio-ethanol production is undertaken by this route. In order to make them commercially viable, efficiency of this process should exceed 10% [20].
(iv) Photoelectrochemical (PEC) reactions: This approach combines the electrocatalytic reduction with the renewable source (solar) that drives selective and highly reduced $CO_2$ reduction products such as methanol, ethanol. However, the instability of semiconductors and the charge carrier separation efficacy limits their potential as a commercially viable process [21–23].
(v) Photocatalytic reactions: These reactions are renewable in nature and drives $CO_2$ conversion to products by exploiting the solar radiation. However, the major challenge lies with the charge-separation efficacy and charge transfer kinetics, as a major fraction of photogenerated electrons are lost by recombination and/or undesired reactions [24, 25].
(vi) Eelectrocatalytic reactions: These are widely studied, and the involved reaction mechanism has been established to a great extent [26, 27] and is discussed in detail, later in this chapter.

In India, the major focus has been on the electrocatalytic and biochemical (by algae) conversion/hydrogenation of $CO_2$ to value-added chemicals/fuel. The biochemical conversion ($CO_2$ microalgae capture and utilization) is discussed in detail in other chapter. The exploitation of renewable sources (solar) to drive these conversion reactions has also been initiated. To get an overview of the current scenario and challenges, a perspective on these approaches is discussed below.

### *3.1 Electrocatalytic Conversion of $CO_2$*

In the electrocatalytic conversion, appropriate catalyst activates and reduces $CO_2$, by participating in an electron transfer reaction and facilitating the acceleration of the desired chemical reaction. A variety of metal catalysts have been explored to convert $CO_2$ that can be categorized into three groups based on product formation: (i) Cu, the only metal capable of reducing $CO_2$ to hydrocarbons (or oxygenated hydrocarbons) at significant rates; (ii) Ag, Au, Zn, Ga, and Pd reduce $CO_2$ to CO as a major product; (iii) Sn, Pb, In, Hg, Cd, Bi, and Tl, mainly produce formate. The metals (Cu, Ag, Au, Zn, Pd, and Ga) are believed to stabilize $*CO_2$ and/or *COOH (*represents a surface adsorption site) on their surfaces, whereas Sn, Pb, In, Hg, Cd, Bi, and Tl produce formate via hydration of non-adsorbing $CO_2$.

The computational studies of the $CO_2$ reduction mechanisms revealed unique property of Cu to catalyze $CO_2$ reduction into hydrocarbons (Fig. 6). The Au and Ag surface binds *CO weakly, thus letting its desorption as CO(g), whereas an intermediate binding strength of *CO on Cu surface enables balance of the barriers for the activation of $CO_2$ and hydrogenation of *CO [28, 29]. Therefore, Cu has received great attention as an electrocatalyst for $CO_2$ reduction than that of other metals. Further investigations on the bulk Cu electrodes with extended surfaces exhibit the dependence of product yield on the electrode potential [30, 31], pressure of $CO_2$ [32–34], electrolyte [35–38], and surface structure [39–42]. Despite of its activity for production of hydrocarbons, it still requires large overpotentials.

A solid oxide fuel cell (SOFC) working in reverse direction can be used as a solid oxide electrolyzer for co-electrolysis of $CO_2$. It works at high temperature

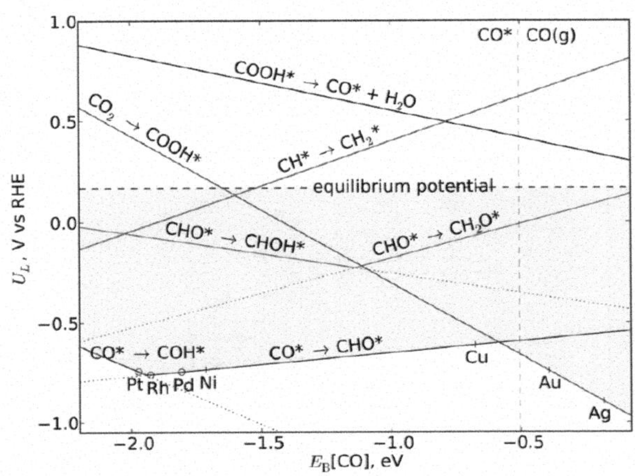


**Fig. 6** DFT calculations illustrating the unique role of Cu among the various transition metals for catalyzing $CO_2$ reduction. The values of EB[CO] on (211) facet of various transition metals are marked to show the most potential-limiting step for a given metal surface. Reprinted with permission from Peterson, A.A and Nørskov, K.J. et al., *J. Phys. Chem. Lett.* 2012, 3, 2, 251–258. Copyright (2012) American Chemical Society

and offers high conversion efficiency and negligible catalyst poisoning. The electrode materials used mainly consist of the cermet of Ni, Pt with yttria-stabilized zirconia (YSZ), $La_{0.9}Sr_{0.1}Ga_{0.8}Mn_{0.2}O_{3-d}$ (LSGM), gadolium-doped ceria [43], $La_{0.6}Sr_{0.4}Co_{0.2}Fe_{0.8}O3_{-d}$ [44] and $La_{1-x}$ $Sr_x$ $Cr_{0.5}$ $Mn_{0.5}$ $O_{3-d}$ [45] for high-temperature $CO_2$ reduction reaction.

Ceria having high oxygen storage capacity has also been extensively studies to drive $CO_2$ reduction. The investigation undertaken suggests important role of the interface of Cu and $CeO_2$ in reducing $CO_2$ to CO and subsequently to methanol. However, whether the extended ceria surface may also activate $CO_2$ to produce methanol remain unclear. The DFT study undertaken by Basu et al., on extended $CeO_2$ (110) surface, examined all suggested mechanisms for the transformation of $CO_2$ to methanol, while considering both carboxyl- and formate (HCOO)-mediated paths. These investigations reveal that the formate intermediate is tightly bound to the ceria surface and thus unlikely to participate methanol formation. On the other hand, the carboxyl-mediated route shows exothermic elementary steps, except the slightly endothermic ($\Delta H = 5$ kJ/mol) dissociation of COOH into CO and OH, and is rate determining as the calculated activation energy is higher (Ea = 126 kJ/mol), Fig. 7 [46]. The progress of major electrocatalyst development in India and their activity are summarized in Table 1.

Recently, enormous interest has grown in designing the transition metal-macrocyclic ligands for $CO_2$ electrocatalytic conversion. Iron 5, 10, 15, 20-tetrakis (2, 6 - dihydroxylphenyl)-porphyrin (Fe TDHPP) shows stable electrocatalytic activity during 4 h, generating CO with a faradic yield above 90%. This is attributed to the high localized proton concentration of phenolic hydroxyl.

Inspired by the metalloenzyme, a cofacial iron (0) tetraphenyl porphyrin dimer reported exhibits highly efficient electroreduction of $CO_2$ [52]. The electrogenerated $Fe^0$ (por) species exhibit excellent CO selectivity with high FE of 95% and a turnover frequency (TOF) of 4300 $s^{-1}$ at an overpotential of 0.66 V. An iridium pincer dihydride electrocatalysts synthesized reduces $CO_2$ to formates

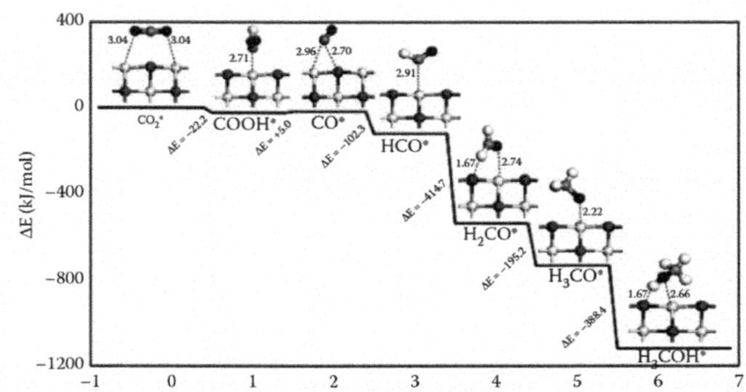


**Fig. 7** Energy diagram of reduction of $CO_2$ to $CH_3OH$ on $CeO_2$ (111) surface. Reprinted with permission from [46] ECS Trans. 68,155. Copyright (2015) The Electrochemical Society

**Table 1** Progress of major electrocatalyst systems designed in India

| Electrode material | Electrolyte | Operating condition | Faradic efficiency (Activity) | Reference |
|---|---|---|---|---|
| *$CO_2$ conversion to CO* | | | | |
| Cu-PDC/LSM | YSZ | 2.5 V | 11% | Kumari et al. [43] |
| Cu-H2L$^{NO2}$ | – | 1.8 V | 25% | Singh et al. [47] |
| Ni-H2L$^{NO2}$ | – | 1.5 V | 74% | Singh et al. [47] |
| 20Sn/ZSM5 | $KHCO_3$ | −2 V | 20.4% | Basu et al. [48] |
| Co-S | | 70 mV | 95%(selectivity) | Dey et al. [49] |
| Fe(0) porphyrin | TsOH | 50 mV | TOFs 1-$10^3$ $s^{-1}$ | Sen et al. [50] |
| *$CO_2$ conversion to $CH_3OH$* | | | | |
| CeO2-X/Cu(II) | – | – | – | Kumari et al. [51] |
| *$CO_2$ conversion to $CH_4$* | | | | |
| 20Sn/$Al_2O_3$ | $KHCO_3$ | −2 V | 12.9 | Basu et al. [48] |

($HCOO^-$) [53], with high selectivity, efficiency, and turnover numbers (TONs ~ 54,200). A series of iron porphyrins that differ only in the distal H-bonding network have been further explored, and these display TOFs ranging from 1.0 $s^{-1}$ to 1000 $s^{-1}$ for converting $CO_2$ to CO [49, 50].

## 3.2 *Photocatalytic Conversion of $CO_2$*

Photocatalysis is considered as a facile approach to activate $CO_2$ and subsequently convert it into useful chemicals, where the adsorption of $CO_2$ onto the surface of photocatalyst lowers its energy barrier. The bent form of $CO_2$ shows high reactivity due to the lowering of the LUMO level of $CO_2$ as the molecule bends. The successive chemical reactions are initiated by one-electron transfer to form the surface-bound $CO_2$ on the photocatalyst. Depending on potential of the charge carriers participating in the chemical reaction, the type of reductant and operating conditions [54], following multiple electron and proton steps determine the various products, Table 2.

In order to drive these reactions, the conduction band edge of the photocatalysts should straddle above the redox potential of the desired reaction. The modification of these photocatalysts with cocatalysts plays vital role in governing the various surface pathways, and thus, the $CO_2$ reduction selectivity also suppresses undesired electron–hole recombination (Fig. 8).

Ever since the first report by Lehn et al. [55] on photocatalytic $CO_2$ reduction using molecular catalyst ruthenium(II) trisbipyridine, $[Ru(bpy)_3]^{2+}$ [55], a variety of strategies are being explored to tune the fundamental processes: activation of $CO_2$ to lower the over potential and improve selectivity, photogenerated electron–hole pair separation and charge transfer kinetics. However, there has been incremental

**Table 2** List of standard reduction potential ($E^0$) of $CO_2$ summarized (at pH = 0)

| Reaction | $E^0$(V vs. NHE) |
|---|---|
| $CO_2 + e^- \rightarrow CO_2^-$ | −1.90 |
| $CO_2 + 2H^+ + 2e^- \rightarrow HCOOH$ | −0.61 |
| $CO_2 + 2H^+ + 2e^- \rightarrow CO + H_2O$ | −0.53 |
| $CO_2 + 4H^+ + 4e^- \rightarrow HCHO + H_2O$ | −0.48 |
| $CO_2 + 6H^+ + 6e^- \rightarrow CH_3OH + H_2O$ | −0.38 |
| $CO_2 + 8H^+ + 8e^- \rightarrow CH_4 + 2H_2O$ | −0.242 |
| $2H_2O + 4H^+ \rightarrow O_2 + 4H^+$ | +0.812 |
| $H^+ + 2e^- \rightarrow H_2$ | −0.42 |

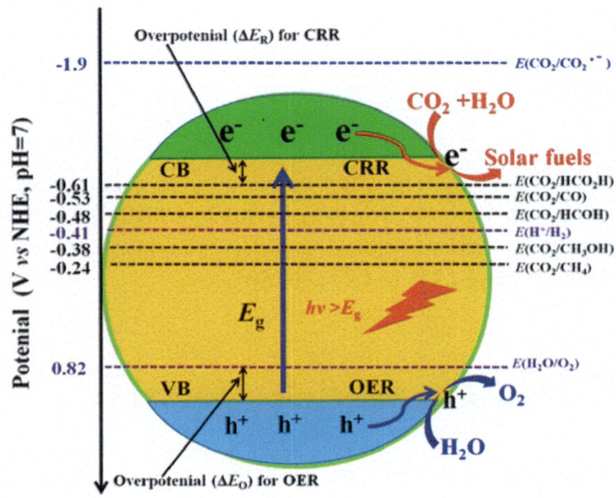


**Fig. 8** Schematic illustration of $CO_2$ photoreduction on a semiconductor. Reprinted with permission from Xin Li, Jiaguo Yu, Mietek Jaroniec, and Xiaobo Chen., *Chemical Reviews*, 2019,119 (6), 3962–4179. Copyright (2019) American Chemical Society.

progress in the photocatalytic $CO_2$ reduction, and still this domain is in embryonic stage.

Li et al. [56] reported the combination of porous $SiO_2$ support and the Cu deposited $TiO_2$ that produces CO and $CH_4$ (60 and 10 μmol $g_{\text{-cat}}^{-1}$ $h^{-1}$, respectively) [56]. Similarly, a variety of photocatalysts such as $RuO_2$ co-catalyst modified mesoporous $ZnGa_2O_4$, MgO-doped $LiNbO_3$, $Zn_{1.7}GeN_{1.8}O$, $HNb_3O_8$ exhibit $CO_2$ reduction to methane [57–61]. The ultrathin $W_{18}O_{49}$ (0.89) nanowires comprising oxygen vacancies exhibit an excellent activity of photochemical reduction of carbon dioxide to methane (2.2 mmol $L^{-1}$ $g^{-1}$ $h^{-1}$) under irradiation with visible light [62]. The porous gallium oxide designed exhibit an efficient photocatalytic conversion of $CO_2$ into $CH_4$, without using any co-particle or sacrificial reagent [62]. The modification of mesoporous ZnO with Cu leads to enhanced activity to convert $CO_2$ to CO, $H_2$, $CH_4$ and $CH_3OH$ as main products [63]. Ag supported on $AgIO_3$ ($Ag/AgIO_3$ particles) display enhanced activity and stability in the photocatalytic conversion of $CO_2$ to $CH_4$ and CO under visible-light irradiation [64]. The hybrid $CuO/Cu_2O$ nanorod arrays drive the photoelectrochemical conversion of $CO_2$ into methanol with the faradic efficiency of 96% at low potential (−0.2 V versus SCE) [65]. Similarly, the $MoS_2$-rods/$TiO_2$-nanotubes together served as photoelectrodes show that their combination can enhance the methanol yield with the recorded faradic efficiency of 111.58% [66].

A variety of molecular catalysts have been designed based on cobalt protoporphyrin perfluorinated cobalt phthalocyanine (CoFPc) to improve selectivity for $CO_2$ conversion to methane [67, 68]. The iron porphyrin molecular catalyst drives the $CO_2$ to CO and $CH_4$ with TONs of 140 and 29 respectively, under visible light radiation (>435 nm). These catalytic systems persistently produce methane even after prolonged irradiation up to four days [69].

Biomimetic or enzyme-based photocatalysis, which is another emerging way to reduce $CO_2$ photocatalytically, has gained interest in recent times [70–73]. Artero and co-workers reported a series of $CO_2$ reducing molecular catalysts comprising Earth abundant elements, which are highly selective to produce formic acid in DMF/water mixtures (FE of 90 ± 10%) at moderate overpotentials (500–700 mV) [74].

There are only a few groups in India involved in undertaking research in the area of photocatalytic $CO_2$ reduction. The progress of major photocatalysts development in India is summarized in Table 3. The N, S, and Fe co-doped strontium titanates show improvement in the photocatalytic reduction of $CO_2$ to C1-C2 hydrocarbons than that of bare strontium titanate. However, they form a mixture of C1-C2 hydrocarbons [75]. A ruthenium trinuclear polyazine complex immobilized through complexation to a graphene oxide support shows selective photocatalytic reduction of $CO_2$ to methanol, with a yield of 3.9 mmol $g_{cat}^{-1}$ in 48 h [76].

The partial reduction of $Ce^{4+}$ to $Ce^{3+}$ in $CeO_2$ leads to the generation of oxygen vacancies [83]. Hazem et al. introduced oxygen vacancies in $CeO_2$ by two different approaches: sunlight-assisted combustion (SAC) and conventional solution combustion (CSC). The SAC-$CeO_2$ (0.702 $\mu$mol $h^{-1}$ $g^{-1}$) shows enhanced photocatalytic $CO_2$ conversion to $CH_3OH$ than that of CSC-$CeO_2$ (0.397 $\mu$mol $h^{-1}$ $g^{-1}$). Such improvement in the photocatalytic reduction of $CO_2$ using SAC- $CeO_2$ is explained by photothermochemical cycle [77].

A metal-free catalyst g-$C_3N_4$ developed with various modifications using rationally chosen precursor—cyanamide—exhibits $CO_2$ photoreduction to $CH_4$ along with trace amount of ethanol and formic acid. The introduction of surface N-deficiency leads to enhanced $CO_2$ conversion activity. Further modification of the nitrogen-deficient g-$C_3Nx_{\text{-C}}$ surface with Cu(0) enables highly selective $CO_2$ photoreduction to only $CH_4$ (3.12 $\mu$molh-1 g-1). The improvement in the activity has been ascribed to the improved light harvesting and charge separation, as indicated by shorter average lifetime of 3.52 ns and higher value of Stern–Volmer quenching constant (11.2 $M^{-1}$) [81].

Unlike most of the transition metal-based molecular catalysts where the reduction of the metal to its zero-valent state is required, Co complex bearing a dithiolate ligand synthesized reduces $CO_2$ in its formal +1 state, enabling it more energy efficient. This molecular catalyst reduces $CO_2$ to CO with ~95% selectivity with a low overpotential of 70 mV. The increased electron density at the Co(I) center due to covalent donation

**Table 3** Progress of photocatalyst systems designed for $CO_2$ conversion in India

| Photocatalyst | Cocatalyst | Light source | Activity | Reference |
|---|---|---|---|---|
| *$CO_2$ conversion to $CH_3OH$* | | | | |
| $CeO_{2-x}$ | – | Xe lamp | 0.702 $\mu molg^{-1}h^{-1}$ | Hezam et al. [77] |
| La-modified $NaTaO_3$ | Co-complex | Hg lamp | 36.2 $\mu molg^{-1}h^{-1}$ | Jeyalakshmi et al. [78] |
| Heteroleptic Ru(II) | | Visible light | $\phi = 0.18$ | Kumar et al. [79] |
| $InTaO_4$ | | Visible light | $\phi = 1.01$ | Singhal and Kumar [80] |
| Ni/$InTaO_4$ | | Visible light | $\phi = 1.85$ | Singhal and Kumar [80] |
| GO-Phen | | Visible light | 3977.57 ± 5.60 $\mu molg^{-1}$ | Kumar et al. [76] |
| $SrTiO_3$ | – | | 9.10 $\mu molg^{-1}h^{-1}$ | Jeyalakshmi et al. [75] |
| $Sr_3Ti_{(2-x-y)}Fe_xS_yO_{(7-z)}N_z$ | Fe doped | | 60.1 $\mu molg^{-1}h^{-1}$ | Jeyalakshmi et al. [75] |
| GO-Ru | $H_2O$ +DMF sat. | Visible light | 482 $\mu molg^{-1}$ | Kumar et al. [79] |
| *$CO_2$ conversion to $CH_4$* | | | | |
| $Cu^{(0)}$decorate dg-$C_3N_{x-C}$ | – | Visible light | 3.12 $\mu molg^{-1}h^{-1}$ | Mishra et al. [81] |
| *$CO_2$ conversion to CO* | | | | |
| CdS | CODH | Visible light | – | Chaudhary et al. [82] |
| *$CO_2$ conversion to $C_2H_5OH$* | | | | |
| $SrTiO_3$ | | | 5.10 $\mu molg^{-1}h^{-1}$ | Jeyalakshmi et al. [75] |

from one thiolate enables it to activate $CO_2$ and the protonation of the bound thiolate plays a vital role in lowering the overpotential that facilitate facile transfer of proton to the bound $CO_2$ [49].

Enzyme (carbon monoxide dehydrogenase, CODH drive $CO_2$ reduction to CO) being known for their extraordinary catalytic activity was also coupled with the light harvesting CdS quantum dots and nanorods in an electroactive manner. The CODH-CdS nanorod integrated system shows highly selective and efficient reduction of $CO_2$ to CO under visible light with the TONs of the order of 23,000 in 5 h, which is better than that of enzyme active site mimics being designed, but inferior than that of activity what these enzymes exhibit in natural systems [82].

## 4 Scale-Up Issues

The above discussion summarizes the progress of the $CO_2$ conversion by both the electrocatalytic and photocatalytic pathways. Particularly for electrocatalytic conversion pathway, an enormous amount of work has been undertaken globally and in India on the electrode materials, their surface modification to have better activation of $CO_2$ and enhanced conversion efficiency [84]. However, the scale-up of this process remains elusive, particularly the stability of the cathode catalyst [85]. There are various issues that are yet to be understood fundamentally, addressed, and optimized. One is that $CO_2$ conversion involves kinetically sluggish proton-coupled multi-electron transfer that inevitably requires high energy, and thus leads to low energy efficiency of the overall conversion process. It may partially be addressed by combining the electrocatalytic reduction with the renewable energy resources. For instance, in photoelectrocatalytic/photocatalytic, the sunlight is used as an energy source to drive these conversions, but the multi-interfaces come in pictures that require further understanding of surface chemistry and their optimization. Other major challenge is the selectivity of $CO_2$ reduction product. Despite optimization by using various co-catalysts that provide selectivity on smaller area of the catalysts, the selectivity at the higher length scale of extended surface of the catalysts for $CO_2$ activation is yet to be optimized and addressed. Furthermore, the selectivity being achieved so far is limited to lower hydrocarbons, and it is still challenging to selectively convert $CO_2$ to chemicals having higher commercial value, for example, C2 or longer chain chemicals. Moreover, the degradation of activity over a period of time is also a serious problem. When scaling-up the catalyst area, it is bound to have increased number of defects, grain boundaries that further promote the degradation of the catalyst surface at a faster rate. Further, the $H_2$ evolution is competing reaction, and it becomes predominant at high applied potential; the gas bubbles formed because of this eventually reduces the active surface of the cathode and also leads to its degradation.

## 5 Conclusions and Outlook

The electrocatalytic reduction has several advantages over the photocatalytic $CO_2$ reduction: control of the conversion process by adjusting the applied potentials and reaction temperature; the chemical consumption can be recycled by recycling the electrolytes; the setup is simple, compact, and easy to scale up. However, to realize the commercial potential, the electrocatalytic reduction of $CO_2$ needs to be integrated with renewable energy resource, so that overall improved energy efficiency of the process is achieved. Despite significant progress in designing the electrocatalytst, photocatalyst, and mechanistic studies, there is ample room to make efforts on the following aspects:

(i) Design of electrocatalysts/photocatalysts based on Earth abundant elements, which have high conversion efficiency, faster charge separation and transfer kinetics, and the photocatalysts that have wide light harvesting ability.
(ii) Optimization of the $CO_2$ interaction with the catalyst/cocatalysts surface in order to facilitate efficient electron transfer and consequently enhancement in the quantum efficiency and faradic efficiency. The coupling interaction of the co-catalysts on the catalyst surface should be optimized such that high selectivity for $CO_2$ conversion is achieved.
(iii) The design of fuel cell with integrated cathode and anode, where an appropriate semiconductor as a photoanode for water oxidation, may provide electrons to drive the reduction on cathode efficiently.
(iv) Another vital aspect is the stability of these catalysts (electrocatalysts/photocatalysts); until now, the catalysts designed can operate up to few hours. However, it is very challenging, nevertheless need to device technologies for their prolong life, to operate for months without significant loss in their activity.

Moreover, it is difficult to get insight into the underlying surface reaction chemistry and mechanism. To understand these aspects, the design of some novel and simple configuration of materials (such as nanosheets, single-atom-based catalysts) is to be explored to gain the mechanistic insight into the surface reaction chemistry that will eventually lead to the development of the practically viable catalysts for $CO_2$ conversion to chemicals and fuel.

## References

1. IPCC (2006) Guideline for national green house gas inventories. (https://www.ipcc-nggip.iges.or.jp/support/Primer_2006GLs.pdf)
2. UNFCCC (2018) India second biennial update report to the united nations framework convention on climate change
3. PTI (2017) PM launches saubhagya scheme to provide power to all, Bus. Line, (http://www.thehindubusinessline.com/news/pm-launches-saubhagya-scheme-toprovide-power-toall/article9872678.ece). (Accessed 6 Jan 2018)
4. BP Statistical Review of World Energy 2019 (2017), 68th edition
5. Kyoto Protocol to the United Nations Framework Convention on Climate Change at united nation (1998)
6. Central Electricity Authority (CEA) (2018) National electricity plan. CEA, New Delhi
7. Kaya Y (1990) Impact of carbon dioxide emission control on GNP growth: interpretation of proposed scenarios
8. Sharma C, Singh R (2012) Estimates of emissions from coal fired thermal power plants in India. In: Proceedings of 2012 international emission inventory conference, Florida, August 13–16, 1–22
9. Sethia VK, Vyas S (2017) An innovative approach for carbon capture and sequestration on a thermal power plant through conversion to multi-purpose fuels—A feasibility study in indian context. Energy Procedia, 114:1288–1296
10. Al-Mamoori A, Krishnamurthy A, Rownaghi AA, Rezaei F (2017) Carbon capture and utilization update. Energy Technol 5:834–849

11. Finn C, Schnittger S, Yellowlees LJ, Love JB (2012) Molecular approaches to the electrochemical reduction of carbondioxide. Chem Commun 48:1392–1399
12. Usman M, Wan Daud WMA (2015) Recent advances in the methanol synthesis via methane reforming processes. RSC Adv 5, 21945–21972
13. Gerard P, Van Der Laan, Beenackers AACM (1999) Beenackers, kinetics and selectivity of the fischer–tropsch synthesis: A literature review. Catal Rev 41:255–318
14. Coates GW, Moore DR (2004) Discrete metal-based catalysts for the copolymerization of $CO_2$ and epoxides: Discovery, reactivity, optimization, and mechanism. Angew Chem Int Ed 43:6618–6639
15. Darensbourg DJ, Ganguly P, Choi W (2006) Metal salen derivatives as catalysts for the alternating copolymerization of oxetanes and carbon dioxide to afford polycarbonates. Inorg Chem 45:3831–3833
16. Qin Z, Thomas CM, Lee S, Coates GW (2003) Cobalt-based complexes for the copolymerization of propylene oxide and $CO_2$: active and selective catalysts for polycarbonate synthesis. Angew Chem Int Ed 42:5484–5487
17. Le GA, Abanades S, Flamant G (2011) $CO_2$ and $H_2O$ splitting for thermochemical production of solar fuels using non stoichiometric ceria and ceria/zirconia solid solutions. Energy Fuels 25:4836–4845
18. Loutzenhiser PG, Steinfeld A (2011) Solar syngas production from $CO_2$ and $H_2O$ in a two-step thermochemical cycle via Zn/ZnO redox reactions: Thermodynamic cycle analysis. Int J Hydrogen Energy 36:12141–12147
19. Smestad GP, Steinfeld A (2012) Review: photochemical and thermochemical production of solarfuels from $H_2O$ and $CO_2$ using metaloxide catalysts. Ind Eng Chem Res 51:11828–11840
20. Acién Fernández FG, González-López CV, Fernández Sevilla JM, Molina Grima E (2012) Conversion of $CO_2$ into biomass by microalgae: How realistic a contribution may it be to significant $CO_2$ removal. Appl Microbiol Biotechnol 96:577–586
21. Barton EE, Rampulla DM, Bocarsly AB. (2008) Selective solar-driven reduction of $CO_2$ to methanol using a catalyzed p-GaP based photoelectrochemical. Cell J Am Chem Soc 6342–6344
22. Kumar B, Llorente M, Froehlich J, Dang T, Sathrum A, Kubiak CP (2012) Photo- Chemical and photoelectrochemical reduction of $CO_2$. Annu Rev Phys Chem 63:541–569
23. Zhao J, Wang X, Xu Z, Loo JSC (2014) Hybrid catalysts for photoelectrochemical reduction of carbondioxide: a prospective review on semiconductor/metal complexo-catalystsystems, J Mater Chem A 15228–15233
24. Das S, Wan Daud WMA (2014) Photocatalytic $CO_2$ transformation into fuel: are view on advances in photocatalyst and photo reactor. Renew Sustain Energy Rev 39:765–805
25. Yin G, Nishikawa M, Nosaka Y, Srinivasan N, Atarashi D, Sakai E, Miyauchi M (2015) Photocatalytic carbondioxide reduction by copperoxide nanocluster-grafted niobate nanosheets. ACS Nano 9:2111–2119
26. Parkinson BA, Weaver PF (1984) Photoelectrochemical pumping of enzymic carbon dioxide reduction. Nature 309:148–149
27. Zhang S, Kang P, Meyer TJ (2014) Nanostructured tin catalysts for selective electrochemical reduction of carbon dioxide to formate. J Am Chem Soc 136:1734–1737
28. Kuhl KP, Hatsukade T, Cave ER, Abram DN, Kibsgaard J, Jaramillo TF (2014) Electrocatalytic conversion of carbon dioxide to methane and methanol on transition metal surfaces. So J Am Chem c., 136:14107–14113
29. Peterson AA, Norskov JK (2012) Activity descriptors for $CO_2$ electroreduction to methane on transition-metal catalysts. J Phys Chem Lett 3:251–258
30. Hori Y, Murata A, Takahashi R (1989) Formation of hydrocarbons in the electrochemical reduction of carbon-dioxide at a copper electrode in aqueous-solution. J Chem Soc - Farad Trans I, 85:2309–2326
31. Kuhl KP, Cave ER, Abram DN, Jaramillo TF (2012) New insights into the electrochemical reduction of carbon dioxide on metallic copper surfaces. Energy Environ Sci 5:7050–7059

32. Hara K, Tsuneto A, Kudo A, Sakata T (1994) Electrochemical reduction of $CO_2$ on a Cu electrode under high-pressure—factors that determine the product selectivity. J Electrochem Soc 141:2097–2103
33. Kas R, Kortlever R, Yilmaz H, Koper MTM, Mul G (2015) Manipulating the hydrocarbon selectivity of copper nanoparticles in $CO_2$ electroreduction by process conditions. Chemelectrochem 2:354–358
34. Kyriacou GZ, Anagnostopoulos AK (1993) Influence of $CO_2$ partial-pressure and the supporting electrolyte cation on the product distribution in $CO_2$ electroreduction. J Appl Electrochem 23:483–486
35. Gupta N, Gattrell, MacDougall BM (2006) Calculation for the cathode surface concentrations in the electrochemical reduction of $CO_2$ in $KHCO_3$ solutions. J Appl Electrochem 36:161–172
36. Murata A, Hori Y (1991) Product selectivity affected by cationic species in electrochemical reduction of $CO_2$ and CO at a Cu electrode. Bull Chem Soc Jpn 64:123–127
37. Ooka H, Figueiredo MC, Koper MTM (2017) Competition between hydrogen evolution and carbon dioxide reduction on copper electrodes in mildly acidic media. Langmuir 33:9307–9313
38. Schouten KJP, Kwon Y, van der Ham CJM, Qin Z, Koper MTM (2011) A new mechanism for the selectivity to C-1 and C-2 species in the electrochemical reduction of carbon dioxide on copper electrodes. Chem Sci 2:1902–1909
39. Hori Y, Wakebe H, Tsukamoto T, Koga O (1995) Adsorption of CO accompanied with simultaneous charge-transfer on copper single-crystal electrodes related with electrochemical reduction of $CO_2$ to hydrocarbons. Surf Sci 335:258–263
40. Hori Y, Takahashi I, Koga O, Hoshi N (2003) Electrochemical reduction of carbon dioxide at various series of copper single crystal electrodes. J Mol Catal A: Chem 199:39–47
41. Huang Y, Handoko AD, Hirunsit P, Yeo BS (2017) Electrochemical reduction of $CO_2$ using copper single-crystal surfaces: effects of CO* coverage on the selective formation of ethylene. ACS Catal 7:1749–1756
42. Takahashi I, Koga O, Hoshi N, Hori Y (2002) Electrochemical reduction of $CO_2$ at copper single crystal Cu(S)-[n (111) x (111)] and Cu(S)-[n (110) x (100)] electrodes. J Electroanal Chem 533:135–143
43. Kumari N, Haider MA, Tiwari PK, Basu S (2019) Carbon dioxide reduction on the composite of copper and praseodymium-doped ceria electrode in a solid oxide electrolysis cells. Ionics, https://doi.org/10.1007/s11581-019-02837-5
44. Kumari N, Haider MA, Basu S (2017) Reduction of $CO_2$ to CO in presence of $H_2$ on strontium doped lanthanum manganite cathode in solid oxide electrolysis cell. J Chem Sci 0–5
45. Steven MP, Connor PA, Tao S, Irvine JTS (2005) Electronic transport in the novel SOFC anode material $La_{1-x}$ $Sr_xCr_{0.5}$ $Mn_{0.5}$ O $_{3\pm\delta}$. Solid state ionics 177:2005–2008
46. Kumari N, Haider MA, Tiwari PK, Basu S (2015) Density functional theory study of $CO_2$ adsorption and reductionon stoichiometric and doped Ceria. ECS Trans 68(3):155–166
47. Singh S, Phukan B, Mukherjee C, Verma A (2015) Salen ligand complexes as electrocatalysts for direct electrochemical reduction of gaseous carbon dioxide to value added products. RSC Adv 5:3581–3589
48. Basu S, Shegokar A., Biswal D (2017) Synthesis and characterization of supported Sn/G-$Al_2O_3$ and Sn/ZSM5 catalysts for $CO_2$ reduction in electrochemical cell. J $CO_2$ Utilization 18:80–88
49. Dey S, Ahmed MdE, Dey A (2018) Activation of Co(I) state in a cobalt-dithiolato catalyst for selective and efficient $CO_2$ reduction to CO. Inorg Chem 57:5939–5947
50. Sen P, Mondal B, Saha D, Rana A, Dey A (2019) Role of 2nd sphere H-bonding residues in tuning the kinetics of $CO_2$ reduction to CO by Iron porphyrin complexes. Dalton Trans. https://doi.org/10.1039/c8dt03850c
51. Kumari N, Haider MA, Agarwal M, Sinha N, Basu S, (2016) Role of reduced CeO (110) surface for CO reduction to CO and Methanol. J Phys Chem C, J Phys Chem C, 120:16626–16635
52. Mohamed EA, Zahran ZN, Naruta Y (2015) Efficient electrocatalytic $CO_2$ reduction with a molecular cofacial iron porphyrin dimer. Chem Commun 51:16900–16903
53. Kang P, Zhang S, Meyer TJ, Brookhart M (2014) Rapid selective electrocatalytic reduction of carbon dioxide to formate by an iridium pincer catalyst immobilized on carbon nanotube electrodes. Angew Chem Int Ed 53:8709–8713

54. Yahaya AH, Gondal MA, Hameed A (2004) Selective laser enhanced photocatalytic conversion of $CO_2$ into methanol. Chem Phys Lett 400:206–212
55. Lehn JM and Ziessel R (1982) Photochemical generation of carbon monoxide and hydrogen by reduction of carbon dioxide and water under visible light irradiation. Proc Nat Acad Sci U.S.A. 279:701–704
56. Li Y, Wang W-N, Zhan Z, Woo M-H, Wu C-Y, Biswas P (2010) Photocatalytic reduction of CO2 with H2O on mesoporous silica supported Cu/TiO2 catalysts. Appl Catal B 100:386–392
57. Liu Q, Zhou Y, Kou J, Chen X, Tian Z, Gao J, Yan S, Zou Z (2010) High-yield synthesis of ultralong and ultrathin Zn2Ge$O_4$ nanoribbons toward improved photocatalytic reduction of $CO_2$ into renewable hydrocarbon fuel. J Am Chem Soc 132:14385–14387
58. Liu Q, Zhou Y, Tian Z, Chen X, Gao J, Zou Z (2012) $Zn_2GeO_4$ crystal splitting toward sheaf-like, hyperbranched nanostructures and photocatalytic reduction of $CO_2$ into $CH_4$ under visible light after nitridation. J Mater Chem 22:2033–2038
59. Nath RK, Zain M (2015) Artificial photosynthesis in concrete surface by using $LiNbO_3$. Adv Environ Biol 9:1–9
60. Li X, Li W, Zhuang Z, Zhong Y, Li Qing, Wang Liya (2012) Photocatalytic reduction of carbon dioxide to methane over $SiO_2$-Pillared $HNb_3O_8$. J Phys Chem C 116(30):16047–16053
61. Yang W-C, Rodriguez BJ, Gruverman A, Nemanich R (2004) Polarization-dependent electron affinity of $LiNbO_3$ surfaces. Appl Phys Lett 85:2316–2318
62. Xi G, Ouyang S, Li P, Ye J, Ma Q, Su N, Bai H, Wang C (2012) Ultrathin $W_{18}O_{49}$ nanowires with diameters below 1 nm: synthesis, near-infrared absorption, photoluminescence, and photochemical reduction of carbon dioxide. Angew Chem Int Ed 51:2395–2399
63. Julio N, de la Peña O'Shea, Víctor A, Coronado JP, Serrano JM, David P (2012) Effect of copper on the performance of ZnO and $ZnO_{1-x}$ $N_x$ oxides as $CO_2$ photoreduction catalysts. Catal Today, 209:21–27
64. He Z, Wang D, Fang H, Chen J, Song S (2014) Highly efficient and stable Ag/Ag$IO_3$ particles for photocatalytic reduction of $CO_2$ under visible light. Nanoscale 6:10540–10544
65. Ghadimkhani G, de Tacconi NR, Chanmanee W, Janaky C, Rajeshwar K (2013) Efficient solar photoelectrosynthesis of methanol from carbon dioxide using hybrid CuO-$Cu_2O$ semiconductor nanorod arrays. Chem Commun 49:1297–1299
66. Li P, Hu H, Xu J, Jing H, Peng H, Lu J, Wu C, Ai S (2014) New Insights into the photo-enhanced electrocatalytic reduction of carbon dioxide on $MoS_2$-Rods/$TiO_2$ NTs with unmatched energy band. Appl Catal B 147:912–919
67. Dong B-X, Qian SL, Bu F-Y, Wu YC, Feng L-G, Teng Y-L, Liu W-L, Li ZW (2018) Electrochemical reduction of $CO_2$ to CO by a heterogeneous catalyst of fe-porphyrin-based metal-organic framework. ACS Appl Energy Matter 1(9):4662–4669
68. Zhang X, Wu Z, Zhang X, Li L, Li Y, Xu H, Li X, Yu1 X, Zhang Z, Liang Y, Hailiang W (2017) Highly selective and active $CO_2$ reduction electrocatalysts based on cobalt phthalocyanine/carbon nanotube hybrid structures. Nature Commun 8:14675
69. Rao H, Lim C-H, Bonin J, Miyake GM, Robert M (2018) Visible-light-driven conversion of $CO_2$ to $CH_4$ with an organic sensitizer and an iron porphyrin catalyst. J Am Chem Soc 140(51):17830–17834
70. El-Zahab B, Donnelly D, Wang P (2008) Particle-tethered NADH for production of methanol from $CO_2$ catalyzed by coimmobilized enzymes. Biotechnol Bioeng 99:508−514
71. Sun Q, Jiang Y, Jiang Z, Zhang L, Sun X, Li J (2009) Greenand efficient conversion of $CO_2$ to methanol biomimetic coimmobilization of three dehydrogenases in protamine-templated titania. Ind Eng Chem Res 48:4210–4215
72. Woolerton TW, Sheard S, Chaudhary YS, Armstrong FA (2012) Enzymes and bio-inspired electrocatalysts in solar fuel devices", Energy Env Sci, 5:7470–7490
73. Xu S-W, Lu Y, Li J, Jiang Z-Y, Wu H (2004) Efficient conversion of $CO_2$ to methanol catalyzed by three dehydrogenases co-encapsulated in an alginate—Silica (ALG − SiO2) Hybrid Gel. Ind Eng Chem Res 45:4567–4573
74. Huan TN, Simon P, Rousse G, Génois I, Artero V, Fontecave M (2017) Porous dendritic copper: an electrocatalyst for highly selective $CO_2$ reduction to formate in water/ionic liquid electrolyte. Chem Sci 8:742–747

75. Jeyalakshmi V, Mahalakshmy R, Krishnamurthya KR, Viswanathan B (2018) Strontium titanates with perovskite structure as photo catalysts for reduction of $CO_2$ by water: influence of Co-doping with N, S & Fe. Catal Today 300:152–159
76. Kumar P, Sain B, Jain SL (2014) Photocatalytic reduction of carbon dioxide to methanol using a ruthenium trinuclear polyazine complex immobilized on graphene oxide under visible light irradiation. J Mater Chem A 2:11246–11253
77. Hezam A, Namratha K, Drmosh AQ, Ponnamma D, Wang J, Prasad S, Ahamed M, Cheng C, Byrappa K (2020) $CeO_2$ nanostructures enriched with oxygenvacancies for photocatalytic $CO_2$ reduction. ACS Appl Nano Mater 3(1):138–148
78. Jeyalakshmi V, Tamilmani S, Mahalakshmy R, Bhyrappa P, Krishnamurthy KR, Viswanathan B (2016) Sensitization of La Modified $NaTaO_3$ with cobalt tetra phenyl porphyrin for photo catalyticreduction of $CO_2$ by water with Uv-Visible light. J Mol Catal A: Chem 420:200–207
79. Kumar P, Bansiwal A, Labhsetwarb N, Jain SL (2015) Visible light assisted photocatalytic reduction of $CO_2$ using graphene oxide supported heteroleptic ruthenium complex. Green Chem 17:1605–1609
80. Singhal N, Kumar U (2012) Fabrication of visible-light-responsive $InVO_4$ for photoreduction of $CO_2$. Chem Select 2:3534–3537
81. Mishra B, Mishra S, Satpati B, Chaudhary YS (2019) Engineering the surface of a polymeric photocatalyst for stable solar-to-chemical fuel conversion from seawater. ChemSusChem 12:3383–3389
82. Chaudhary YS, Woolerton TW, Allen CS, Warner JH, Pierce E, Ragsdalec SW, Armstrong FA (2012) Visible light-driven $CO_2$ reduction by enzyme coupled CdS nanocrystals. Chem Commun 48:58–60
83. Liyanage AD, Perera SD, Tan K, Chabal Y, Balkus KJ (2014). Synthesis, characterization, and photocatalytic activity of Y-Doped $CeO_2$ Nanorods ACS catalysis 4(2):577–584
84. Malik K, Singh S, Verma A, Suddhasatwa B (2016) Electrochemical reduction of $CO_2$ for synthesis of green fuel. WIRES: Energy & Environ 6(4):1–17
85. Gag Garima, Basu Suddhasatwa (2015) Studies on degradation of copper nano particles in cathode for $CO_2$ electrolysis to organic compounds. Electrochim Acta 177:359–365

# Thermodynamic Analysis of Ionic Liquids for $CO_2$ Capture, Regeneration and Conversion

**Indrajit Das, V. Ramkumar, and Ramesh L. Gardas**

**Abstract** Global warming due to emission of greenhouse gases, especially carbon dioxide ($CO_2$), has a major impact on climate change. A large number of absorbents were tested for $CO_2$ capturing, and the literature reports suggested that ionic liquids (ILs) showed better performance than conventional absorbents. In recent years, several researchers studied experimental and theoretical methods to understand the role of ILs in $CO_2$ capture, regeneration and conversion. ILs have potential to replace conventional absorbents for $CO_2$ capture due to the combination of unique properties and good solvating capability. The physical and chemical properties of ILs could be modified by tuning their ionic mobilities to achieve specific application requirements. Further, $CO_2$ solubility in ILs can be attributed to free volume effect and Lewis acid–base interactions. The significant enhancement in viscosity of IL after $CO_2$ absorption can be controlled by the addition of water as co-solvent, and the presence of water has the potential to compete with the $CO_2$ for absorption. After $CO_2$ absorption, the recycling of ILs is important, which can be performed by heating. From the viewpoint of thermodynamics, the enthalpy change is the main driving force for $CO_2$ absorption for ILs, and other thermodynamic parameters can be derived for this absorption process.

**Keywords** Ionic liquid · $CO_2$ capture · Absorption · Solubility · Separation · Free volume · Enthalpy change

## Abbreviations

| | |
|---|---|
| ILs | Ionic liquids |
| $CO_2$ | Carbon dioxide |
| Z | Solubility |
| T | Temperature |

I. Das · V. Ramkumar · R. L. Gardas (✉)
Department of Chemistry, Indian Institute of Technology Madras, Chennai 600036, India
e-mail: gardas@iitm.ac.in
URL: http://www.iitm.ac.in/info/fac/gardas



M. Goel et al. (eds.), *Climate Change and Green Chemistry of $CO_2$ Sequestration*, Green Energy and Technology, https://doi.org/10.1007/978-981-16-0029-6_9

| | |
|---|---|
| K | Kelvin |
| $K^{\theta}$ | Equilibrium constant |
| Mol | Mole |
| MPa | Mega Pascal |
| $P_{CO_2}$ | $CO_2$ partial pressure (kPa) |
| $H_m$ | Henry's constant (kPa) |
| $C_3$ | The molar ratio of active ionic liquids to the total ionic liquids |
| R | Universal gas constant |
| MEA | Monoethanolamine |
| MOFs | Metal–organic frameworks |
| VOCs | Volatile organic compounds |
| DES | Deep eutectic solvents |
| TFSI | Bistriflimide |
| $PF_6$ | Hexafluorophosphate |
| $\Delta H^{\theta}$ | Enthalpy change |
| $\Delta S^{\theta}$ | Entropy change |
| $\Delta G$ | Gibbs free energy change |
| $C_p$ | Heat capacity at constant pressure |
| $Q$ | Heat |
| ATR-IR | Attenuated total reflection-infra red |
| VSA | Vacuum swing adsorption |
| PSA | Pressure swing adsorption |
| TSA | Temperature swing adsorption |
| Me | Methyl |
| MeO | Methyl oxide |
| [hmim][FEP] | 1-hexyl-3-methylimidazolium tris(pentafluoroethyl) trifluorophosphate |
| [bmpyr][FEP] | 1-butyl-1-methylpyrrolidinium tris(pentafluoroethyl) trifluorophosphate |
| [bmim][$PF_6$] | 1–butyl-3-methylimidazolium hexafluorophosphate |
| [bmim][$BF_4$] | 1-butyl-3-methylimidazolium tetrafluoroborate |
| [$C_8$mim][$PF_6$] | 1-octyl-3-methylimidazolium hexafluorophosphate |
| [$C_8$mim][$BF_4$] | 1-octyl-3-methylimidazolium tetrafluoroborate |
| [bmim][$NO_3$] | 1-butyl-3-methylimidazolium nitrate |
| [emim][$EtSO_4$] | 1-ethyl-3-methylimidazolium ethyl sulfate |
| [N-bupy][$BF_4$] | N-butylpyridinium tetrafluoroborate |
| [$C_2$-mim][$Tf_2N$] | 1-ethyl-3-methylimidazolium bis(trifluoromethylsulfonyl) imide |
| [$C_4$-mim][$Tf_2N$] | 1-butyl-3-methylimidazolium bis(trifluoromethylsulfonyl) imide |
| [$C_6$-mim][$Tf_2N$] | 1-hexyl-3-methylimidazolium bis(trifluoromethylsulfonyl) imide |
| [$C_8$-mim][$Tf_2N$] | 1-octyl-3-methylimidazolium bis(trifluoromethylsulfonyl) imide |
| [bmim][DCA] | 1-butyl-3-methylimidazolium dicyanamide |

| | |
|---|---|
| [bmim][$NO_3$] | 1-butyl-3-methylimidazolium nitrate |
| [bmim][$BF_4$] | 1-butyl-3-methylimidazolium tetrafluoroborate |
| [bmim][$PF_6$] | 1-butyl-3-methylimidazolium hexafluorophosphate |
| [bmim][TfO] | 1-butyl-3-methylimidazolium trifluoromethanesulfonate |
| [bmim][$Tf_2N$] | 1-butyl-3-methylimidazolium bis(trifluoromethylsulfonyl) imide |
| [bmim][methide] | 1-butyl-3-methylimidazolium methide |
| [Emim][$B(CN)_4$] | 1-ethyl-3-methylimidazolium tetracyanoborate |
| [Emim][$Tf_2N$] | 1-ethyl-3-methylimidazolium bis (trifluoromethylsulfonyl) imide |
| [$P_{66614}$][Met] | Trihexyl(tetradecyl)phosphonium methionate |
| [$P_{66614}$][Pro] | Trihexyl(tetradecyl) phosphonium prolinate |
| [$P_{66614}$][4-Me-PhO] | Trihexyl(tetradecyl) phosphonium p-methylphenolate |
| [$P_{66614}$][4-Cl-PhO] | Trihexyl(tetradecyl) phosphonium m p-chlorophenolate |
| [$NH_2$p—bim][$BF_4$] | 1-butyl-2-propylamine imidazolium tetrafluoroborate |

## 1 Introduction

One of the major green house gases responsible for climate change is $CO_2$. Large amounts of $CO_2$ are emitted into the atmosphere in the process of burning fossil fuels to produce electricity and heat. The increased $CO_2$ concentration in the atmosphere has been a debate worldwide, aiming to reduce the emission of $CO_2$ and other greenhouse gases [1]. The exponential increase in $CO_2$ emissions into the earth's atmosphere since industrial growth has resulted in a rise in $CO_2$ concentration in the atmosphere last five decades. India is stated to be the second or the third-largest producer of $CO_2$ emission due to the development of thermal power stations, transport and urbanization [2]. Power stations use fuel for combustion to generate electricity, and the emission rate of $CO_2$ from fuel (coal) is comparatively high.

The $CO_2$ emission will increase and continue to grow in the future if we do not find a solution to reduce the emission or to capture the emission. In this regard, $CO_2$ capture will be considered as a temporary solution to the problem. Moreover, the energy gas resources, such as biogas, shale gas, natural gas and syngas, contain $CO_2$ as an impurity, which lowers the heating value and gas quality. Therefore, one crucial solution to green house gas control, gas upgradation and purification of fuel is $CO_2$ separation. The capture of $CO_2$ is a challenging process, and technically, many obstacles have to be overcome to commercialize these processes [3]. Commercialization of the $CO_2$ capturing process at large scale is cost-ineffective, which is a major barrier, and the available technology for $CO_2$ capture is based on amine solvents (monoethanolamine, MEA), and this process involves a chemical reaction with a large enthalpy of reaction [4]. Therefore, a great quantity of heat is needed to free the captured $CO_2$ in the regeneration step. The existing methods for $CO_2$ capture are not cost-effective and not suitable for large-scale applications.

Thus, extended research has pushed to produce new methods/patterns/materials for $CO_2$ capture, which include $CO_2$ capture by chemical/physical absorption, solid adsorbents, membranes and biomimetic approaches [5].

Thither is a need to get more efficient with low price choices for efficient absorbing and subsequently regenerating the $CO_2$ for further use or sequestration. Absorption of $CO_2$ in liquid salts is one of the most promising alternatives. The technology essentially uses custom-designed salts, which are liquid at room temperature, as well known as ionic liquids, to soak up the carbon dioxide (Fig. 1).

New and emerging materials for $CO_2$ capture were metal–organic frameworks (MOFs), and a special class of absorption materials, namely ILs, was offered as an alternative to volatile commercial solvents [6]. The particular focus is on the applicable interest of ionic liquids as a $CO_2$ capture. Thus, the objective is to emphasize the recent developments in designing suitable ionic liquids for $CO_2$ capture from various sources.

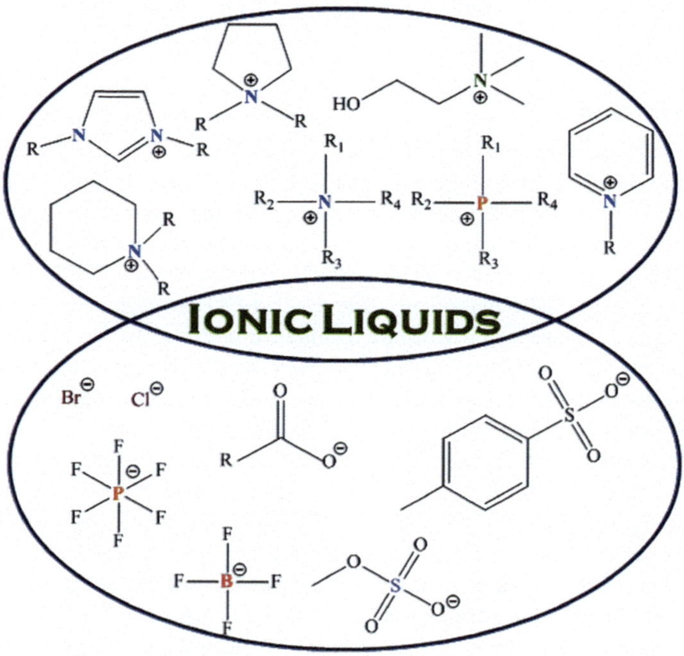


**Fig. 1** Common cation and anion constituents of ionic liquids

## 2 Ionic Liquids and $CO_2$ Absorption

ILs are described as a novel class of molten salts consisting of ions, with a melting point less than 100 °C [7]. The uses of ionic liquids were limited until Freemantle [8] described the likely role of ionic liquids as solvents for green chemistry [9]. Soon after this, researchers showed interest in ionic liquids and in their use for new applications and used in the field of organic synthesis [10], biochemistry [11], catalysis [12], electrochemistry [13], separation technology [14] and fluid engineering [15].

ILs have remarkable properties, such as negligible vapor pressure, high thermal stability, non-flammable, tunability, solvation properties and high $CO_2$ solubility. These unique properties were the driving force for researchers to replace the traditional industrial solvents called the volatile organic compounds (VOCs) by ionic liquids. Non-volatile ionic liquids were substituted by the conventional volatile solvents to prevent environmental contamination. Furthermore, ILs are called as "designer solvents" due to their tunable properties, which provide an extra degree of freedom for tuning the physicochemical properties of the aimed application by structural modifications of cations and anions or by changing the combination of cations and anions [16].

A large number of absorbents were tested for $CO_2$ capture. However, liquid salt-based absorbents (e.g., ILs and deep eutectic solvents, [DES]) show better performance. In recent years, the solubility of carbon dioxide in different liquid salts (e.g., ILs and DES) is studied by many researchers using experimental and theoretical methods. Blanchard et al. [17] were the first to report high $CO_2$ solubility in imidazolium-based ILs reaching a $CO_2$ mol fraction of 0.6 at 8 MPa and 298.15 K and also showed that the bistriflimide anion has high affinity for $CO_2$ capture compared to hexafluorophosphate anion for ILs with same cation [18]. Amine-containing imidazole functionalized ionic liquid was for the first time used to absorb $CO_2$, indicating that the ionic liquid is capable of absorbing about 0.5 mol carbon dioxide per mole of ionic liquid at atmospheric pressure [19]. Later, many researchers have developed other quaternary phosphorous ionic liquids with amino acid anions with which the absorption of 1 mol of carbon dioxide was achieved, for example, Song et al. reported $CO_2$ solubility in various aqueous amino acid salt solutions, where maximum $CO_2$ absorption was achieved with arginine (1 mol $CO_2$/mol solvent) [20]. Although these ILs showed a good $CO_2$ absorption capacity, their high viscosity after $CO_2$ absorption made them unsuitable for industrial application [21].

The presence of water has been shown to affect the physical properties of the IL and also has the potential to compete with the $CO_2$ for absorption [22]. Joan Brennecke (University of Notre Dame, USA), James Davis (University of South Alabama, USA), Chris Hardacre (QUILL, Queen's University Belfast, UK), Mert Atilhan (Qatar University, Qatar), Jin-Won Park (Yonsei University, Seoul, South Korea) and few more research groups are involved researching the likely use of few ionic liquids for $CO_2$ capture [23]. Nevertheless, the systematic approach to understanding the $CO_2$-superbase ionic liquid interaction mechanism and comprehensive work of phase equilibrium and thermodynamic information for designing industrial

**Table 1** $CO_2$ solubility experimental data for imidazolium-based ionic liquids

| Ionic liquid | T (K) | P (bar) | Max solubility (mol $CO_2$/mol IL) | References |
|---|---|---|---|---|
| [hmim][FEP]<br>[bmpyr][FEP] | 283.15–298.6 | 0.297–18.1 | 0.517<br>0.498 | Zhang et al. [52] |
| [bmim][$PF_6$] | 293–393 | 1–97 | 0.555 | Perez-Salado Kamps et al. [53] |
| [bmim][$PF_6$]<br>[bmim][$BF_4$] | 283–348 | 0.096–19.99 | 0.379<br>0.364 | Shiflett and Yokozeki [54] |
| [bmim][$PF_6$]<br>[$C_8$mim][$PF_6$]<br>[$C_8$mim][$BF_4$]<br>[bmim][$NO_3$]<br>[emim][$EtSO_4$]<br>[N-bupy][$BF_4$] | 313.15, 323.15,333.15 | 0.97–92 | 0.729<br>0.755<br>0.513<br>0.708<br>0.579<br>0.423 | Blanchard et al. [55] |
| [$C_2$-mim][$Tf_2N$]<br>[$C_4$-mim][$Tf_2N$]<br>[$C_6$-mim][$Tf_2N$]<br>[$C_8$-mim][$Tf_2N$] | 301–344.4 | 10–43 | 0.761<br>0.803<br>0.833<br>0.845 | Shin et al. [56] |
| [bmim][DCA]<br>[bmim][$NO_3$]<br>[bmim][$BF_4$]<br>[bmim][$PF_6$]<br>[bmim][TfO]<br>[bmim][$Tf_2N$]<br>[bmim][methide] | 298.15, 313.15, 333.15 | 1–120 | 0.550<br>0.475<br>0.520<br>0.640<br>0.641<br>0.681<br>0.761 | Aki et al. [45] |

processes are scarce. Table 1 lists some example of imidazolium-based ILs and their $CO_2$ solubility.

In the Indian scenario, $CO_2$ absorption studies in various aminate ILs and aminates with different types of amine sites (primary and secondary amines) was studied by Ashok Khanna [24]. However, several research groups used molecular dynamics simulations to understand the interactions of $CO_2$ with couple of ionic liquids [25–27]. The experimental measurements of $CO_2$ solubility in amines and aqueous solution of amines as a function of temperature and pressure are vastly studied by Mondol and Bajpai [28], Kundu and Kumar [29]. Gardas et al. studied the absorption of $CO_2$ in guanidinium-based carboxylate ionic liquids and showed that the increase in alkyl chain length of the carboxylate anion increases the $CO_2$ absorption capacity [30]. Researchers are also engaged in ionic liquid and in their synthesis, simulation and experimental thermodynamics to explore their potential use as an absorbent for $CO_2$ capture and regeneration.

## 3 Thermodynamics of $CO_2$ Absorption

Thermodynamic properties of functionalized IL systems are essential to lead the pattern of new functionalized ILs for $CO_2$ capture. Nowadays, great attempts have been dedicated to the fields on the thermodynamic properties of $CO_2$ chemical absorption by functionalized ILs. A good example model is Brennecke, and his group [31] proposed "deactivated IL" model and "two-reaction" model to analyze the thermodynamic properties of $CO_2$ in amine-functionalized ILs. Hu, with his co-workers [32], studied the thermodynamic properties of $CO_2$ capture in low-viscous fluorine-substituted phenolic ILs through the "deactivated IL" model. Thus, it is very important to expand the thermodynamic studies of different types of functionalized ILs.

For the evaluation of new absorbents in practical applications, thermodynamic properties such as the absorption Gibbs energy, enthalpy and entropy change are of great importance and to understand the thermodynamic driving force for the absorption. Normally, the absorption of $CO_2$ in ILs consists of two parts, the physical absorption and the chemical absorption.

Let us consider the equations for physical and chemical, $CO_2$ capture expressed as follows:

$$CO_2(g) \rightarrow CO_2(l) \tag{1}$$

$$CO_2(g) + IL\ (l) \rightarrow CO_2 - IL(l) \tag{2}$$

In Eq. (1), $CO_2$ is in gaseous and liquid states, while in Eq. (2), IL(l) before and $CO_2$-IL (l) represents after $CO_2$ capture. Since our ILs are basic, they interact chemically with acid ($CO_2$) to form complex IL-$CO_2$ (l). Brennecke's model assumes that only 1:1 reaction takes place, and the maximum quantity of the ionic liquids is allowed to react with $CO_2$. The "deactivated IL" model can be expressed as Eq. (3)

$$Z = \frac{P_{CO_2}/H_m}{1 - P_{CO_2}/H_m} + \frac{K^{\theta}.P_{CO_2}.C_3}{1 + K^{\theta}.P_{CO_2}} \tag{3}$$

where $Z$ denotes solubility (mol $CO_2$/mol IL); $K^{\theta}$, $P_{CO2}$ and $H_m$ denote the equilibrium constant; $CO_2$ partial pressure (kPa) and the Henry's constant (kPa), respectively. $C_3$ stands for the molar ratio of active ionic liquids to the total ionic liquids. The first term denotes the contribution from physical absorption, while the second term is the contribution from chemical absorption. Generally, if the pressure is less than 100 kPa, physical adsorption is negligible compared to chemical adsorption [32]. Considering the fact that pressure is less than 100 kPa, the $CO_2$ adsorption system can be treated as an ideal chemical reaction in the studied experimental temperature range, and we can write the Eq. (4) as,

$$Z = \frac{K^{\theta}.P_{CO_2}.C_3}{1 + K^{\theta}.P_{CO_2}} \quad (4)$$

If we fit the experimental solubility of $CO_2$ data with different temperatures in Eq. (2), we will get equilibrium constant value at different temperatures. By using the van't Hoff equation, the reaction enthalpy and reaction entropy for the $CO_2$ absorption by the ILs could be determined using Eq. (5)

$$\ln K^{\theta} = -\frac{\Delta H^{\theta}}{RT} + \frac{\Delta S^{\theta}}{R} \quad (5)$$

where $\Delta H^{\theta}$ and $\Delta S^{\theta}$ represent the enthalpy (J $mol^{-1}$) and entropy (J $mol^{-1}$ $K^{-1}$) change, respectively, for the chemical adsorption of $CO_2$. T is the thermodynamic temperature in K, R denotes the universal gas constant ($R = 8.314$ J $mol^{-1}$ $K^{-1}$), and the Gibbs free energy could be calculated using the following Eq. (6).

$$\Delta G = -\text{RT} \ln K^{\theta} \quad (6)$$

$\Delta G^{\theta}$ value suggests the interaction between ILs and $CO_2$, and we can compare among different ILs, which one forming stronger bonds with $CO_2$. Negative $\Delta S^{\theta}$ value indicates that the degree of disorder of the system becomes smaller due to the strong interaction of the IL with $CO_2$ molecules.

## 4 Thermodynamics of $CO_2$ Separation

Thermodynamic analysis is important for theoretical evaluation of $CO_2$ separation. The $CO_2$ separation process is studied based on the first and second thermodynamic laws. The theoretical energy penalty for $CO_2$ separation, the energy used and the quantity of absorbent needed are considered with Gibbs free energy change ($\Delta G$). For screening $CO_2$ absorbents, the main screening options are based on thermodynamic properties, energy use and Gibbs free energy change.

Thermodynamic analysis is usually conducted on the basis of the Gibbs free energy change for $CO_2$ separation with the ILs. The operational condition, the amount of IL needed and the energy use are investigated theoretically for minimum Gibbs free energy change in a reversible process, etc. The thermodynamic results are in a relationship with the operating condition and the thermodynamic properties using the criteria of the amount of ILs needed and the energy use. Zang et al. have designed a framework for conducting thermodynamic analysis for $CO_2$ separation with IL and have given an illustration, as shown in Fig. 2.

The gas mixture containing $CO_2$ and other gases is separated into pure $CO_2$ at a temperature and pressure (298.15 K, 1 bar) in system. By thermodynamics principle, the Gibbs free energy change of the separation process is positive ($\Delta G_1 > 0$), and to make the overall system spontaneous, which comprises system and surrounding,

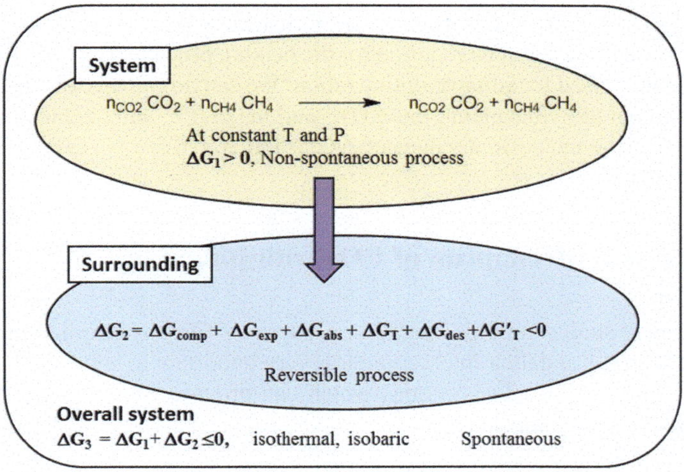


**Fig. 2** Schematic representation of thermodynamic analysis for $CO_2$ separation of gas mixture

Gibbs free energy change is used as the criterion of spontaneous system at a constant temperature and pressure with $\Delta G_3 \leq 0$. Therefore, surrounding should provide extra energy ($\Delta G_2$), which should be satisfied with negative contribution of $\Delta G_2$. The overall system can be designed reversibly at constant temperature and pressure by including system and surroundings, and the Gibbs free energy change ($\Delta G$) can be used as the criterion to minimize the work input.

The thermodynamics of the absorption process will involve absorption at a lower temperature ($T_L$) followed by desorption of the $CO_2$ by heating to a higher temperature ($T_H$) and regeneration of the solvent, followed by cooling back to $T_L$ to repeat the cycle. Therefore, in this respect, there are three elements that affect the thermodynamics, and these are $\Delta H^{abs}$ (enthalpy of absorption, also desorption), the $\Delta G^{ads}$ (Gibbs energy of adsorption) and the energy required to change the chamber temperature, $Q = mC_P\,(T_H - T_L)$, where $m$ is the mass and $C_P$ the heat capacity of the chamber. Thus, the equilibrium constant for binding (K), at $T_L$ is $K > 1$ and at $T_H$ is $K < 1$. $K$ is related to $\Delta G^{ads}$ by Eq. (7) given below.

$$K = \exp\left[\frac{-\Delta G^{ads}}{RT}\right] \tag{7}$$

$\Delta H^{ads}$ will determine the heat needed for regeneration, but also the rate of change with temperature of $\Delta G^{ads}$, as shown by the Gibbs–Helmholtz relationship in Eq. (8),

$$\frac{\Delta H^{ads}}{RT} = \left[\frac{-d(\Delta G^{ads}/RT)}{dT}\right] \tag{8}$$

If $\Delta H^{ads}$ is low, then for desorption, very little heat will be required at $T_H$, but for $T_H \gg T_L$, more energy will be required for the heating process. Thus, the thermodynamic analysis is used for suitable optimization. We can do calculations to minimize the energy required by balancing the $\Delta G^{ads}$ and the $\Delta H^{ads}$ and chemically modify solvents to achieve the desired values of $\Delta G^{ads}$ and $\Delta H^{ads}$.

## 5 Interaction Mechanism of $CO_2$ with Ionic Liquids

There are many studies on the absorption capacity of $CO_2$ with ionic liquids, $CO_2$ separation from ILs and their interaction. It is very important to know the interaction mechanism between IL-$CO_2$ systems, which can provide valuable information to design novel IL-based solvent.

(a) **Lewis Acid and Base interaction**

$CO_2$ solubility with [$BF_4$] and [$PF_6$] anions by attenuated total reflectance (ATR)-IR spectroscopy were studied by Kazarian et al., and they found weak acid–base interaction existing between $CO_2$ with [$BF_4$] and [$PF_6$] anion [33]. From Raman spectroscopy and DFT study, it was proved that there was a very weak charge-transfer interaction between oxygen atom of $COO^-$ group and the carbon atom of $CO_2$ molecule [34]. Generally, strength of Lewis acid–base interaction is directly proportional to anion basicity, and solubility of $CO_2$ is inversely proportional to the binding energy between $CO_2$ and ILs. In case of halides, the interaction strength decreases with the increasing size of anions [35]. But, there are some reports which show that not only Lewis acid–base interaction plays a role for $CO_2$ solubility, there are some other effects like free volume effects, ILs containing [$BF_4^-$] ion Lewis acid–base interaction is less compared to ILs containing [$PF_6^-$] ion although [$BF_4^-$] is stronger base than [$PF_6$]$^-$ ion [36].

(b) **Free Volume**

Using molecular simulation study, it was found that at 8 bar and 298 K, $CO_2$ solubility of 1-hexyl-3-methylimidazolium hexafluorophosphate ([Hmim][$PF_6$]) is 70% lower than that of 1-hexyl-3-methylimidazolium tris(pentafluoroethyl)trifluorophosphate ([Hmim][FEP]). This phenomenon can be explained by free volume effect. [FEP] anion contains six additional $CF_2$/$CF_3$ groups, which lead to a larger free volume than [$PF_6$] anion [36, 37]. Generally, larger free volume of ILs accommodates more $CO_2$ molecule (Fig. 3).

There is a good relationship between the interaction energy of cation and anion and $CO_2$ solubility. For example, 1-ethyl-3-methylimidazolium tetracyanoborate ([Emim][$B(CN)_4$]) shows higher $CO_2$ solubility than 1-ethyl-3-methylimidazolium bis (trifluoromethylsulfonyl)imide ([Emim][$Tf_2N$]). This can be explained by free volume effect. In 1-ethyl-3-methylimidazolium tetracyanoborate ([Emim][$B(CN)_4$]), interaction between cation and anion is weaker compared to bis

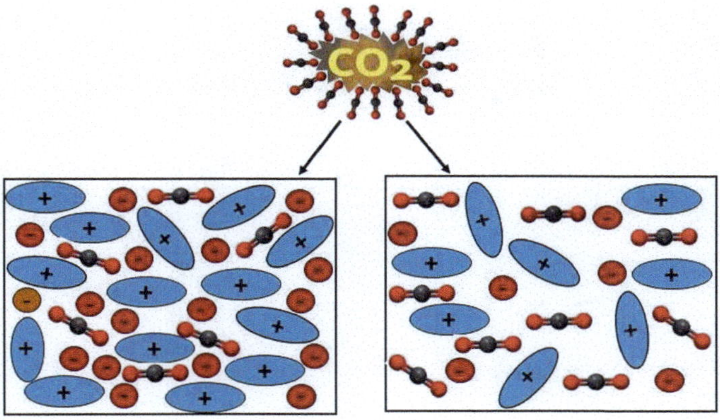


**Fig. 3** Illustration showing the free volume of ionic liquid and accommodation of $CO_2$ [37]

(trifluoromethylsulfonyl)imide ([Emim][$Tf_2N$]), which permits the lattice expansion easier, and more number of $CO_2$ will insert into the interstitial space (or free volume) of ILs. Moreover, it was found that at low concentration, $CO_2$ filled small voids located in the interior space of ILs, and at higher concentration, all the small voids formed larger voids, and more $CO_2$ was accommodated [38].

(c) **Rearrangements**

Huang et al. studied the behavior of $CO_2$ on 1-butyl-3-methylimidazolium hexa fluorophosphate ([bmim][$PF_6^-$]) and confirmed that the size of $CO_2$ molecule is large enough compared to the cavities of ILs, so $CO_2$ molecule will not be accommodated inside the cavities of ILs, but after small angular rearrangement of [$PF_6^-$] anion, large cavities formed and $CO_2$ molecules will be accommodated in that position [39] (Fig. 4).

(d) **Amine-based IL -$CO_2$ chemical Interaction**

Amine groups connected with either cation or anions of ILs, commonly react with through various mechanisms. In case of single and double functionalized amines, $CO_2$ reacts to form the product by carbamate mechanism which is very similar to amine reacts in aqueous medium [19, 40] (Fig. 5).

In the case of 1,2-dimethyl-(3-aminoethyl)imidazolium-based ILs as a cation and fluoride, chloride, bromide and iodide as anion, they capture $CO_2$ by molar ratio 1:1 and form carbamic acid [41]. It was observed that after adding -$NH_2$ groups into the imidazolium group, $CO_2$ absorption capacity increased significantly [42]. Similarly, anion-functionalized amine in ILs are, for example, trihexyl(tetradecyl)phosphonium methionate [$P_{66614}$][Met] and trihexyl(tetradecyl) phosphonium prolinate [$P_{66614}$][Pro] which showed equimolar $CO_2$ absorption capacity, and it favors the carbamic acid formation [21].

(e) **Non-amine-based IL-$CO_2$ chemical interaction**

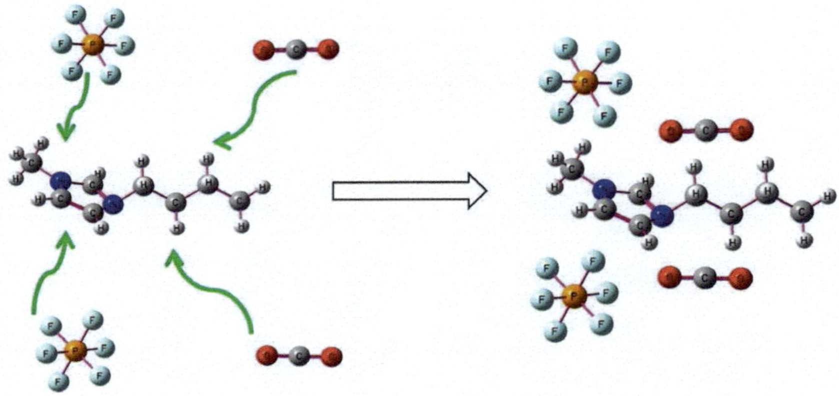

Fig. 4 Illustration showing the rearrangement of [Bmim][$PF_6^-$] with $CO_2$ [39]

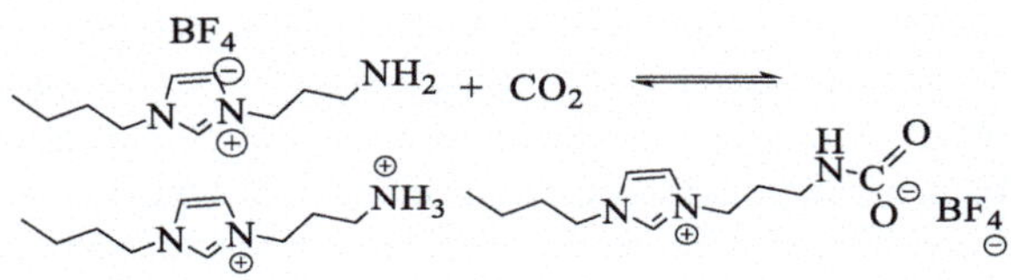

Fig. 5 Proposed reaction mechanism of $CO_2$ with [$NH_2$p-bim][$BF_4$]

The main disadvantage of amine-based ILs is that its viscosity is too high, and we cannot reuse it for further application. Moreover, the interaction between $CO_2$ and ILs is very strong for amine-based ILs, which causes difficulty for the regeneration of $CO_2$. Among all non-amine-based ILs, phenolic ILs play a major role for the reversible and efficient capture of $CO_2$. In phenolic ILs, phenol acts as an anion, and its physicochemical and $CO_2$ absorption capacity depend on electron-donating ability, electron-withdrawing ability and position of the substituents.

Wang et al. studied phenol-based ILs and their performances for $CO_2$ absorption. Generally, for phenolic-based ILs, $CO_2$ absorption performances depend on the number of the substituents and electron-donating, withdrawing effect of the substituent. They showed that [$P_{66614}$][4-Me-PhO] and [$P_{66614}$][4-Cl-PhO] had higher absorption capacity of 0.91 and 0.95 mol $IL^{-1}$ [43, 44] (Fig. 6).

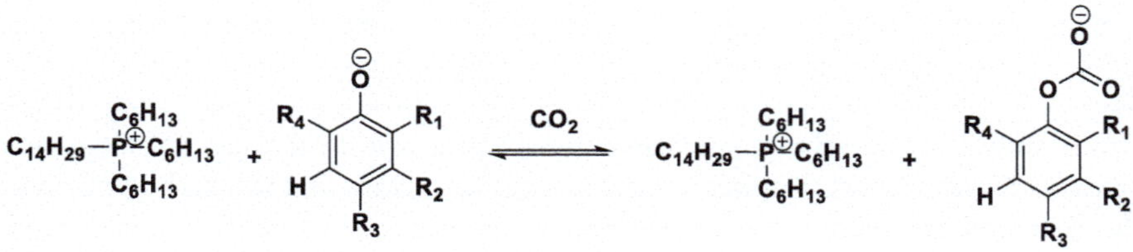

Fig. 6 $CO_2$ absorption mechanism by phenolic ILs

## 6 Structural Effects

The most significant feature of ILs compared to conventional organic solvents is by changing its structural units (cations/anions) and functional groups. We can change its property. There are millions of cations, anions and functional groups that can be tailored not only to the physicochemical properties of ILs but also to $CO_2$ absorption capacity. Here, we will discuss the effect of cations, anions and functional groups for solubility of $CO_2$ on physisorption and chemisorption of ILs.

(a) **Cationic and anionic effect on Physisorption of ILs**

ILs can capture $CO_2$ through the physisorption process between cation, anion and $CO_2$. Generally, $CO_2$ solubility depends on the types of constituent ions. First, $CO_2$ solubility by conventional ILs was reported by Blanchard et al. They reported that at 25 °C and 8.30 MPa, $CO_2$ solubility in conventional IL, 1-butyl-3-methylimidazoliu hexafluorophosphate ([Bmim][$PF_6$]) was 0.75 mol.$IL^{-1}$ [17]. $CO_2$ solubility of imidazolium-based ILs is common, and in the literature, a huge number of data available [6, 18, 45]. Anthony et al. studied various types of cations, e.g., imidazolium, ammonium, pyrrolidinium, and phosphonium and anions bis(trifluoromethylsulfonyl)imide, hexafluorophosphate, and tetrafluoro borate) and their effects on $CO_2$ solubility and concluded that there was no major change in $CO_2$ solubility for cationic change when anion is same. In contrast, anionic effect is more prominent. Among all the anions, bis(trifluoromethylsulfonyl) imide shows strongest affinity for $CO_2$ solubility, regardless of whether the ILs cation is imidazolium, ammonium, pyrrolidinium and phosphonium [18]. Another group of researcher found that at 25 °C, for imidazolium-based ILs, $CO_2$ solubility decreased in the following order: (tris(trifluoromethylsulfonyl)methide) > [$Tf_2N$] > [TfO] > [$BF_4$] > [DCA] > [$NO_3$]. It was likewise observed that ILs containing fluorine groups on the anions have a higher $CO_2$ solubility than nonfluorinated anions, and with the increase in number of fluorine groups, solubility of $CO_2$ will increase, e.g., the $CO_2$ solubility trend followed the order: [FEP] > [methide] > [$Tf_2N$] > [$PF_6$] > [$BF_4$]. Although cation's contribution is less effective, we still cannot ignore it. Generally, with an increase in cation chain length, solubility of $CO_2$ in ILs will increase, for example, $CO_2$ solubility increased in the following order on the imidazolium ring [Omim] > [Hmim] > [Bmim]. This could be explained by free volume effect. The densities of imidazolium-based ILs decrease with an increase in alkyl chain length, and as a consequence, free volume of ILs will also increase to accommodate more $CO_2$ gas, which is a favorable condition for $CO_2$ solubility [39, 46].

(b) **Cationic and anionic effect on Chemisorption of ILs**

Although ILs is a promising solvent for $CO_2$ capture, still physisorption of ILs cannot compete with chemisorption of ILs because $CO_2$ absorption capacity is lower for physisorption compared to chemisorption. Therefore, chemisorption of ILs with $CO_2$ is very important, which can be done by functionalized ILs. In case of chemisorption, interaction between cation and anion is strong, resulting in ILs with difficult for desorption and high energy demand for regeneration. Higher $CO_2$ capacity is achieved at the expense of increasing the enthalpy and energy consumption. Wang et al. prepared phenolic-based ILs and studied their cost-effective $CO_2$ absorption capacity by tuning anions of ILs [43, 44]. The experimental result shows different behaviors for electron-donating and electron-withdrawing groups. The electron donation groups (e.g., MeO and Me) on anion increased $CO_2$ absorption capacity, and electron-withdrawing groups (e.g., Cl, $CF_3$ and $NO_2$) showed lower absorption capacity. The reason stated is the relation between the quantitative relationship between the basicity of ILs, its absorption enthalpy and $CO_2$ absorption capacity.

(c) **Regeneration of ILs from (IL + $CO_2$) mixture**

The main disadvantage of using ILs is its higher viscosity and higher production cost. To overcome these viscosity issues, researcher is investigating binary mixture of ILs with various solvents, and by separating $CO_2$ from ILs and by reusing the ILs, we can minimize the production cost of ILs. Moreover, directly, ILs will not be useful due to the presence of other components. Among all the regeneration techniques, vacuum swing adsorption (VSA), pressure swing adsorption (PSA) and temperature swing adsorption (TSA) are very important. VSA and PSA are similar; in both the cases, adsorption pressure is higher and desorption pressure is lower. In VSA, adsorption operates at atmospheric pressure and desorption operates at lower pressure, but for PSA, adsorption pressure is higher than atmospheric pressure and desorption pressure is atmospheric pressure. By passing hot air or steam, the regeneration is done in TSA [47]. Huang et al. studied the recycling of amido-based ILs with different anions for $CO_2$ absorption and concluded that up to six cycles, rapid $CO_2$ absorption rates remained which conclude that amido-functionalized ILs are highly recyclable [48].

## 7 Conversion of Carbon Dioxide into Value-Added Chemicals

The captured $CO_2$ from the air or any other sources can be stored or can be utilized for the production of value-added products. Among various value-added products, methanol and formic acid have high interest because of their industrial use as a fuel or to generate $H_2$ gas and methanol, which is the source for formaldehyde, dimethyl ether, light olefins and gasoline [49].

The abundant and easily available carbon resource—$CO_2$—is environmentally friendly gas for making commercial chemicals and materials [50]. Chemical transformation of $CO_2$ into useful compounds by developing a catalytic process is important for green chemistry. However, $CO_2$ is used as raw material by a few industrial processes, because $CO_2$ is the most thermodynamically stable molecule. In this context, the biggest obstacle for establishing industrial processes for $CO_2$ conversion would be due to its low energy level and its inherent thermodynamic stability and kinetic inertness hindering the development of efficient catalysts that achieve $CO_2$ activation and its subsequent functionalization. Only by understanding the underlying principles of $CO_2$ activation, we can effectively use $CO_2$ as an environmentally friendly and economically feasible source of carbon.

## 8 Conclusions and Outlook

The combination of unique properties of ILs made them potential candidates to replace conventional absorbents for $CO_2$ capture, for both flue gases and gas sweetening purposes. Tuning of ILs property by altering its cation and anion makes it a special solvent compared to other organic solvents. The effect of cation and anion of ILs on $CO_2$ solubility and selectivity for chemisorption and physisorption processes is discussed in detail. Chemical equilibrium constants of $CO_2$ with ILs are derived at different temperatures from the "deactivated IL" model. From equilibrium data, a thermodynamic parameter such as Gibbs energy, reaction enthalpy and reaction entropy in the absorption of $CO_2$ is also discussed. Based on the Gibbs free energy change and the amount of IL needed, the energy use of $CO_2$ separation from biogas is also discussed theoretically. Detailed understanding of regeneration of ILs and the conversion of $CO_2$ into value-added products will be a crucial step for commercial applications. There are many optimistic indications that ILs are on their way to becoming a commercial success story [51], and now, ILs moved from academic curiosity to industrial process and products.

**Acknowledgements** The authors are grateful to the Department of Science & Technology (DST), India, for financial support through two projects: EMR/2016/005810 and DST/INT/Portugal/P-01/2017. Indrajit Das would like to thank the Council of Scientific and Industrial Research (CSIR), India, for doctoral fellowship: 09/084(0748)/2019-EMR-I. V. Ramkumar would like to thank the Department of Chemistry, IIT Madras, for permission to pursue a Ph.D. under staff category.

## References

1. Haszeldine RS (2009) Carbon capture and storage: how green can black be? Science 325:1647–1651
2. Ahmad S, Baiocchi G, Creutzig F (2015) $CO_2$ Emissions from direct energy use of urban households in india. Environ Sci Technol 49:11312–11320

3. Gough C (2008) State of the art in carbon dioxide capture and storage in the UK: An experts' review. Int J Greenhouse Gas Control, 2:155–168
4. Rochelle GT (2009) Amine scrubbing for $CO_2$ capture. Science 325:1652–1654
5. MacDowell N, Florin N, Buchard A, Hallett J, Galindo A, Jackson G, Adjiman CS, Williams C, Shah N, Fennell P (2010) An overview of $CO_2$ capture technologies. Energy Environ Sci 3:1645–1669
6. Anthony JL, Maginn EJ, Brennecke JF (2002) Solubilities and thermodynamic properties of gases in the ionic liquid 1-n-butyl-3-methylimidazolium hexafluorophosphate. J Phys Chem B 106:7315–7320
7. Freemantle M (2010) An introduction to ionic liquids. RSC Publishing, London
8. Freemantle M (1998) Designer solvents: ionic liquids may boost clean technology development. Chem Eng News 76:32–37
9. Seddon KR (1997) Ionic liquids for clean technology. J Chem Tech Biotechnol 68:351–356
10. Earle MJ, Seddon KR (2000) Ionic liquids. Green solvents for the future. Pure Appl Chem 72:1391–1398
11. Brennecke JF, Maginn EJ (2001) Ionic liquids: innovative fluids for chemical processing. AIChE J 47:2384–2389
12. Zhao D, Wu M, Kou Y, Min E (2002) Ionic liquids: applications in catalysis. Catal Today 74:157–189
13. Buzzeo MC, Evans RG, Compton RG (2004) Non-haloaluminate room-temperature ionic liquids in electrochemistry-a review. Chem Phys Chem 5:1106–1120
14. Han X, Armstrong DW (2007) Ionic liquids in Separations. Acc Chem Res 40:1079–1086
15. Zhao H (2006) Innovative applications of ionic liquids as "Green" engineering liquids. Chem Eng Commun 193:1660–1677
16. Rogers RD, Seddon K (2003) Ionic liquids-solvents of the future? Science 302:792–793
17. Blanchard LA, Hancu D, Beckman EJ, Brennecke JF (1999) Green processing using ionic liquids and $CO_2$. Nature 399:28–29
18. Anthony JL, Anderson JL, Maginn EJ, Brennecke JF (2005) Anion effects on gas solubility in ionic liquids. J Phys Chem B 109:6366–6374
19. Bates ED, Mayton RD, Ntai I, Davis JH (2002) $CO_2$ capture by task-specific ionic liquids. J Am Chem Soc 124:926–927
20. Song HJ, Park S, Kim H, Gaur A, Park JW, Lee SJ (2012) Carbon dioxide absorption characteristics of aqueous amino acid salt solutions. Int J Greenhouse Gas Control 11:64–72
21. Gurkan BE, de la Fuente JC, Mindrup EM, Ficke LE, Goodrich BF, Price EA, Schneider WF, Brennecke JF (2010) Equimolar $CO_2$ absorption by anion-functionalized ionic liquids. J Am Chem Soc 132:2116–2117
22. Taylor SFR, McCrellis C, McStay C, Jacquemin J, Hardacre C, Mercy M, Bell RG, de Leeuw NH (2015) $CO_2$ capture in wet and dry superbase ionic liquids. J Solution Chem 44:511–527
23. Lee TB, Oh S, Gohndrone TR, Morales-Collazo O, Seo S, Brenneke JF, Schneider WF (2016) $CO_2$ chemistry of phenolate-based ionic liquids. J Phys Chem B 120(8):1509–1517
24. Khanna AK, Sistla YS (2015) $CO_2$ absorption studies in amino acid-anion based ionic liquids. Chem Eng J 273:268–276
25. Bhargava BL, Krishna AC, Balasubramanian S (2008) Molecular dynamics simulation studies of $CO_2$-[bmim][PF6] solutions: Effect of $CO_2$ concentration. AIChE J 54:2971–2978
26. Chandran A, Prakash K, Senapati S (2010) Self-assembled inverted micelles stabilize ionic liquid domains in supercritical $CO_2$. J Am Chem Soc 132:12511–12516
27. Rao SS, Gejji SP (2016) 2009; $CO_2$ absorption using fluorine functionalized ionic liquids: interplay of hydrogen and σ-hole interactions. J Phys Chem A 120:1243–1260
28. Mondal MK, Bajpai A (2013) Equilibrium solubility of $CO_2$ in aqueous mixtures of DEA and AEEA. J Chem Eng Data 58:1490–1495
29. Kundu M, Kumar G (2012) Solubility of $CO_2$ in Aqueous blends of (N-Methyl-2-ethanolamine + N-Methyl-diethanolamine) and (N-Methyl-2-ethanolamine + 2-Amino-2-methyl-1-propanol). J Chem Eng Data 57:3203–3209

30. Ramkumar V, Gardas RL (2019) Thermophysical properties and carbon dioxide absorption studies of guanidinium-based carboxylate ionic liquids. J Chem Eng Data 64(11):4844–4855
31. Goodrich BF, de la Fuente, Gurkan JCBE, Lopez ZK, Price EA, Huang Y, Brennecke JF (2011) Effect of water and temperature on absorption of $CO_2$ by amine-functionalized anion-tethered ionic liquids. J Phys Chem B 115:9140–9150
32. Zhang XM, Huang K, Xia S, Chen YL, Wu YT, Hu XB (2015) Low-viscous fluorine-substituted phenolic ionic liquids with high performance for capture of $CO_2$. Chem Eng J 274:30–38
33. Kazarian SG, Briscoe BJ, Welton T (2000) Combining ionic liquids and supercritical fluids: in situ ATR-IR study of $CO_2$ dissolved in two ionic liquids at high pressures. Chem Commun (Cambridge, U. K.), 2047 – 2048
34. Cabaco MI, Besnard M, Danten Y, Coutinho JAP (2011) Solubility of $CO_2$ in 1-Butyl-3-methyl-imidazolium-trifluoro Acetate Ionic Liquid Studied by Raman Spectroscopy and DFT investigations. J Phys Chem B 115:3538–3550
35. Bhargava BL, Balasubramanian S (2007) Probing anion-carbon dioxide interactions in room temperature ionic liquids: Gas phase cluster calculations. Chem Phys Lett 444:242–246
36. Zhang XC, Huo F, Liu ZP, Wang WC, Shi W, Maginn EJ (2009) Absorption of $CO_2$ in the ionic liquid 1-n-Hexyl-3-methylimidazolium tris(pentafluoroethyl) trifluorophosphate ([hmim][FEP]): A molecular view by computer simulations. J Phys Chem B 113:7591–7598
37. Klahn M, Seduraman A (2015) What determines $CO_2$ solubility in ionic liquids? A molecular simulation study. J Phys Chem B 119:10066–10078
38. Babarao R, Dai S, Jiang D (2011) Understanding the high solubility of $CO_2$ in an ionic liquid with the tetracyanoborate anion. J Phys Chem B 115:9789–9794
39. Huang XH, Margulis CJ, Li YH, Berne BJ (2005) Why is the partial molar volume of $CO_2$ so small when dissolved in a room temperature ionic liquid? Structure and dynamics of $CO_2$ Dissolved in [Bmim][$PF_6$]. J Am Chem Soc 127:17842–17851
40. Zhang JZ, Jia C, Dong HF, Wang JQ, Zhang XP, Zhang SJ (2013) A novel dual amino-functionalized cation-tethered ionic liquid for $CO_2$ capture. Ind Eng Chem Res 52:5835–5841
41. Cao BB, Du JY, Liu SY, Zhu X, Sun XJ, Sun HT, Fu H (2016) Carbon dioxide capture by amino-functionalized ionic liquids: DFT based theoretical analysis substantiated by FT-IR investigation. RSC Adv 6:10462–10470
42. Zhang SJ, Yuan XL, Chen YH, Zhang XP (2005) Solubilities of $CO_2$ in 1-butyl-3-methylimidazolium Hexafluorophosphate and 1,1,3,3-tetramethylguanidium Lactate at Elevated Pressures J Chem Eng Data 50:1582–1585
43. Wang CM, Luo HM, Li HR, Zhu X, Yu B, Dai S (2012a) Tuning the physicochemical properties of diverse phenolic ionic liquids for equimolar $CO_2$ capture by the substituent on the anion. Chem Eur J 18:2153–2160
44. Wang CM, Luo HM, Li HR, Zhu X, Yu B Dai S (2012b) Tuning the physicochemical properties of diverse phenolic ionic liquids for equimolar $CO_2$ capture by the substituent on the anion. Chem Eur J 18:2153–2160
45. Aki S, Mellein BR, Saurer EM, Brennecke JF (2004) High-pressure phase behavior of carbon dioxide with imidazolium-based ionic liquids. J Phys Chem B 108:20355–20365
46. Muldoon MJ, Aki SNVK, Anderson JL, Dixon JK, Brennecke JF (2007) Improving carbon dioxide solubility in ionic liquids. J Phys Chem B 111:9001–9009
47. Cheng-Tung C, Chao-Yuh C (2004) Carbon dioxide recovery by vacuum swing adsorption. Sep Purif Technol 39:51–65
48. Huang Y, Cui G, Wang H, Li Z, Wang J (2019) Absorption and thermodynamic properties of $CO_2$ by amido-containing anion-functionalized ionic liquids. RSC Adv. 9:1882–1888
49. Kar S, Goeppert A, Surya Prakash GK (2019) Integrated $CO_2$ capture and conversion to formate and methanol: Connecting two threads. Acc Chem Res 52, 10:2892–2903
50. Arakawa H, Aresta M, Armor JN, Barteau MA, Beckman EJ, Bell AT, Bercaw JE, Creutz C, Dinjus E, Dixon DA, Domen K, DuBois DL, Eckert J, Fujita E, Gibson DH, Goddard WA, Goodman DW, Keller J, Kubas GJ, Kung HH, Lyons JE, Manzer LE, Marks TJ, Morokuma K, Nicholas KM, Periana R, Que L, Rostrup-Nielson J, Sachtler WM, Schmidt LD, Sen A, Somorial GA, Stair PC, Stults BR Tumas W (2001) Catalysis research of relevance to carbon management: progress, challenges, and opportunities. Chem Rev 101:953–996

51. Shiflett MB (2020) Commercial applications of ionic liquids. Springer Nature Switzerland AG
52. Zhang X, Liu Z, Wang W (2008) Screening of ionic liquids to capture $CO_2$ by COSMO-RS and experiments. AIChE J 54:2717–28
53. Perez-Salado Kamps A, Tuma D, Xia J, Maurer G (2003) Solubility of $CO_2$ in the ionic liquid [bmim][$PF_6$]. J Chem Eng Data 48:746–749
54. Shiflett MB, Yokozeki A (2005) Solubilities and diffusivities of carbon dioxide in ionic liquids: [bmim][$PF_6$] and [bmim][$BF_4$]. Ind Eng Chem Res 44:4453–4464
55. Blanchard LA, Gu Z, Brennecke JF (2001) High-pressure phase behavior of ionic liquid/$CO_2$ systems. J Phys Chem B 105:2437–44
56. Shin E-K, Lee B-C, Lim JS (2008) High-pressure solubilities of carbon dioxide in ionic liquids: 1-alkyl-3-methylimidazolium bis (trifluoromethylsulfonyl) imide. J Supercrit Fluids 45:282–292

# Heterogeneous Catalysis for Chemical Fixation of $CO_2$ via Carbonylation Reactions

**Ganapati V. Shanbhag, Nagendra Kulal, and B. J. Vaishnavi**

**Abstract** One of the major constituents of greenhouse gases is carbon dioxide whose concentration in the atmosphere is increasing at an alarming rate due to the disorganized human activities. Therefore, capturing $CO_2$ and utilizing it to make various value-added chemicals is regarded as a green transformation, which thereby balances the carbon footprint. In this chapter, the advances in $CO_2$ as the $C_1$ source by utilizing it as a carbonylating agent for important organic transformations for the synthesis of cyclic carbonate, substituted urea, cyclic urea, carbamates, glycerol carbonate and dimethyl carbonate are discussed. These products have a wide range of applications as specialty solvents, starting materials/intermediates for polymer, paint, agrochemicals, pesticides, herbicides and pharmaceutical industries. Also, a few products like dimethyl carbonate have potential applications as fuel additives. Although $CO_2$ is thermodynamically and kinetically stable molecule, it can be activated by basic sites in the catalyst due to the electron deficiency of the carbonyl carbon at appropriate reaction condition. Various catalyst systems such as metal oxides, mixed metal oxides, supported metal oxides, metal-organic framework, bifunctional catalysts and importance of solvent have been highlighted and discussed in detail for the efficient production of these commercially important chemicals from $CO_2$.

**Keywords** $CO_2$ · Carbonylation · Heterogeneous catalyst · Organic carbonate · Substituted urea

## Abbreviations

| | |
|---|---|
| ACN | Acetonitrile |
| BBA | N-benzylidenebenzylamine |
| BDC | 1,4-Benzenedicarboxylicacid |
| CDI | 1,1'-Carbonyldiimidazole |

G. V. Shanbhag (✉) · N. Kulal · B. J. Vaishnavi
Materials Science and Catalysis Division, Poornaprajna Institute of Scientific Research (PPISR), Bidalur Post, Devanahalli, Bengaluru 562164, India
e-mail: shanbhag@poornaprajna.org

 
M. Goel et al. (eds.), *Climate Change and Green Chemistry of $CO_2$ Sequestration*, Green Energy and Technology, https://doi.org/10.1007/978-981-16-0029-6_10

| | |
|---|---|
| CLZ | Conventional co-precipitation method |
| $COCl_2$ | Phosgene |
| 2-Cypy | 2-Cyanopyridine |
| DBU | Dibenzylurea |
| DMC | Dimethyl carbonate |
| DMF | N,N-Dimethylformamide |
| DMSO | Dimethyl sulfoxide |
| DPU | Diphenyl urea |
| ECH | Epichlorohydrin |
| GC | Glycerol carbonate |
| MBA | N-methylbenzylamine |
| MBC | Methyl benzyl carbamate |
| MDI | 4,4'-Diphenylmethane diisocyanate |
| MeOH | Methanol |
| MOF | Metal organic framework |
| MPC | Methyl-N-phenylcarbamate |
| MS | Molecular sieve |
| NLZ | Novel ethylene glycol combustion method |
| PC | Propylene carbonate |
| TBAB | Tetra-n-butylammonium bromide |
| TOF | Turnover frequency |
| TPD | Temperature programmed desorption |
| XDI | *p*-Xylylene diisocyanate |
| ZnGly | Zinc glycerolate |

# 1 Introduction

$CO_2$ activation and valorization into value-added products is an emerging area of research considering the increasing environmental concerns caused by the emission of $CO_2$ into the atmosphere by various modes resulting in air pollution leading to health hazards [1]. Especially, air pollution due to $CO_2$ emission is increasing in countries like China and India, where energy-intensive productive activities based on coal are largely expanding thanks to local large reserve of this form of fossil carbon. The best way to address this problem is by preventing the emission of $CO_2$ into the atmosphere. However, it is very difficult to prevent the $CO_2$ emission totally because of the dependency of human life on chemicals and fuels which form $CO_2$ during their synthesis, consumption as well as their disposal. Hence, conversion of $CO_2$ into chemicals is today considered with much attention as it could contribute to cycling of carbon and mimicking nature that makes thousands of compounds from atmospheric $CO_2$ [2]. Such conversions can be done either directly with the out-gases rich in $CO_2$ from chemical industries or the captured $CO_2$ from various sources. $CO_2$ can be converted into various chemicals such as cyclic carbonates, carbamates, substituted urea and urethanes by reacting with raw materials like epoxides, amines, alcohols and alkyl halides, respectively, via carbonylation reaction [1]. Chemicals like alcohols, ethers, olefins, paraffins, formaldehyde and formic acid can be obtained by catalytic hydrogenation of $CO_2$ [3]. Synthesis of carboxylic acid, esters and isocyanates is other important products which can be obtained by $CO_2$ reactions. It is challenging for researchers to activate $CO_2$ and convert to chemicals because it is chemically inert in nature. Several homogeneous catalysts were reported for the catalytic conversion of $CO_2$ in liquid phase batch reactions for the synthesis of organic carbonates, cyclic urea, substituted urea, carbamates, urethanes, carboxylic acids, etc. [1, 2]. However, homogeneous catalysts suffer from drawbacks such as tedious and expensive work-up for catalyst recovery and recycle [4]. The small amount of the catalyst may remain in the product as impurity causing the issues for further applications. The homogeneous catalysts also suffer from low thermal stability and difficulty in handling and storage as many of them are moisture and air sensitive in nature. They can be easily deactivated due to trace impurities. Hence, heterogeneous catalyst without any such drawbacks is preferred which activates $CO_2$, enhances reaction rates and can be separated and recycled easily after the reaction.

$CO_2$ carbonylation products have several important commercial applications reported in the literature. Cyclic carbonates synthesized by cycloaddition of $CO_2$ to epoxides are widely used as electrolyte components in lithium batteries, polar solvents and chemical intermediates and also a monomer for polycarbonate synthesis [5, 6]. Presently, a poisonous gas, phosgene ($COCl_2$) is used for this purpose. Organic carbamates are industrially important raw materials in terms of sustainable chemistry as they can be thermally decomposed into isocyanates such as 4,4'-diphenylmethane diisocyanate (MDI), *p*-xylylene diisocyanate (XDI) and 1,4-bis(isocyanatemethyl)cyclohexane which are well-known as high-value chemicals mainly used as starting materials for polyurethane synthesis. This polymer has several applications for making coatings, foams, adhesives, fibres and plastics [7]. Additionally, they are also used as intermediates of agrochemicals, pesticides, herbicides, pharmaceutical agents and the protection of functional groups, such as an amine. Substituted ureas are synthesized from carbonylation of amines with $CO_2$. These are useful as chemical intermediates in the synthesis of pharmaceuticals, agricultural chemicals, and in dye chemistry [8]. Glycerol carbonate is an industrially important product having wide applications in various fields like the pharmaceutical, polymer and paint industries [9]. It is obtained from carbonylation of $CO_2$ with glycerol, a biorefinery by-product from biodiesel synthesis. On the other hand, dimethyl carbonate formed by methanol and $CO_2$ reaction is a fast evaporating polar solvent. DMC can be used as carbonylating and methylating agent in organic synthesis [10]. It can also be used as a fuel additive due to its high oxygen content.

In this chapter, the design of solid catalysts for the utilization of $CO_2$ by converting into commercially important products via carbonylation reactions is discussed. These are mainly acid-base bifunctional catalysts which are invariably required as $CO_2$ is activated by basic active sites, whereas the substrate is activated by the acidic active centres present in the catalyst. However, the amount, proximity, nature and strength of active sites play an important role in getting high efficiency in terms of conversion of the raw material and selectivity of the product. Other parameters are surface area, particle size and pore size of the catalyst which also may play important part in enhancing the efficiency of the catalyst.

# 2 $CO_2$ as Carbonylating Agent for Commercially Important Reactions

## *2.1 Carbonylation (Cycloaddition) of $CO_2$ with Epoxide to Produce Cyclic Carbonate*

As a potential pathway for the effective utilization of $CO_2$, the synthesis of cyclic carbonate from cycloaddition of $CO_2$ to epoxides has received much attention in terms of green chemistry and atom economy as there is no formation of any by-products [11, 12]. Currently, there is considerable interest in cyclic carbonate synthesis because it can be used in many areas of applications. Cyclic carbonates have been used as intermediate for polycarbonate [13], polyurethanes [14] and dimethyl carbonate [15], solvent in lithium ion batteries [16], paint, personal care and cosmetics products [6, 17]. Cyclic carbonate synthesis has been successfully performed from epoxide and $CO_2$ in the industry [11]. The reaction of $CO_2$ oxidative carboxylation of olefin and $CO_2$ reaction with glycol are the other two routes for the synthesis of cyclic carbonates [11]. Studies have revealed that the synthesis of cyclic carbonate is thermodynamically favoured over the polymerization process of polycarbonate [18]. Both homogenous and heterogeneous catalysts systems have been developed for cyclic carbonate production. Homogenous catalysts such as quaternary ammonium salt, metal halide with melamine, metal halide with ionic liquid, salen metal complex and ionic liquid have been reported as catalysts for $CO_2$ and epoxide reaction [12]. Heterogeneous catalysts such as polymer, MOF, supported metal complexes, metal supported on polymer, supported quaternary ammonium salt, amine functionalized silica, supported ionic liquids, halide supported mesoporous carbon nitride, and individual and mixed oxide have been reported [19].

$CO_2$ is thermodynamically and kinetically stable molecule. However, due to the electron deficiency of the carbonyl carbon, $CO_2$ has a strong affinity towards nucleophiles and electron donating groups [1]. In other words, $CO_2$ gets activated on the basic sites of the catalyst, whereas the acid sites on the catalyst surface help to activate the epoxide. The C–O bond of the epoxide is loosened and electron-rich oxygen attacks the carbonyl carbon of the $CO_2$ to form cyclic carbonate. Thus, a catalyst with acid-base bifunctional property favours the cycloaddition of $CO_2$ with epoxide. Two recent works, one on metal organic framework catalyst and the other on mixed metal oxide catalyst are discussed in detail in the chapter.

Metal organic framework (MOF), a class of crystalline organic-inorganic porous material composed of metal ion and ligands, has been used as catalyst for different applications [20]. Their crystalline nature, tunable structure, high surface area and thermal stability due to their diverse functional groups show distinct properties. One such application is cycloaddition of $CO_2$ with epoxide to cyclic carbonate synthesis. Multi-variate MOFs by incorporating different functionalized ligands into a single framework have been reported [21]. They functionalized 1,4-benzenedicarboxylicacid (BDC) to make 2-aminoterephthalicacid and 2-hydroxyterephthalicacid which were combined to form a different set of multi-variate MOFs. Multi-variate of MOF-5 consisting of $Zn_4O$ secondary building units and benzene-dicarboxylate type linkers [22]. The 3-D framework of MOF-5 consists of $[Zn_4O]^{6+}$ clusters in octahedral subunits connected by BDC linkers. The number of acidic and basic sites of the catalysts was listed in Table 1 (Scheme 1) [21].

$NH_3$ and $CO_2$ temperature programme desorption (TPD) was used mainly to quantify the acidic and basic sites. Chemically functionalized MOFs like MOF-5-$NH_2$ and MOF-5-MIX shows a higher number of basic sites and acidic sites compared to the other synthesized catalysts. Basicity from $CO_2$ chemisorption technique for MOF-5-$NH_2$ and MOF-5-MIX was estimated to be 0.6 and 5.8 mmol/g, respectively. $NH_2$ functionalized MOF framework is expected to be responsible for $CO_2$ adsorption, whereas metal (Zn atom) in the framework is responsible for the acidic sites to activate the epoxide.

Cycloaddition of $CO_2$ with epoxide reaction was performed in a stainless-steel reactor, and reaction parameters were varied to achieve the best performance. Under catalyst-free condition, cycloaddition reaction did not convert epichlorohydrin (ECH) even after 6 h of reaction time (Table 2, Entry 1). Homogeneous catalyst precursors

**Table 1** Concentration of acidic and basic sites in the catalysts [21] Copyright (2020) Elsevier

| Entry | Catalyst | Acidic sites $NH_3$-TPD (mmol/g) | Basic sites $CO_2$-TPD (mmol/g) |
|---|---|---|---|
| 1 | MOF-5 | 2.34 | 0.38 |
| 2 | MOF-5-$NH_2$ | 2.34 | 0.60 |
| 3 | MOF-5-OH | 4.20 | 0.52 |
| 4 | MOF-5-MIX | 5.90 | 0.58 |

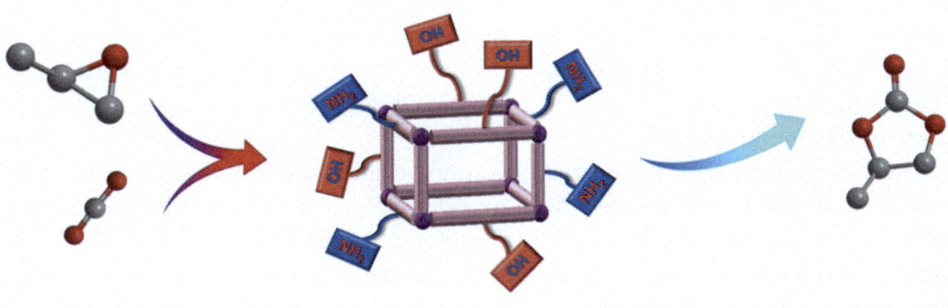


**Scheme 1** Schematic representation of the synthesis of cyclic carbonates from epoxides and $CO_2$. Reprinted with permission from Ref. [21]. Copyright (2020) Elsevier

**Table 2** Catalytic activity comparison of all the catalysts for the cycloaddition reaction of ECH and $CO_2$ [21] Copyright (2020) Elsevier

| Entry | Catalyst | Conversion[d] (%) | Selectivity[d] (%) |
|---|---|---|---|
| 1 | None | – | – |
| 2 | $Zn(NO_3)_2 \cdot 6H_2O$ | 19 | 98 |
| 3 | $H_2BDC$ | 9 | 95 |
| 4 | $H_2BDC$-$NH_2$ | 15 | 98 |
| 5 | $H_2BDC$-OH | 12 | 96 |
| 6 | MOF-5/TBAB | 40 | >99 |
| 7[a] | MOF-5/TBAB | 95 | >99 |
| 8 | MOF-5-$NH_2$/TBAB | 81 | >99 |
| 9 | MOF-5-OH/TBAB | 69 | >99 |
| 10 | MOF-5-MIX/TBAB | 98 | >99 |
| 11[b] | MOF-5-MIX | 94 | 97 |
| 12[c] | MOF-5-MIX | 99 | 98 |

Reaction condition: Epichlorohydrin (ECH) = 25 mmol, Catalyst = 0.5 mol%, TBAB = 0.5 mol%, Pressure = 1.2 MPa $CO_2$, Temperature = 50 °C, Time = 6 h, semi-batch reaction
[a]Temperature = 80 °C, catalyst = 1.0 mol% TBAB = 1.0 mol%
[b]Catalyst = 0.7 mol%, no TBAB, Pressure = 1.2 MPa $CO_2$, Temperature = 110 °C, Time = 8 h
[c]Catalyst = 0.7 mol%, no TBAB, Pressure = 1.2 MPa $CO_2$, Temperature = 120 °C, Time = 8 h
[d]Determined by GC

tested for their individual catalytic performance and ECH carbonate yields were much lower than MOF-5. MOF-5 in the presence of tetra-n-butylammonium bromide (TBAB) as a co-catalyst showed 40% conversion of ECH with 99% of selectivity for ECH carbonate under the same reaction condition. Notably, the conversion of ECH increased when functionalized MOF-5 catalyst was applied under the optimized condition. MOF-5-OH catalyst showed lower catalytic activity than the MOF-5-$NH_2$ due to its lower $CO_2$ chemisorption capacity as well as the lower number of basic active sites. The multi-functionalized MOF-5-MIX catalyst showed 98% of ECH conversion and 99% of selectivity for ECH carbonate. Both functionalized –$NH_2$ and –OH groups accelerated the rate of the cycloaddition of $CO_2$ with epoxide reaction which has been computationally and experimentally proved by previous reports [23, 24]. The presence of –OH functional group in the framework could activate the ring opening of epoxide, whereas –$NH_2$ functional group could help to polarize $CO_2$ in the reaction [25].

Kulal et al. reported the mixed metal oxide as an acid-base bifunctional catalyst for cycloaddition of $CO_2$ with propylene oxide which gives mainly propylene carbonate and a few minor side products such as acetone, propionaldehyde and 2-ethyl-4-methyl-1,3-dioxolane [19]. The absence of the catalyst and solvent gave a trace amount of propylene carbonate yield (~0.2). Under the same reaction condition, the addition of solvent, N,N-dimethylformamide (DMF) increased the yield to 10% which could be attributed to activation of $CO_2$ by DMF due to its basic nature. The blank reaction studies revealed that there was a need to design a suitable catalyst to further increase the product yield. Authors synthesized different mono and binary mixed metal oxides which contained both acidic and basic bifunctional nature but differ in amount, nature and strength of these active sites. Binary mixed oxides were chosen based on the combination of transition metals (Mn, Fe, Co, Ni, Cu and Zn) with alkaline earth metal (Ba) and post transition metal (Sn). Mixed oxide catalyst with 1:1 metal ratio was synthesized by co-precipitation method, and the reaction was carried out under 135 °C, 40 bar $CO_2$ pressure, 1.8 g propylene oxide and DMF as solvent (10 ml). The activity of the synthesized binary oxides was found to have a good correlation with a total number of both weak and medium active sites as shown in Fig. 1. As the amount of weak and medium acid + base sites increased, propylene carbonate yield increased irrespective of the type of metal oxide present in the catalyst. Further, there was no correlation of catalytic activity found for any other catalyst properties like only acid sites, only basic sites, nature of metal oxide, active site density, etc. This is an important finding which shows that the combined number of acid and base sites is decisive parameter to get better catalytic activity for this reaction.

Among the screened mixed oxides, Sn-Ni oxide gave the highest propylene carbonate (PC) yield of 73.8% followed by Mn–Ba oxide (69.2%). Figure 1 shows that cycloaddition of $CO_2$ with propylene oxide could be mainly driven by the amount of acid and base sites in the mixed oxides. When the molar ratio of metals in the mixed oxide was varied, properties of mixed oxides changed which led to the difference in catalytic activity. Hence, the active sites could be tuned by changing the molar ratio of metals in the composition of mixed oxide as well as by calcination temperature. For Mn–Ba mixed oxide, Mn to Ba metal molar ratio was varied to obtain Mn:Ba ratios from 0.8 to 7.4 and catalytic activity was tested for propylene carbonate synthesis. Mn–Ba(4.3:1) oxide catalyst containing similar amount of acidity (92 μmol/g) and

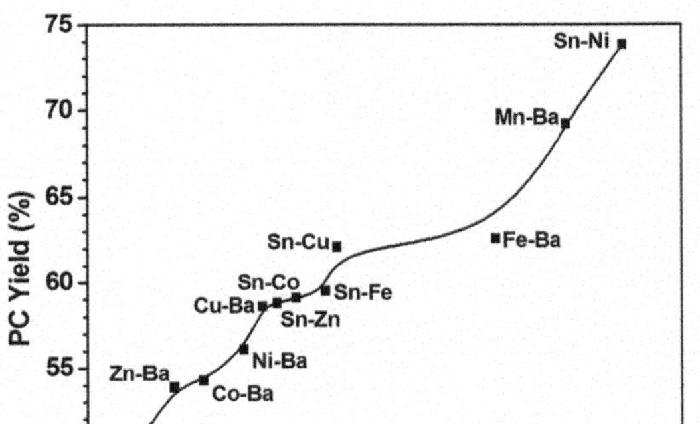


**Fig. 1** Correlation of active sites with catalytic activity for mixed oxides calcined at 600 °C Reaction condition: 1.8 g propylene oxide, 40 bar $CO_2$ pressure, 11 ml DMF, 135 °C, 10 h, 20 wt% w. r. t. propylene oxide. Reprinted with permission from Ref. [19]. Copyright (2019) Elsevier

basicity (111 μmol/g) showed a better activity for cycloaddition of $CO_2$ with propylene oxide reaction. Similarly, Sn–Ni oxides with different metal compositions, Sn–Ni(1.5:1) oxide showed the best performance with 97.6% propylene oxide conversion and 90.2% PC yield under optimized reaction condition. Mn–Ba(4.3:1) and Sn-Ni(1.5:1) oxide catalysts were tested for different epoxide and $CO_2$ reactions (Table 3). For different epoxides, corresponding carbonate yield decreased in the order; propylene oxide > epichlorohydrin > styrene oxide > cyclohexene oxide.

Propylene oxide and epichlorohydrin gave high cyclic carbonate yield and catalytic activity decreased with an increase of steric hindrance caused by side chain substituent of epoxides [26]. Also, electron donating $-CH_3$ group seems to have a positive effect on the epoxide conversion, whereas electron withdrawing –Cl and phenyl groups decreased the epoxide conversion. On the other hand, cyclohexene oxide (33–37% yield) is distinctively different from other epoxides. It is a bicyclic compound where epoxide ring is fused with 6-membered ring and therefore ring opening is more difficult compared to other epoxides.

**Table 3** Cycloaddition of reaction of $CO_2$ with various epoxides catalysed by Mn–Ba (4.3:1) and Sn-Ni (1.5:1) oxides calcined at 600 °C. Ref. [19]. Copyright (2019) Elsevier

| Reactant | Product | Catalyst | Time (h) | Carbon balance (%) | Conv. % | Yield % |
|---|---|---|---|---|---|---|
| 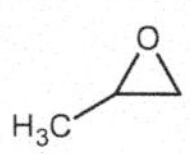 | 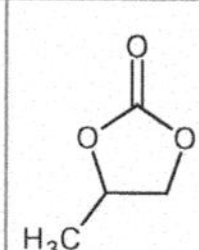 | Blank | 6 | 96.5 | 20.5 | 19.1 |
| | | Mn–Ba(4.3:1) | 6 | 95.3 | 98.1 | 96.0 |
| | | Sn-Ni(1.5:1) | 8 | 97.6 | 96.9 | 90.2 |
| 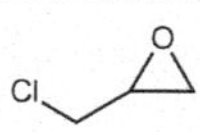 | 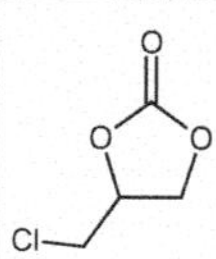 | Blank | 6 | 98.3 | 11.9 | 11.6 |
| | | Mn–Ba(4.3:1) | 6 | 96.1 | 94.2 | 87.7 |
| | | Sn-Ni(1.5:1) | 8 | 96.9 | 93.5 | 85.4 |
| 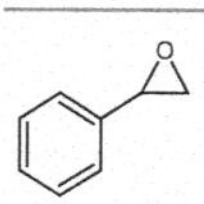 | 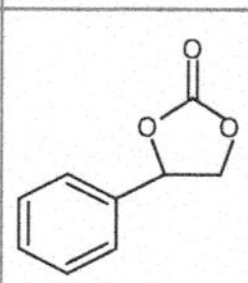 | Blank | 6 | 97.7 | 9.5 | 9.3 |
| | | Mn–Ba(4.3:1) | 6 | 95.2 | 89.8 | 82.3 |
| | | Sn-Ni(1.5:1) | 8 | 96.0 | 87.4 | 75.3 |
| 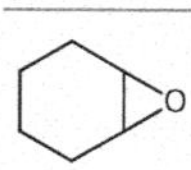 | 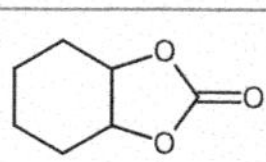 | Blank | 6 | 99.1 | Trace | 0.0 |
| | | Mn–Ba(4.3:1) | 6 | 94.7 | 41.7 | 36.5 |
| | | Sn-Ni(1.5:1) | 8 | 96.1 | 38.1 | 33.3 |

Reaction condition: 30.1 mmol of epoxides, 160 °C, 25 bar pressure, 11 ml DMF (solvent), 20 wt% of catalyst (w.r.t. epoxide)

## 2.2 Synthesis of Substituted Urea and Cyclic Urea from Carbonylation of Amine with $CO_2$

Disubstituted urea and cyclic urea are significant chemical intermediates and industrial raw materials. They are widely used in the synthesis of pharmaceuticals, agricultural chemicals, dyes as well as antioxidants in gasoline and additives in plastics. Especially cyclic urea, a heterocyclic motif, is frequently observed in biologically active molecules [27]. Urea derivatives have been traditionally synthesized with phosgene route. The phosgene process has significant drawbacks such as toxicity of phosgene and produces a large amount of salts as a by-product by the neutralization. Urea derivatives also synthesized from various methods such as reductive carbonylation of amines with CO, oxidative carbonylation of amines with CO, transamidation of ureas with amine, reaction of carbonates with amines and reaction of $CO_2$ and amines [28, 29]. Among these methods, direct synthesis of urea derivatives from amine and $CO_2$ is promising, because water is the only by-product without generating any toxic side products. Heterogeneous catalyst and non-catalytic routes are more desirable than the homogeneous catalysts from the environmental, industrial and economical viewpoints. Only a few heterogeneous catalysts were reported for dialkylurea synthesis such as polymer supported Au nanoparticles [30], KOH/PEG [31] and Mg–Al hydrotalcite [32]. As for cyclic urea synthesis, two effective catalysts were reported such as polyethylene-glycol supported potassium hydroxide (KOH/PEG1000) [31] and nanoparticulate $CeO_2$ prepared by a biopolymer-template process [33]. Among the heterogeneous catalysts, Tomishige and co-workers reported pure $CeO_2$ as the most efficient catalyst for both dialkylurea and cyclic urea synthesis [8, 27]. $CeO_2$ gave better yield than other tested individual oxides like ZnO, $ZrO_2$, $Sm_2O_3$, $La_2O_3$, $Pr_6O_{11}$, $\gamma$-$Al_2O_3$, $Y_2O_3$, MgO, $TiO_2$ and $SiO_2$. Various metal oxides were tested as catalyst for the synthesis of dialkylurea and cyclic urea from amine and $CO_2$. The results of TOF, conversion of amine and selectivity for desired product are show in Tables 4 and 5 (Scheme 2).

The metal oxide such as activated forms of ZnO and $CeO_2$ gave higher conversion, selectivity and turn over frequency (TOF) than the other metal oxides. In the case of ZnO, the insoluble solid material was observed in the reaction mixture, particularly on the surface of ZnO which implies that the polymerization reaction had occurred. The produced polymer may cover the active sites of ZnO leading to its deactivation [27]. Pure $CeO_2$ is therefore the most efficient catalyst for dialkylurea and cyclic urea synthesis among the various metal oxides tested. It is also observed that the solvent plays an important role in the dialkylurea synthesis. To study the solvent effect, the reactions were carried out using various solvents in the absence and presence of $CeO_2$. Solvent DMSO (Dimethyl sulfoxide) showed the highest conversion with moderate selectivity in the absence of $CeO_2$ catalyst followed by N-methylpyrrolidone (NMP). Other solvents such as THF, DME, benzene, 2-propanol, $CH_3CN$, methanol and sulfolane showed relatively lower yields. The rate of formation of dibutylurea in the case of DMSO was the highest among the solvents. In the presence of $CeO_2$, DMSO also showed higher conversion than the other solvents used.

**Table 4** Catalyst screening for the synthesis of dibutylurea from $CO_2$ and n-butylamine [8] Copyright (2016) Elsevier

| Entry | Metal oxide | $S^a(m^2g^{-1})$ | $MO_x$ amount (mmol) | t (h) | Conv. (%) | Selec. (%) | $V_1$ (mmol $h^{-1}$ $g^{-1}$) | $V_2$ (mmol $h^{-1}$ $m^{-2}$) |
|---|---|---|---|---|---|---|---|---|
| 1 | $CeO_2$ | 88 | 0.1 | 1 | 24 | 94 | 77 | 0.87 |
| 2[b] | | | 2 | 4 | 73 | 99 | – | – |
| 3 | ZnO | 66 | 0.1 | 1 | 10 | 86 | 33 | 0.50 |
| 4[b] | | | 2 | 4 | 42 | 97 | – | – |
| 5 | $ZrO_2$ | 46 | 0.1 | 1 | 8 | 86 | 12 | 0.25 |
| 6 | | | 2 | 4 | 23 | 98 | – | – |
| 7 | $Sm_2O_3$ | 38 | 0.1 | 1 | 9 | 77 | 9 | 0.23 |
| 8[b] | | | 2 | 4 | 31 | 94 | – | – |
| 9 | $La_2O_3$ | 28 | 0.1 | 1 | 9 | 70 | 2.8 | 0.10 |
| 10 | $Pr_6O_{11}$ | 30 | 0.1 | 1 | 7 | 79 | 0.2 | 0.01 |
| 11 | $\gamma$-$Al_2O_3$ | 97 | 0.1 | 1 | 8 | 66 | <0.1 | <0.01 |
| 12 | $Y_2O_3$ | 70 | 0.1 | 1 | 7 | 70 | <0.1 | <0.01 |
| 13 | MgO | 46 | 0.1 | 1 | 6 | 76 | <0.1 | <0.01 |
| 14 | $TiO_2$ | 55 | 0.1 | 1 | 5 | 70 | <0.1 | <0.01 |
| 15 | $SiO_2$ | 535 | 0.1 | 1 | 8 | 68 | <0.1 | <0.01 |
| 16 | None | – | – | 1 | 8 | 72 | – | – |

Reaction condition: metal oxide 0.1 mmol (based on metal), n-butylamine 20 mmol, N-methylpyrrolidone solvent 8 ml, Ar 0.1 MPa, 160 °C, $CO_2$ 5 MPa (at 160 °C)

[a]Specific surface area determined by BET method

[b]$CO_2$ 5 MPa (at RT)

**Table 5** Reaction of ethylenediamine and $CO_2$ in methanol to synthesize cyclic urea over various metal oxides[a]. Ref. [27]. Copyright (2013) The Royal Society of Chemistry

| Entry | Catalyst | SA ($m^2g^{-1}$) | Conversion (%) | Selectivity (%) | | TOF[b] ($h^{-1}$) |
|---|---|---|---|---|---|---|
| | | | | 1 | others | |
| 1 | Pure $CeO_2$ | 78 | 22 | 93 | 7 | 4.0 |
| 2[c] | Pure $CeO_2$ | 78 | 98 | 96 | 4 | – |
| 3 | ZnO | 5.8 | 40 | 68 | 32 | 5.4 |
| 4[d] | ZnO | 5.8 | 42 | 7 | 93 | – |
| 5 | CaO | 12 | 3.1 | 32 | 68 | 0.20 |
| 6 | $La_2O_3$ | 7.0 | 5.0 | 16 | 84 | 0.16 |
| 7 | $TiO_2$ | 48 | 2.0 | 95 | 5 | 0.38 |
| 8 | MgO | 37 | 2.6 | 23 | 77 | 0.12 |
| 9 | $ZrO_2$ | 91 | 2.1 | 90 | 10 | 0.38 |
| 10 | $Pr_6O_{11}$ | 33 | 5.5 | 13 | 87 | 0.14 |
| 11 | $Al_2O_3$ | 182 | 0.8 | 88 | 12 | 0.14 |
| 12 | None | – | 0.8 | 87 | 13 | – |

[a]Reaction condition: ethylenediamine (10 mmol), methanol (6.4 g), metal oxide (metal: 0.2 mmol), $PCO_2 = 0.5$ MPa, T = 160 °C, t = 1 h. [b]Molar ratio of the converted substrate to the total amount of the metal atom per hour. [c]12 h. [d]4 h

**Scheme 2** Substituted urea and cyclic urea synthesis from carbonylation of amine with $CO_2$

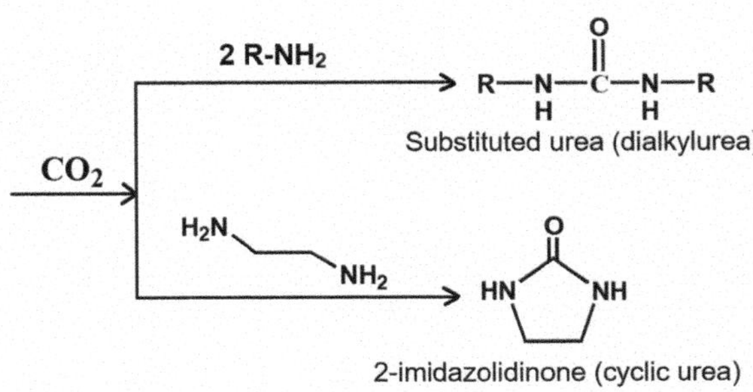


The total rate of formation of dibutylurea with $CeO_2$ catalyst under different solvents and blank reaction with solvent was plotted v/s dielectric constant of the solvent as shown in Fig. 2. The effect of solvent study for dibutylurea synthesis indicates that the solvent with too high polarity is not preferable due to the deactivation of $CeO_2$ by adsorption of the solvent on the surface of the catalyst. The difference in the rates v/s dielectric constant plot shows that the combination of NMP solvent and $CeO_2$ catalyst is better for dibutylurea synthesis (Fig. 2c).

The solvent study for the synthesis of 2-imidazolidinone from ethylenediamine and $CO_2$ is given in Fig. 3 [34]. For this reaction, isopropyl alcohol (IPA) solvent showed higher conversion and selectivity, whereas $CH_3CN$ and NMP exhibited lower selectivity for the desired product. The formation of side products in presence IPA

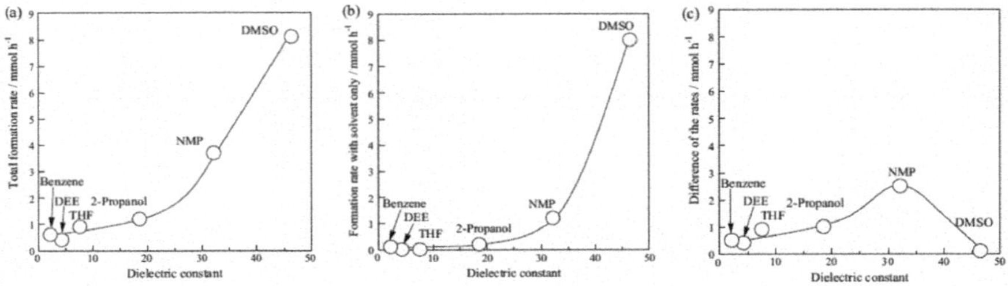


**Fig. 2** Correlations between the formation rates and the dielectric constant of solvents: **a** Total formation rate is the rate measured using $CeO_2$ and **b** The formation rate with solvent only without a catalyst **c** Difference of the rates v/s dielectric constant. Reprinted with permission from Ref. [8]. Copyright (2016) Elsevier

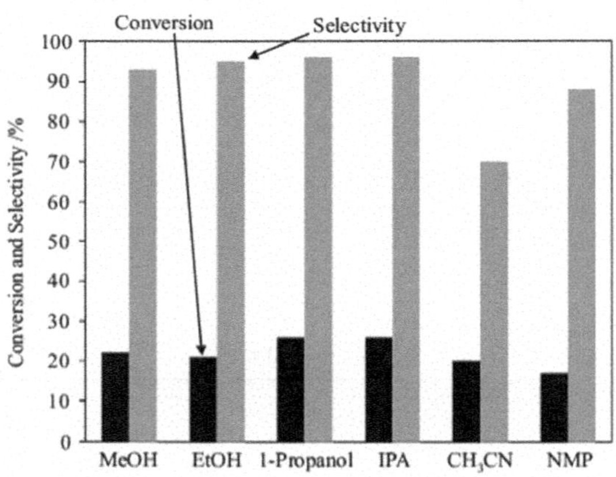


**Fig. 3** Solvent effect for the synthesis of 2-imidazolidinone from ethylenediamine and $CO_2$ over $CeO_2$. Reaction condition: 10 mmol ethylenediamine, 200 mmol solvent, 0.086 g $CeO_2$, 0.5 MPa of $CO_2$ pressure, 160 °C, 1 h. Reprinted with permission from Ref. [27]. Copyright (2013) The Royal Society of Chemistry

is lower compared with other solvents. It is also shown that the branched alcohol like IPA performs better than linear alcohols due to steric hindrance which inhibits the alcohol and $CO_2$ side reaction. It is interesting to note that the carbonylation of monoamine with $CO_2$ to form dialkylurea requires polar aprotic solvents, whereas the formation of cyclic urea from diamine and $CO_2$ reaction is promoted by polar protic solvents. Among various catalysts, thermally stable and reusable $CeO_2$ with a simple method of preparation was a highly effective catalyst for carbonylation of mono and diamine with $CO_2$.

## 2.3 *Direct Catalytic Synthesis of Carbamates from $CO_2$, Anilines and Alcohols*

Carbamates are attractive chemicals as their applications range from agrochemicals, protecting groups, and pharmaceuticals to the synthesis of polyurethanes. Traditionally they are produced using toxic reagents such as phosgene or carbon monoxide. Other alternative routes are aminolysis of carbonates, coupling reactions of dimethyl carbonate and N, N'-substituted urea, alcoholysis of urea, etc. which are eco-friendly [7]. Usage of $CO_2$ as carbonyl source has influenced a lot of researchers as it encourages carbon balance in the environment. Variety of carbamates have been synthesized using various heterogeneous catalyst systems such as ionic liquids [35], polymer- and carbon nanotubes supported catalysts [36], metal oxides such as $CeO_2$, CuO, TiO, $Y_2O$, $\gamma$-$Al_2O_3$, $\alpha$-$Al_2O_3$, $La_2O_3$, $Cs_2CO_3$, $Pr_6O_{11}$ [7] and supported metal catalysts [37], etc. The general scheme of this transformation is as shown in Scheme 3.

In a study, it was demonstrated $CeO_2$ as an effective catalyst for the synthesis of methyl benzylcarbamate (Scheme 4) by three component coupling of benzylamine, $CO_2$ and methanol in a liquid phase batch reactor under pressure [38]. The catalyst gave 99% conversion and 92% selectivity in the absence of dehydrating agents under optimized conditions. In addition to the desired methyl benzylcarbamate (MBC), dibenzylurea (DBU), N-benzylidenebenzylamine (BBA), N-methylbenzylamine (MBA) and dimethyl carbonate(DMC) were detected as side products. All these side products are obtained by side reactions as shown in Scheme 5.

The amount of MBC and DMC increased gradually with an increase in methanol amount, and the amount of DBU and BBA decreased. Therefore, considering the high conversion and selectivity, an excess amount of methanol is thought to be optimum as a large amount of methanol can suppress the side reactions as well as shift the equilibrium to the product side. $CO_2$ pressure has a positive effect on the desired product selectivity. The amount of MBC increased with increase in pressure up to 3 MPa and then remained constant with further increase. The products like MBC,

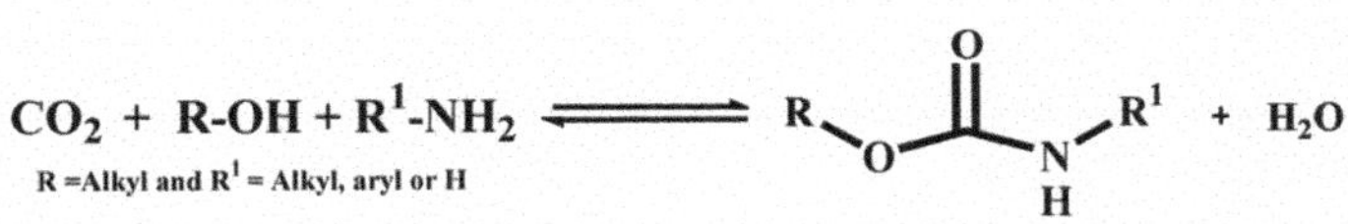


**Scheme 3** Synthesis of carbamate from $CO_2$, alcohol and amine

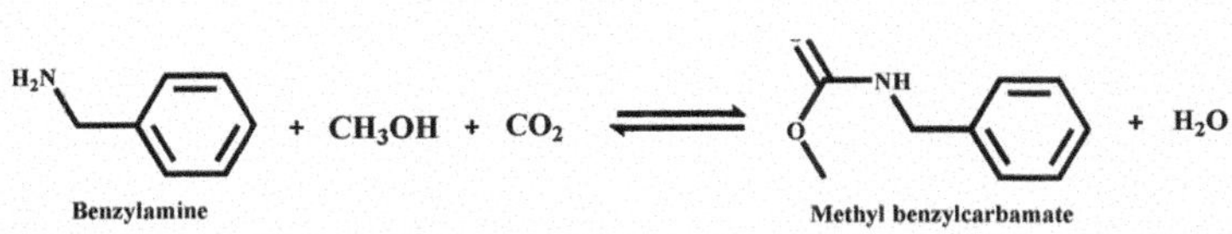


**Scheme 4** One-pot methyl benzylcarbamate synthesis from $CO_2$, amine and methanol

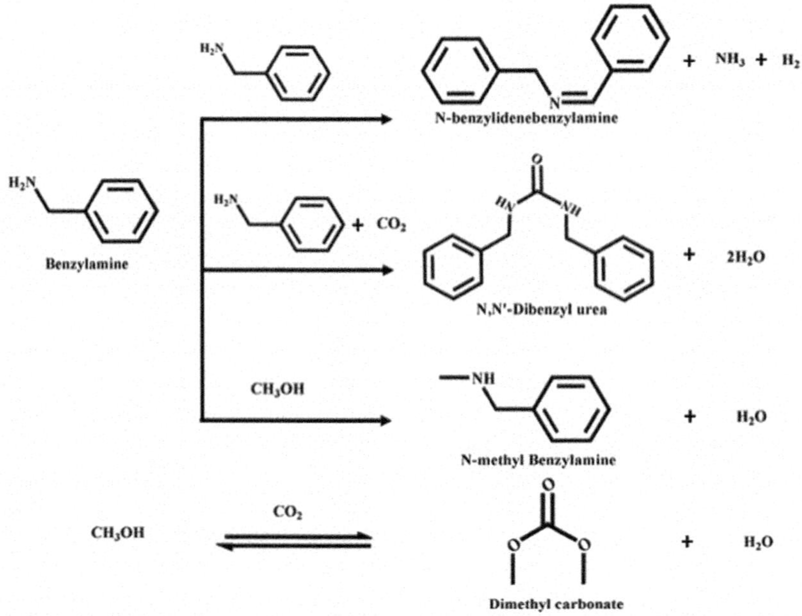


**Scheme 5** Possible side reactions in carbamate synthesis

DBU and DMC increased with increase in $CO_2$ pressure, whereas BBA and N-methylamine formations were not sensitive to $CO_2$ pressure. One possible reason is that these products were not derived from $CO_2$ and the acidic site on $CeO_2$ is helpful for the formation of these two products independent of $CO_2$ pressure and $CO_2$ coverage. As the acidic $CO_2$ is adsorbed on the basic sites, the acidic sites of $CeO_2$ are still available for independent amine side reactions. In this study, it was found that the formation of DBU, MBA, DMC is catalysed by $CeO_2$, and that of BBA proceeds in the liquid phase without $CeO_2$. The main formation route of the target product MBC is suggested to be the catalytic reaction of benzylamine with dimethyl carbonate or the precursor of dimethyl carbonate formed from methanol and $CO_2$ over $CeO_2$ catalyst.

Recently, the transformation of aniline with $CO_2$ and methanol was studied to produce methyl-N-phenylcarbamate (MPC) by the combination of $CeO_2$ catalyst and 2-cyanopyridine as a dehydrating agent in a stirred batch reactor [39]. Among the various carbamates, N-arylcarbamates are the most useful carbamates. However, the transformation itself suffers from few drawbacks such as equilibrium limitation, the inertness of $CO_2$, side reactions and low reactivity of arylamines (low basicity and nucleophilicity). The reaction scheme and side products like carbonates (from $CO_2$ and alcohols) and N,N′-disubstituted urea (from $CO_2$ and amines) are shown in Scheme 6 [39].

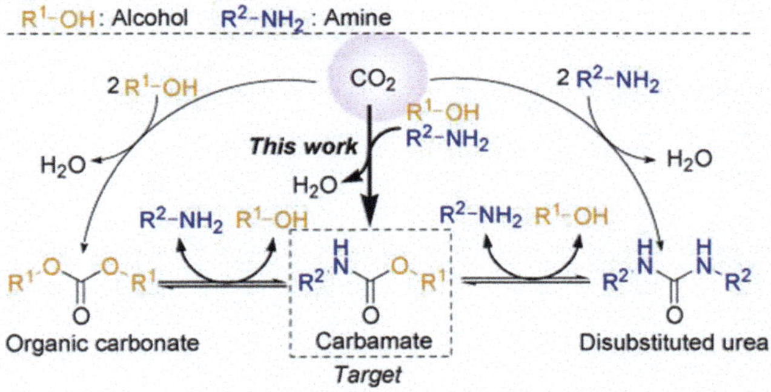


**Scheme 6** Direct and indirect routes for carbamate formation from $CO_2$, amine and alcohol. Reprinted with permission from Ref. [39]. Copyright (2018) Wiley

During the reaction over $CeO_2$ catalyst and 2-cyanopyridine, the selectivity to diphenyl urea (DPU) was very high (~80%) at the initial hour and then it decreased as the reaction progressed. In contrast, the MPC selectivity increased with time to reach 98% at 6 h. These results indicate that DPU is the intermediate for the desired product in this reaction. The maximum MPC yield was 97% in 8 h, which was much higher compared with other reported heterogeneous catalysts. MPC is formed by two consecutive reactions (Scheme 7) [39]: (1) DPU formation from $CO_2$ and aniline and (2) MPC formation from DPU and a methoxy source.

Among the various metal oxide catalysts tested such as $CeO_2$, $SiO_2$-$Al_2O_3$, $SmO_2$, ZnO, $Nb_2O_5$, $Y_2O_3$, $TiO_2$, the catalyst that performed the best was $CeO_2$ (Fig. 4).

The removal of $H_2O$ from the reaction media by the assistance of dehydrating agents helps to overcome the severe reaction equilibrium to some extent (Scheme 8) [39]. Among the various dehydrating agents tested, the performance of 2-cyanopyridine was prominent and it can be recovered by dehydration of the produced 2-picolinamide over alkali metal-supported $SiO_2$ catalysts (Fig. 5).

$CeO_2$ catalyst alone showed low conversion for this reaction, whereas 2-cyanopyridine alone hardly converted the reactant (Fig. 6). N-phenyl-pyridine-2-

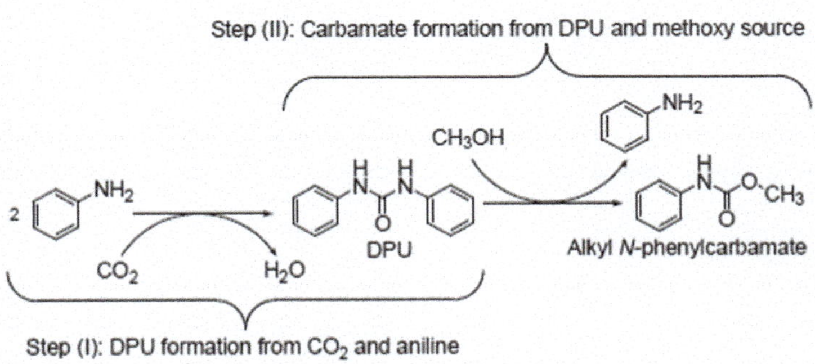


**Scheme 7** Formation of MPC from $CO_2$, aniline and $CH_3OH$. Reprinted with permission from Ref. [39]. Copyright (2018) Wiley

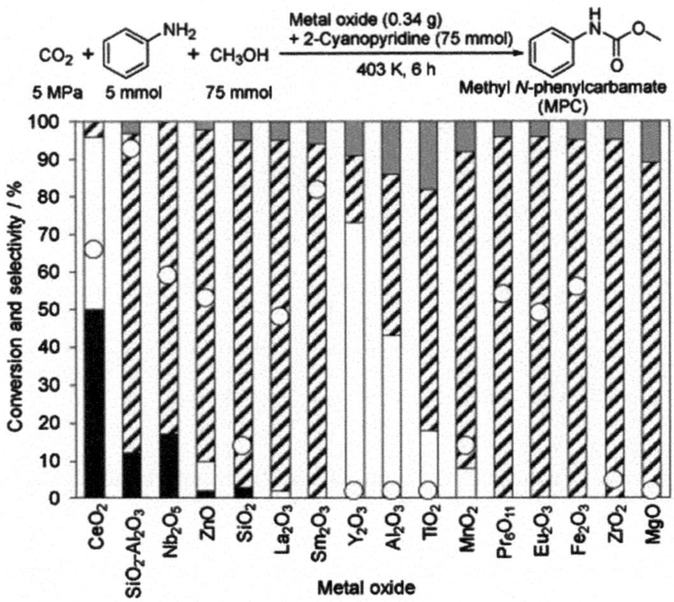


**Fig. 4** Screening of metal oxides with 2-cyanopyridine in MPC formation from $CO_2$, aniline and methanol. (O) aniline conversion. The bars represent selectivity. Black bars: MPC, white bars: DPU, striped bars: amidine, grey bars: others. Others include *N*-methylaniline, *N,N*-dimethylaniline and methyl-*N*-phenyl-pyridine-2- carboximidate. Reprinted with permission from Ref. [39]. Copyright (2018) Wiley

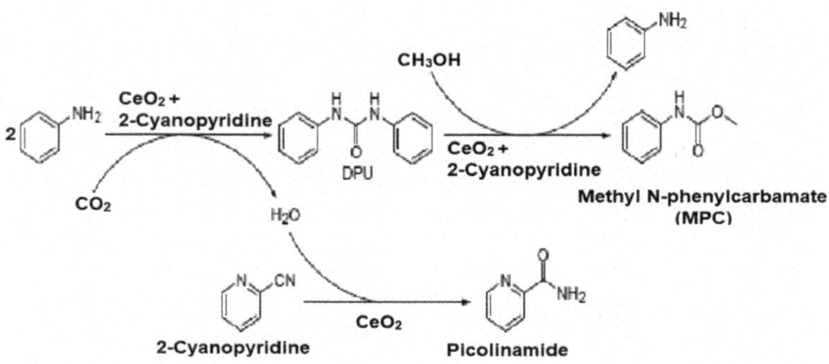


**Scheme 8** Proposed reaction route of MPC formation from $CO_2$, aniline and methanol using $CeO_2$ and 2-cyanopyridine. Reprinted with permission from Ref. [39]. Copyright (2018) Wiley

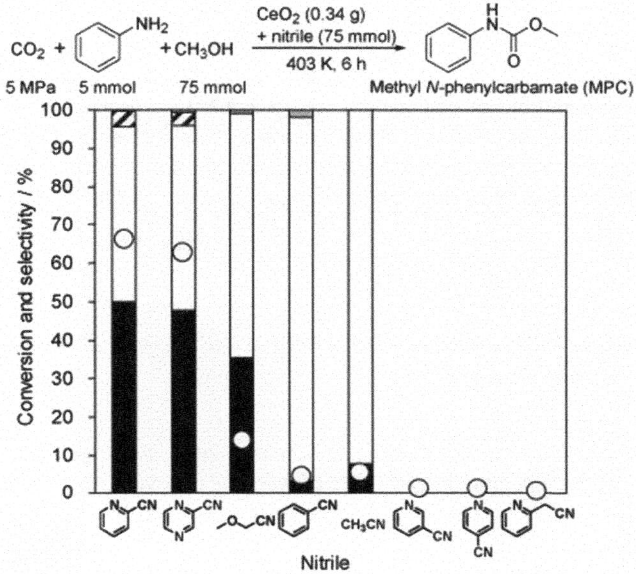


**Fig. 5** Screening of nitriles with $CeO_2$ in MPC formation from $CO_2$, aniline and methanol. (O) aniline conversion. The bars represent selectivity. black: MPC, white: DPU, stripe: amidine, grey: others. Others include *N*-methylaniline, *N,N*-dimethylaniline and methyl-*N*-phenyl-pyridine-2-carboximidate. Reprinted with permission from Ref. [39]. Copyright (2018) Wiley

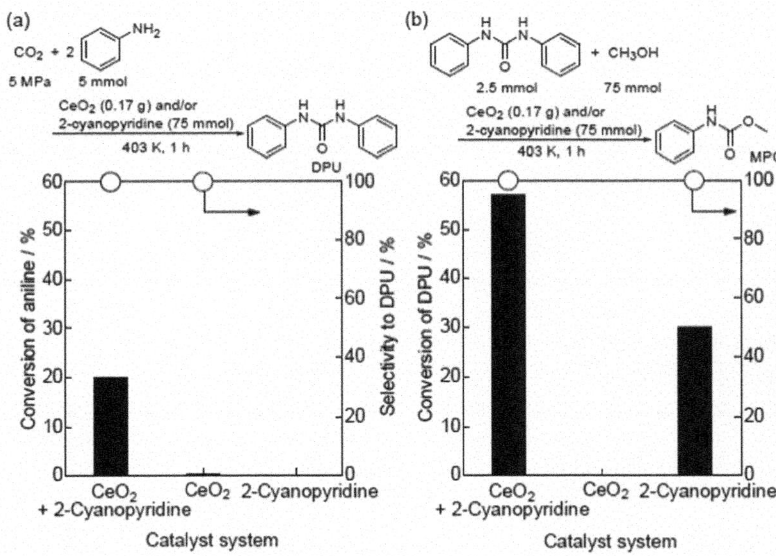


**Fig. 6** Comparison of catalysts. **a** reaction of aniline with $CO_2$. **b** reaction of DPU with $CH_3OH$ in the presence of $CO_2$. Reprinted with permission from Ref. [39]. Copyright (2018) Wiley

carboxamidine (amidine) was produced from 2-cyanopyridine and aniline. Hence, these results indicate the necessity of the combination of $CeO_2$ and 2-cyanpyridine for the reaction.

Therefore, this method directs a sequential formation of C-N and C-O bonds by the drastic shift in the reaction equilibrium towards N,N′-substituted urea formation and a remarkable catalytic function of $CeO_2$ in combination with 2-cyanopyridine resulting in the formation of various carbamates including N-arylcarbamates with high selectivities.

## 2.4 *Direct Carbonylation of Glycerol with $CO_2$ to Produce Glycerol Carbonate*

Glycerol carbonate, a five-membered cyclic carbonate, is one among the various chemicals derived from glycerol that can be used as a novel component in several fields such as batteries, industrial solvents, cosmetics, synthesis of polymers and chemical intermediates [9]. As the applications of glycerol carbonate (GC) grew, there is an urge for the development of greener route, as the conventional routes were polluting the ecosystem. There are a few transformations reported for the GC synthesis from glycerol and carbonylating agents such as dimethyl carbonate, phosgene, urea, ethylene carbonate, 1,1′-carbonyldiimidazole (CDI), carbon monoxide and carbon dioxide [40, 41] as shown in Scheme 9. Apart from the bio-based starting material glycerol, epichlorohydrin, 3-Choloro-1,2-propanediol are also reported for the same [42].

Among the routes mentioned in the Scheme 9, synthesis using $CO_2$ as carbonyl source serves as the best as $CO_2$ is an abundant renewable $C_1$ source, non-toxic, non-flammable and inexpensive. On the other hand, glycerol is a biorefinery by-product obtained from transesterification of vegetable oils or animal fat to produce biodiesel. Therefore, this carbonylation of glycerol with $CO_2$ is the transformation which uses two inexpensive and abundant by-products to make glycerol carbonate [9] and this makes the reaction highly economical and eco-friendly in nature.

Various heterogeneous catalysts are reported for the direct carbonylation of glycerol with $CO_2$. Aresta et al. were the first to report metal alkoxides (n-$Bu_2Sn(OMe)_2$ and n-$Bu_2SnO$) [43] as catalysts for this reaction. Some important catalysts reported for the direct carbonylation of glycerol are summarized in Table 6. It is to be noted that this transformation suffers from thermodynamic limitation. To overcome this loophole, dehydrating agents (also act as coupling agents) such as acetonitrile, 2-cyanopyridine and adiponitrile are used to push the reaction equilibrium towards the GC side [44–45].

A metal-free catalytic transformation was reported, where they proved that 2-cyanopyridine acted as a promoter in activating carbonyl bond of $CO_2$ in addition to its role as a dehydrating agent [55]. In the absence of the catalyst and additives, 2-cyanopyridine alone gave 11.4% yield (31.1% conversion) at 170 °C and 10 Mpa

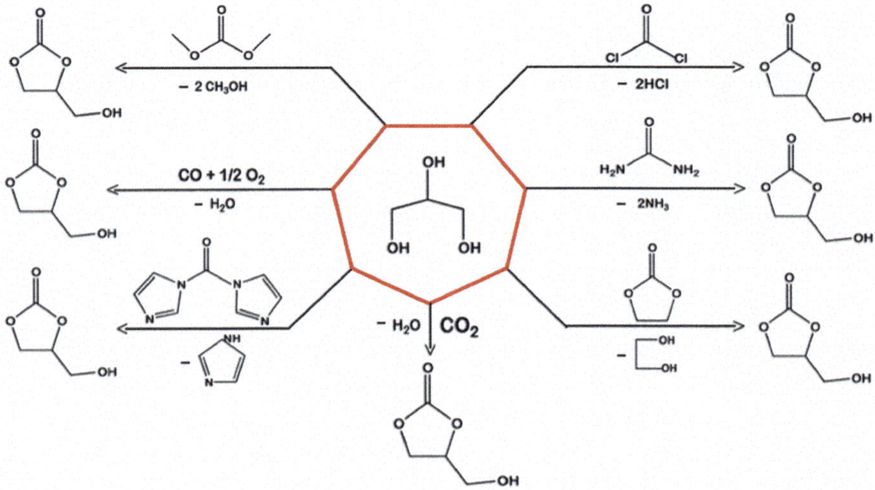


**Scheme 9** Carbonylation of glycerol to produce glycerol carbonate

which was close to the catalyst employed with the best performance ($CeO_2$ with 35.5% conversion and 14.2% yield).

They also studied the reaction with different dehydrating agents and found that the activity decreased in the following order: 2-cyanopyridine > 3-cyanopyridine > 4-cyanopyridine > acetonitrile. The theoretical simulation yielded three structures and three binding energy values ($E_b$) for 2-cyanopyridine, which states that there are three possible interactions of $CO_2$ with nitrogen viz. site-a (N in pyridine), site-b (N in cyano group), and both simultaneous interactions referred to as site-c as shown in Table 7. The third interaction (site-c: 2c-T) led to the formation of five-membered ring intermediate with the lowest adsorption energy 57.71 kcal/mol which makes the structure 'c-T' to form easily over the rest (a-T and b-T). This was confirmed by comparing the calculated IR spectrum of 2c-T with the experimentally obtained spectrum (Fig. 7). It confirms that the five-membered ring structure intermediate was formed due to the strong interaction existed simultaneously between $CO_2$ and 'N' of both cyano group and pyridine of 2-cyanopyridine. In the case of 3-cyanopyridine and 4-cyanopyridine, due to their stereo positions of 'N' (i.e. unavailability of both the N atoms in the close proximity for the $CO_2$ interaction), only two structures and $E_b$ values were obtained which implies that there are only two possible ways of interaction with $CO_2$ (site-a and site-b).

On the basis of the experimental and theoretical calculations, Scheme 10 [55] shows the proposed reaction route, where five-membered ring intermediate was obtained due to the activation of $CO_2$ by 2-cyanopyridine in the first step followed by the formation of 2-picolinamide. This shows that the dehydrating agent has a very important role in shifting the thermodynamic limited equilibrium towards the product side.

**Table 6** Literature studies on the synthesis of glycerol carbonate from glycerol and $CO_2$

| SL | Catalyst | T (°C) | P (Mpa) | Glycerol conv. (%) | Glycerol carbonate (%) | | Dehydrating agent | Solvent | References |
|---|---|---|---|---|---|---|---|---|---|
| | | | | | Selectivity | Yield | | | |
| 1. | $Ce_{0.98}Zr_{0.02}O_2$ | 150 | 4 | 40.9 | 88.7 | 36.3 | 2-Cypy | DMF | Liu et al. [46] |
| 2. | $nBu_2SnO$ | 80 | 13.8 | – | – | 35 | 13x zeolite | MeOH | George et al. [47] |
| 3. | $CeO_2$ | 150 | 4 | 38 | 87.6 | 33.3 | 2-Cypy | DMF | Liu et al. [45] |
| 4. | Zn/Al/La/Li | 170 | 4 | 39.5 | 47.3 | 18.7 | ACN | – | Li et al. [48] |
| 5. | La modified Zn-Al hydrotalcite | 170 | 4 | 35.5 | 45.2 | 16 | ACN | – | Li et al. [49] |
| 6. | $Cu/La_2O_3$ | 150 | 7 | 33.4 | 45.4 | 15.2 | ACN | – | Zhang and He [50] |
| 7. | $La_2O_2CO_3/ZnO$ (Novel method) | 170 | 4 | 24.3 | 59.2 | 14.4 | ACN | – | Park et al. [51] |
| 8. | $La_2O_2CO_3/ZnO$ | 170 | 4 | 30.3 | 47.2 | 14.3 | ACN | – | Li et al. [52] |
| 9. | ZnO | 180 | 15 | – | – | 8.1 | – | – | Ozorio and Mota [53] |
| 10. | $ZnWO_4$-ZnO | 150 | 5 | 6.5 | 100 | 6.5 | – | DMF | Ozorio and Mota [53] |
| 11. | Zn-Zeolite Y | 180 | 10 | – | – | 5.8 | – | – | Ozorio et al. [54] |
| 12. | $CeO_2$ | 170 | 10 | 35.5 | 14.2 | 5.0 | 2-Cypy | – | Su at al. [55] |
| 13. | Cu/Mg–Al-Zr | 150 | 7 | 11.5 | 20.9 | 2.4 | ACN | – | Zhang and He [56] |

(continued)

**Table 6** (continued)

| SL | Catalyst | T (°C) | P (Mpa) | Glycerol conv. (%) | Glycerol carbonate (%) | | Dehydrating agent | Solvent | References |
|---|---|---|---|---|---|---|---|---|---|
| | | | | | Selectivity | Yield | | | |
| 14. | n-$Bu_2Sn(Ome)_2$ and $Bu_2SnO$ | 180 | 5 | 6.8 | – | – | MS | – | Aresta et al. [43] |

2-Cypy = 2-cyanopyridine, ACN = acetonitrile, DMF = N,N-dimethylformamide, MeOH = methanol, MS = Molecular sieves

**Table 7** Theoretical calculations of the structures and binding energies of $CO_2$ connecting with different cyanopyridines

| | Possible connection structure | | |
|---|---|---|---|
| | a-T | b-T | c-T |
| 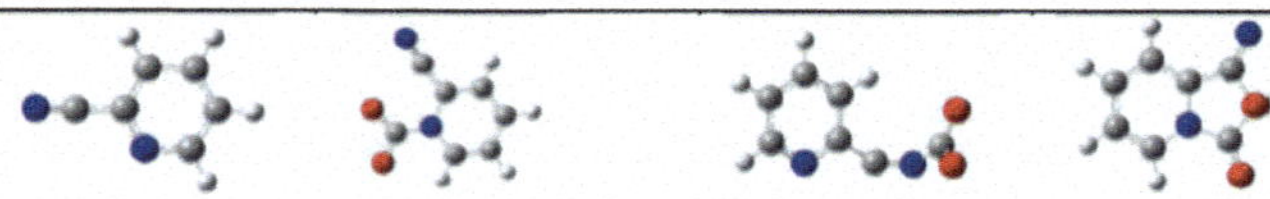 | | | |
| 2-cyanopyridine | $E_b$ = 74.69 | $E_b$ = 92.26 | $E_b$ = 57.71 |
| 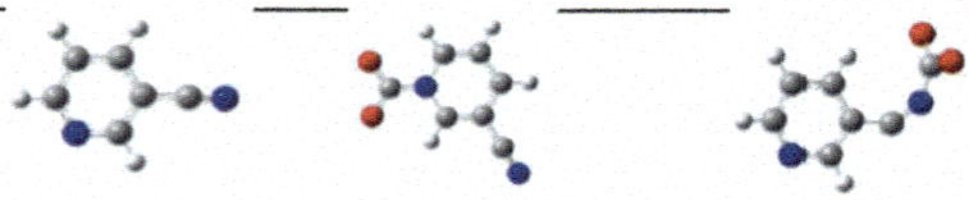 | | | |
| 3-cyanopyridine | $E_b$ = 72.83 | $E_b$ = 92.63 | |
| 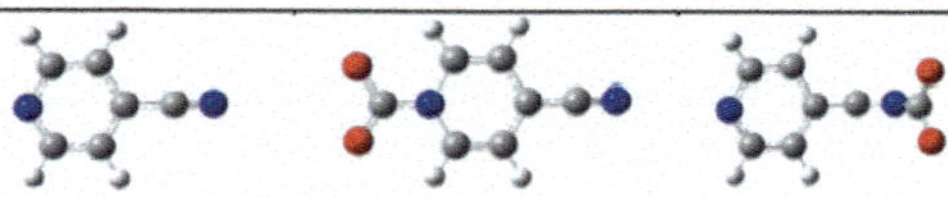 | | | |
| 4-cyanopyridine | $E_b$ = 67.39 | $E_b$ = 92.57 | |

The unit of energy is kcal/mol. a-T indicates $CO_2$ connect to N in pyridine; b-T indicates $CO_2$ connect to N in C ≡ N; and c-T indicates $CO_2$ connect to both N in pyridine and N in C ≡ N. Reprinted with permission from Ref. [55]. Copyright (2017) The Royal Society of Chemistry

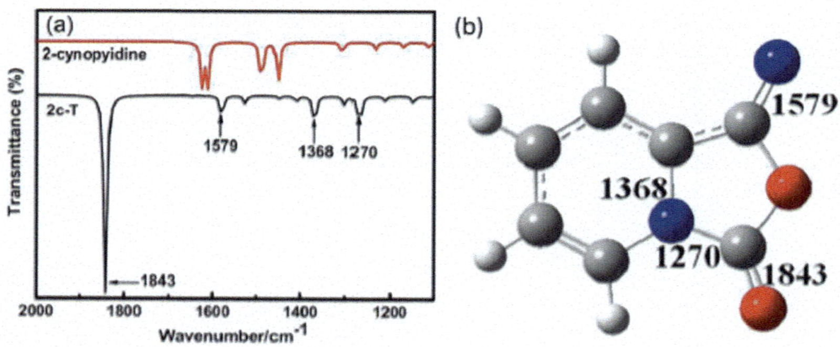


**Fig. 7** **a** Calculated IR spectrum of 2c-T (in black) and 2-cyanopyridine (in red). **b** The optimized structures of 2-cyanopyridine and $CO_2$ which can form a five-membered ring. Reprinted with permission from Ref. [55]. Copyright (2017) The Royal Society of Chemistry

In another work, [51], the importance of crystal phases was illustrated for $La_2O_2CO_3$/ZnO catalyst for the synthesis of GC using $CO_2$ as carbonyl source. The study was regarding the synthesis of $La_2O_2CO_3$/ZnO catalyst by two different methodologies resulting in unique catalytic activity. The catalyst was prepared by the conventional co-precipitation method (CLZ) and novel ethylene glycol combustion

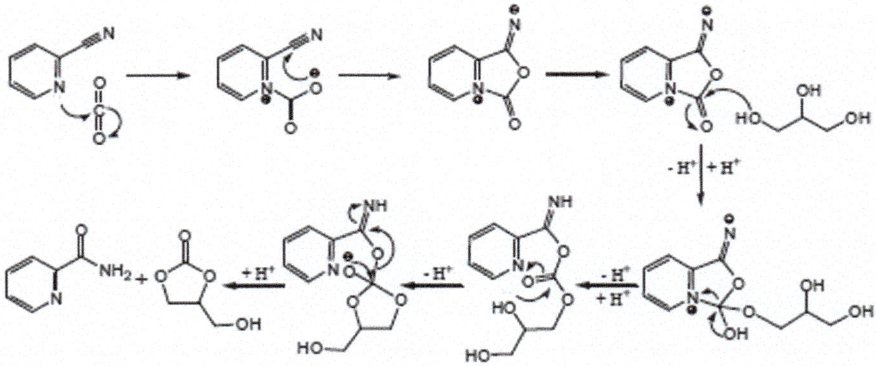

**Scheme 10** Possible reaction route for carbonylation of $CO_2$ with glycerol in the presence of 2-cyanopyridine. Reprinted with permission from Ref. [55]. Copyright (2017). The Royal Society of Chemistry

method (NLZ), which predominantly resulted in a hexagonal and monoclinic crystal structure, respectively.

The reaction pathway involved two steps on each solid phase (Scheme 11) [51] in which zinc glycerolate (ZnGly) is obtained from the reaction of ZnO phase with glycerol giving water as a by-product (Step 1) followed by the insertion of $CO_2$ into ZnGly, over the $La_2O_2CO_3$ phase resulting in GC product (Step 2). The activity of the catalyst synthesized by the NLZ method was better than the catalyst obtained from CLZ method.

According to Table 8, the conversion of glycerol was better in the case of CLZ catalyst, whereas the selectivity was not competitive with the NLZ catalyst. The lower selectivity was due to the strong interaction of the generated ZnGly with the CLZ catalyst in the first step because of which it was retained on the spent catalyst. This activity indicates the inhibition of Step 2 in the CLZ catalyst. However, in the case of NLZ catalyst, though the conversion was low, the selectivity towards the GC was better due to the short-lived ZnGly intermediate which was found to be in less concentration in the spent catalyst. This was confirmed by comparing

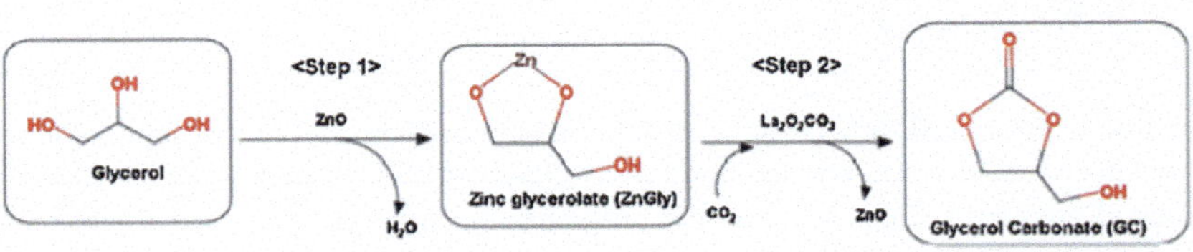


**Scheme 11** Simplified reaction pathways over $La_2O_2CO_3/ZnO$ catalysts for glycerol carbonation with $CO_2$. Reprinted with permission from Ref. [51]. Copyright (2017) Elsevier

**Table 8** Catalytic reaction results of the prepared catalysts for glycerol carbonation with $CO_2$. (Reaction conditions: initial pressure $PCO_2$ = 4.0 MPa, reaction temperature = 170 °C, reaction time = 12 h)

| Samples | Glycerol conversion (%) | GC yield (%) | Selectivity (%) | | | TOF[a] (X $10^{-4}$, $s^{-1}$) |
|---|---|---|---|---|---|---|
| | | | GC | MA | DA | |
| CLZ-0.25 | 28.7 ± 0.3 | 8.9 ± 0.2 | 31.1 ± 1.0 | 50.0 ± 1.6 | 19.9 ± 1.1 | 1.33 |
| CLZ-1 | 38.0 ± 0.5 | 9.1 ± 0.3 | 23.9 ± 2.0 | 46.1 ± 2.2 | 30.0 ± 1.8 | 1.03 |
| NLZ-0.25 | 21.1 ± 0.3 | 10.4 ± 0.4 | 49.3 ± 1.6 | 30.7 ± 0.8 | 20.0 ± 1.1 | 1.55 |
| NLZ-1 | 24.3 ± 0.1 | 14.4 ± 0.3 | 59.2 ± 0.8 | 26.4 ± 2.1 | 14.3 ± 1.3 | 1.63 |

[a]Apparent turnover frequency = mol of GC formed/[(mole of catalyst used) × (reaction time)]. Reprinted with permission from Ref. [51]. Copyright (2017) Elsevier

the characterizations of both fresh and spent catalysts using XRD, FTIR, TGA, HR-TEM techniques which provided clear evidence of the formation of highly crystalline ZnGly nanoparticles over the spent CLZ catalysts.

The difference in the catalytic activity of the catalysts could be closely related to their bulk and surface properties. The unique activity was due to the behaviour of the crystalline structures of the catalysts on the kinetics of the reaction which is illustrated in Scheme 12 [51]. The high reactivity of the monoclinic crystallite structure of $La_2O_2CO_3$ (NLZ) facilitates faster $CO_2$ reaction with ZnGly in Step 2. It is known that the $Ln_2O_2^{2+}$ layer in the monoclinic $La_2O_2CO_3$ phase had oxygen atoms oriented with the central Ln atom in pyramidal form, whereas the same layer in the hexagonal phase consisted of distorted $LaO_8$ coordinated rhombohedra. This indicates that there are fewer neighbouring oxygen atoms of La in the monoclinic $La_2O_2CO_3$ phase than in the hexagonal $La_2O_2CO_3$ phase.

Therefore, the good catalytic activity of La-Zn catalysts was said to be due to the formation of differed crystalline structures of $La_2O_2CO_3$ and the synergy between La and Zn.

## 2.5 *Dimethyl Carbonate Synthesis from Carbonylation of Methanol with $CO_2$*

Dimethyl carbonate (DMC) is a green and environmentally benign reagent owing to its various reactivity and widespread prospective use in the chemical industry [10]. It is used as a substitute for dimethyl sulphate and methyl halides in methylation reactions and as a carbonylating agent to substitute toxic phosgene in the manufacture of polycarbonate and urethane polymers [57]. It has also been assessed for use as a non-aqueous electrolyte in rechargeable lithium batteries. DMC is a powerful contender to assist the petroleum refining industry to fulfil the requirements of the

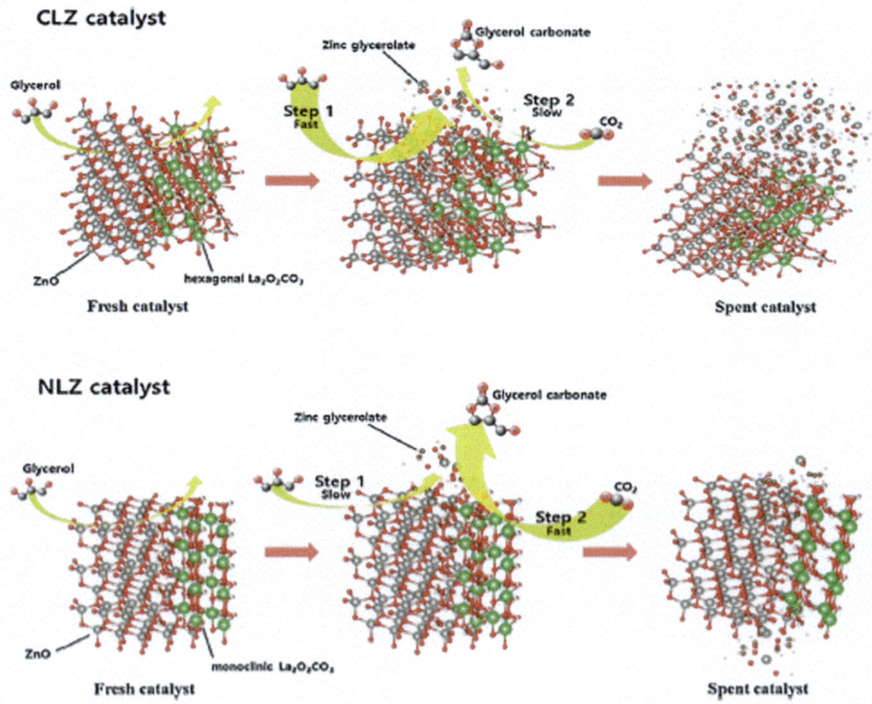


**Scheme 12** Schematic views of the crystalline structures of the catalysts during the reaction. Reprinted with permission from Ref. [51] Copyright (2017) Elsevier

Clean Air Act for oxygen in gasoline. DMC has a good blending octane number and it is both low in toxicity and quickly degradable [58].

The direct synthesis DMC from methanol and $CO_2$ has been acquiring importance because of the abundance of $CO_2$ (Scheme 13), its environmental benign nature and high atom efficiency of the synthesis process [44, 46]. Direct synthesis of DMC from methanol over various heterogeneous catalysts has been investigated such as Cu-Fe bimetal catalyst, heteropolyacid/metal oxide, zirconia, $CeO_2$ and $CeO_2$-$ZrO_2$ [59–60, 61]. Among the reported catalysts, $CeO_2$-based catalysts showed promising conversion and yield for DMC. The thermodynamic parameters calculated for DMC synthesis from methanol and $CO_2$ were $\Delta G = + 26.21$ kJ/mol and $\Delta H = -27.90$ kJ/mol at 25 °C [62]. The yield of DMC without an additive reported so far is very low because the reaction is severely equilibrium limited. Therefore, it is

**Scheme 13** Direct synthesis of dimethyl carbonate from methanol and $CO_2$

necessary to provide an effective way to shift the equilibrium to the product side for the commercialization of DMC from methanol and $CO_2$ route. It has been reported that the water removal during the reaction can shift the equilibrium to the right by adding an effective dehydrating agent and thus can improve the DMC yield. In 2013, Tomishige et al. reported pure $CeO_2$ for DMC synthesis using a 2-cyanopyridine as an organic dehydrating agent and achieved 94% of DMC yield [63]. $CeO_2$ is an efficient catalyst for both DMC synthesis and 2-cyanopyridine hydrolysis to form 2-picolinamide. $CeO_2$ morphology has an influence on the crystal plane and the concentration of their surface oxygen vacancies. Overbury and co-workers synthesized $CeO_2$ of different morphologies; nanorods, nanocubes and nano-octahedra [64]. Among these morphologies, nanorods contained the highest concentration of oxygen vacancy sites because of their enclosed (110) and (100) crystal plane. Wang et al. studied the effect of morphology of $CeO_2$ for DMC synthesis with 2-cyanopyridine as a dehydrating agent and they found that $CeO_2$ nanorods exhibited higher yield than nanocubes and nano-octahedra because of their higher concentration of oxygen vacancy and acid-base sites [65]. Liu et al. reported that Zr-doped $CeO_2$ nanorods possessed higher activities for DMC synthesis compared to pure $CeO_2$ nanorods because of the presence of high concentration of oxygen vacancies on the Zr-doped $CeO_2$ nanorods [44]. The surface oxygen vacancy sites could promote the adsorption and activation of $CO_2$ by filling one oxygen atom of the $CO_2$ molecule to the oxygen vacancy [66]. Liu et al. prepared various Zr content on $CeO_2$ nanorods by hydrothermal method and their activity in DMC synthesis was investigated [44]. It was found that the surface oxygen vacancy and $CO_2$ adsorption varied with Zr content which is crucial to achieve a high yield of DMC.

The oxygen vacancy on the surface of the catalyst could act as Lewis acid site that promotes adsorption of $CO_2$. Thus, authors further investigated the relationship between the surface oxygen vacancy, $CO_2$ adsorption capacity and catalytic activity as shown in Fig. 8. The $Zr_{0.1}Ce$ nanorods showed 1.3 and 6.7 times higher DMC yield than individual oxide $CeO_2$ and $ZrO_2$, respectively. The DMC yield decreased

**Fig. 8** Relationship between catalyst activity, $CO_2$ adsorption amount and the concentration of the surface oxygen vacancy. Reprinted with permission from Ref. (Liu et al. [44, 46]. Copyright (2018) The American Chemical Society

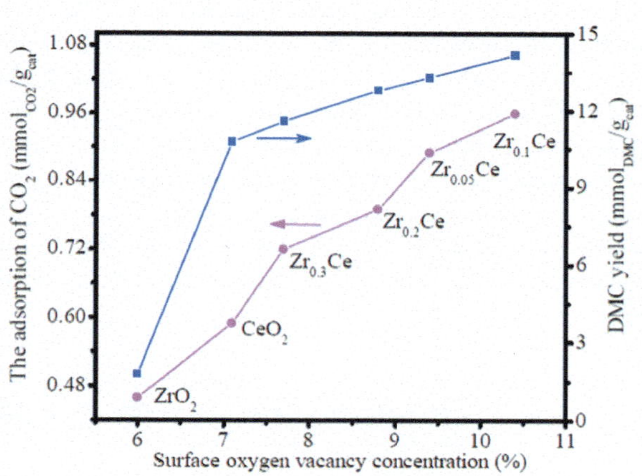

in the order: $Zr_{0.1}Ce > Zr_{0.05}Ce > Zr_{0.2}Ce > Zr_{0.3}Ce > CeO_2 > ZrO_2$, which followed the trend of the surface oxygen vacancy concentrations and the capacity of $CO_2$ adsorption. The findings of this research provide an insight into the impact of Zr-doping (on $CeO_2$) on DMC synthesis and offer a technical foundation for developing a fresh approach for the design of highly efficient $CeO_2$-based catalysts for direct carbonylation of methanol with $CO_2$.

## 3 Conclusions and Outlook

$CO_2$ activation and utilization by converting into value-added products is emerging as an important area of research due to environmental concerns caused by the $CO_2$ emission. Today's production of chemicals, in general, mainly depends on the utilization of diminishing fossil-based carbon sources. So, fixation of $CO_2$ by its insertion into the raw chemicals to make high-valued chemicals is considered as the most effective way of cycling carbon and reducing atmospheric $CO_2$. Moreover, the synthesis of chemicals using $CO_2$ as an abundant renewable, non-toxic, non-flammable and inexpensive $C_1$ source is highly desirable. The $CO_2$ in gaseous form is converted into various chemicals such as cyclic carbonates, carbamates, substituted urea and urethanes by reacting with raw materials like epoxides, amines, alcohols and alkyl halides, respectively, via carbonylation reaction. A poisonous gas, phosgene ($COCl_2$) is currently used as the carbonylating agent but it is unadvisable to use for its obvious health and environmental concerns. Hence, $CO_2$ is a much better option compared to phosgene. However, it is challenging for researchers to activate $CO_2$ and convert to chemicals because of its chemically inert nature but can be activated by suitable catalysts and by employing proper reaction condition. Although there are many studies on the use of homogeneous catalysts for carbonylation reactions using $CO_2$, a heterogeneous catalyst is preferred which activates $CO_2$, enhances reaction rates and can be separated and recycled easily after the reaction unlike homogeneous catalyst. In this chapter, it can be seen that acid-base bifunctional catalysts are mainly required for carbonylation reaction as $CO_2$ is activated by basic sites, whereas the substrate is activated by the acidic active centres present in the catalyst. However, the number, nature and strength of acidic and basic sites could be crucial in getting high efficiency for the formation of carbonylation product in high yields. Currently, the limitation of these $CO_2$ transformations is the requirement of severe conditions like high temperature and high $CO_2$ pressure. Milder conditions can be achieved only by designing more efficient catalysts in future research. Moreover, some reactions like carbonylation of certain alcohols with $CO_2$ are severely limited by chemical equilibrium and need dehydrating agents like 2-cyanopyridine to drive away the by-product water to shift the equilibrium towards the carbonylation product. However, adding such additives in the reaction increases the parallel reactions and undesired side products which need to be avoided. Moreover, it adds to the cost of the whole process which may not be economically viable. Hence, there is a need for the researchers to concentrate on this aspect to find better alternatives to improve the product yield for equilibrium limited $CO_2$ transformations. Probably, we shall witness a lot of vigorous research on chemical transformations of $CO_2$ in this decade and hope that some of the processes will see a light of the day to commercially produce a greater number of $CO_2$-based chemicals.

## References

1. Sakakura T, Choi J-C, Yasuda H (2007) Chem Rev 107:2365–2387
2. Aresta M, Dibenedetto A, Angelini A (2013) Chem Rev 114:1709–1742
3. Galadima A, Muraza O (2019) Renew Sust Energ Rev 115:109333
4. Janssen M, Müller C, Vogt D (2011) Green Chem 13:2247–2257
5. Clements JH (2003) Ind Eng Chem Res 42:663–674
6. Shaikh A-AG, Sivaram S (1996) Chem Rev 96:951–976
7. Tomishige K, Tamura M, Nakagawa Y (2019) Chem Rec 19:1354–1379
8. Tamura M, Ito K, Nakagawa Y, Tomishige K (2016) J Catal 343:75–85
9. de Caro P, Bandres M, Urrutigoïty M, Cecutti C, Thiebaud-Roux S (2019) Front Chem 7
10. Tundo P, Selva M (2002) Acc Chem Res 35:706–716
11. Dai W-L, Luo S-L, Yin S-F, Au C-T (2009) Appl Catal A Gen 366:2–12
12. Xiao L-F, Li F-W, Peng J-J, Xia C-G (2006) J Mol Catal A: Chem 253:265–269
13. Darensbourg DJ, Ulusoy M, Karroonnirum O, Poland RR, Reibenspies JH, Çetinkaya B (2009) Macromolecules 42:6992–6998
14. North M, Pasquale R, Young C (2010) Green Chem 12:1514–1539
15. Yoshida Y, Arai Y, Kado S, Kunimori K, Tomishige K (2006) Catal Today 115:95–101
16. Decortes A (2010) Chem Commun 46:4580–4582
17. Aresta M, Dibenedetto A, Gianfrate L, Pastore C (2003) J Mol Catal A: Chem 204:245–252
18. Pescarmona PP, Taherimehr M (2012) Catal. Sci Technol 2:2169–2187
19. Kulal N, Vasista V, Shanbhag GV (2019) J $CO_2$ Util 33:434–444
20. Du D-Y, Qin J-S, Li S-L, Su Z-M, Lan Y-Q (2014) Chem Soc Rev 43:4615–4632
21. Kurisingal JF, Rachuri Y, Gu Y, Choe Y, Park D-W (2020) Chem Eng J (in press)
22. Song J, Zhang Z, Hu S, Wu T, Jiang T, Han B (2009) Green Chem 11:1031–1036
23. Han L, Choi H-J, Choi S-J, Liu B, Park D-W (2011) Green Chem 13:1023–1028
24. Takahashi T, Watahiki T, Kitazume S, Yasuda H, Sakakura T (2006) Chem Commun 1664–1666
25. Sun J, Wang J, Cheng W, Zhang J, Li X, Zhang S, She Y (2012) Green Chem 14:654–660
26. Liu M, Liu B, Shi L, Wang F, Liang L, Sun J (2015) RSC Adv 5:960–966
27. Tamura M, Noro K, Honda M, Nakagawa Y, Tomishige K (2013) Green Chem 15:1567–1577
28. Bigi F, Maggi R, Sartori G (2000) Green Chem 2:140–148
29. Kreye O, Mutlu H, Meier MA (2013) Green Chem 15:1431–1455
30. Shi F, Zhang Q, Ma Y, He Y, Deng Y (2005) J Am Chem Soc 127:4182–4183
31. Kong D-L, He L-N, Wang J-Q (2010) Synlett 2010:1276–1280
32. Zhang R, Hua L, Guo L, Song B, Chen J, Hou Z (2013) Chin J Chem 31:381–387
33. Primo A, Aguado E, Garcia H (2013) ChemCatChem 5:1020–1023
34. Tamura M, Honda M, Nakagawa Y, Tomishige K (2014) J Chem Technol 89:19–33
35. Song D, Li D, Xiao X, Cheng C, Chaemchuen S, Yuan Y, Verpoort F (2018) J $CO_2$ Util 27:217–222
36. Krawczyk T, Jasiak K, Kokolus A, Baj S (2016) Catal Lett 146:1163–1168
37. Shang J, Guo X, Li Z, Deng Y (2016) Green Chem 18:3082–3088
38. Honda M, Sonehara S, Yasuda H, Nakagawa Y, Tomishige K (2011) Green Chem 13:3406–3413
39. Tamura M, Miura A, Honda M, Gu Y, Nakagawa Y, Tomishige K (2018) ChemCatChem 10:4821–4825
40. Pinto B, Mota CDA, in: Advances in Biorefineries, Elsevier (2014)

41. Teng WK, Ngoh GC, Yusoff R, Aroua MK (2014) Energ Convers Manage 88:484–497
42. Sarode-Dolas R (2018) Inter J Lat Technol Eng Manage Appl Sci 7:39
43. Aresta M, Dibenedetto A, Nocito F, Pastore C (2006) J Mol Catal A: Chem 257:149–153
44. Liu B, Li C, Zhang G, Yao X, Chuang SS, Li Z (2018a) ACS Catal 8:10446–10456
45. Liu J, Li Y, Zhang J, He D (2016) Appl Catal A Gen 513:9–18
46. Liu J, Li Y, Liu H, He D (2018b) Biomass Bioenerg. 118:74–83
47. George J, Patel Y, Pillai SM, Munshi P (2009) J Mol Catal A: Chem 304:1–7
48. Li H, Jiao X, Li L, Zhao N, Xiao F, Wei W, Sun Y, Zhang B (2015) Catal Sci Technol 5:989–1005
49. Li H, Xin C, Jiao X, Zhao N, Xiao F, Li L, Wei W, Sun Y (2015) J Mol Catal A: Chem 402:71–78
50. Zhang J, He D (2014) J Colloid Interface Sci 419:31–38
51. Park C-y, Nguyen-Phu H, Shin EW (2017) Mol Catal 435:99–109
52. Li H, Gao D, Gao P, Wang F, Zhao N, Xiao F, Wei W, Sun Y (2013) Catal Sci Technol 3:2801–2809
53. Ozorio LP, Mota CJ (2017) ChemPhysChem 18:3260–3265
54. Ozorio LP, Pianzolli R, da Cruz Machado L, Miranda JL, Turci CC, Guerra AC, Souza-Aguiar EF, Mota CJ (2015) Appl Catal A Gen 504:187–191
55. Su X, Lin W, Cheng H, Zhang C, Wang Y, Yu X, Wu Z, Zhao F (2017) Green Chem 19:1775–1781
56. Zhang J, He D (2015) J Chem Technol 90:1077–1085
57. Wang F, Zhao N, Xiao F, Wei W, Sun Y (2012) Distill: Adv Model Appl 61–90
58. Ono Y (1997) Catal Today 35:15–25
59. La KW, Jung JC, Kim H, Baeck S-H, Song IK (2007) J Mol Catal A: Chem 269:41–45
60. Zhou Y, Wang S, Xiao M, Han D, Lu Y, Meng Y (2012) RSC Adv 2:6831–6837
61. La KW, Youn MH, Chung JS, Baeck SH, Song IK (2007) Solid state phenom. Trans Tech Publ 119:287–290
62. Cai Q, Lu B, Guo L, Shan Y (2009) Catal Commun 10:605–609
63. Honda M, Tamura M, Nakagawa Y, Sonehara S, Suzuki K, Ki F, Tomishige K (2013) Chemsuschem 6:1341–1344
64. Mann AK, Wu Z, Calaza FC, Overbury SH (2014) ACS Catal 4:2437–2448
65. Wang S-P, Zhou J-J, Zhao S-Y, Zhao Y-J, Ma X-B (2015) Chin Chem Lett 26:1096–1100
66. Xin C, Hu M, Wang K, Wang X (2017) Langmuir 33:6667–6676
67. Tomishige K, Sakaihori T, Ikeda Y, Fujimoto K (1999) Catal Lett 58:225–229
68. Zhang Z-F, Liu Z-T, Liu Z-W, Lu J (2009) Catal Lett 129:428–436

# Application $CO_2$ as Refrigerant for HVAC, Supermarket, and Cold Chain—A Review

**Manju Lata Pal, Ashish Kumar Yadav, and Dileep Kumar Gupta**

**Abstract** Present scenario global warming is a major concern worldwide, due to the continuous increment of $CO_2$ level in the environment. The power plants, chemicals, manufacturing industries, and automobiles applications are the major cause for the same. Nowadays, various protocols and norms have been in the implementation stage to either reduce the $CO_2$ level or to capture the $CO_2$ produced during the process. In this chapter, one of the utilization of captured $CO_2$ has been explored as a natural refrigerant, which has favourable thermophysical properties and having the lowest global warming potential (GWP). Many research is going on worldwide for different application in the HVAC sector, like automobile air conditioning, supermarket, cold chain, industrial application, etc. The cold chain is one of the major sectors where the use of $CO_2$ as refrigerant can be appreciated, especially for the referred vehicles and cold storage for NH3 charge reduction. This chapter summarizes the recent development of the $CO_2$ refrigeration system based on the available literature, for the supermarket and cold chain applications followed by advantages of $CO_2$ as refrigerant, its technical challenges, and opportunities over the existing systems.

**Keywords** $CO_2$ refrigerant · Supermarket · Trans-critical cycle · Cold chain

## Abbreviations

| | |
|---|---|
| COP | Coefficient of Performance |
| DX | Direct Expansion |
| D-X | Direct Expansion |
| GC | Gas Cooler |
| GWP | Global Warming Potential |
| HP | High Pressure |

M. L. Pal · A. K. Yadav · D. K. Gupta (✉)
Mechanical Engineering Department, Institute of Infrastructure, Technology, Research And Management, Ahmedabad, India
e-mail: dileep.vnit@gmail.com

 
M. Goel et al. (eds.), *Climate Change and Green Chemistry of $CO_2$ Sequestration*, Green Energy and Technology, https://doi.org/10.1007/978-981-16-0029-6_11

| | |
|---|---|
| HVAC | Heating, Cooling and Ventilation |
| LT | Low temperature |
| MT | Medium temperature |
| NCCD | National Centre for Cold-chain Development |
| NH3 | Ammonia |
| ODP | Ozone Depletion Potential |
| UNEP | United Nations Environment Programme |

## 1 Introduction

The various research is going on to capture the $CO_2$ from the various sources; however, the storage of the captured $CO_2$ is the major challenge, which again leaves a question whether the capturing of $CO_2$ with storage is a feasible and economical solution than the protection from $CO_2$. In the other hand, the HVAC sectors are in the transition phase due to the enactment of Montreal and Kyoto Protocols to control the ozone depletion and global warming, respectively. Moreover, due to the challenges associated with the storage of $CO_2$ captured from the various sectors, the use of $CO_2$ as a refrigerant in the HVAC applications could be an appreciable solution.

$CO_2$ as a natural substance with its favourable thermophysical properties and the lowest GWP attract to use it in HVAC applications. There are many research going on worldwide for different applications in the HVAC sector. The chapter will give the glimpse of the present and future status of cold chain infrastructure in India, followed by the advantages of $CO_2$ as refrigerant, its technical challenges and opportunities over the existing systems, and the development of $CO_2$ refrigeration systems for the above-mentioned application.

## 2 Indian Cold Chain Infrastructure

Cold chain is an environment controlled logistics chain, ensuring continuous care from source-to-user, containing only of storage and distribution-related needs in which the inventory is maintained within pre-set ambient considerations. The cold chain does not modify the essential features of the harvest or product handled. As per the data from the National Centre for Cold-chain Development (NCCD) India, the status of Cold chain Infrastructure Gap is shown in Table 1. The data indicate the huge scope and demand of HVAC for Infrastructure development of India.

1. Why $CO_2$ as refrigerant.

$CO_2$ has favourable and excellent thermophysical properties and it is most suitable as it has the lowest GWP. The detailed comparison of properties with different refrigerants is shown in Table 2. It is being already in use as a refrigerant in the colder climate

**Table 1** Infrastructure needed in Indian cold chain[a]

| Types | Capacity requirement | Available | Gap |
|---|---|---|---|
| Reefer vehicles | 61,826 | 9,000 | 52,826 |
| Cold storage (Bulk) | 341,64,411 MT* | 318,23,700 MT* | 32,76,962 MT* |
| Cold storage (Hub) | 9,36,251 MT* | | |
| Packing-house | 70,080 | 249 | 69,831 |
| Ripening chamber | 9,131 nos | 812 nos | 8,319 nos |

[a]National centre for cold-chain development (NCCD) report 2014, MT* = Metric Tonnes

**Table 2** Properties of the refrigerants[a] [10]

| Refrigerant | GWP | ODP | Critical Point | Safety | Refrigerant | GWP | ODP | Critical Point | Safety |
|---|---|---|---|---|---|---|---|---|---|
| R22 | 1780 | 0.04 | 96.1 | A1 | R407C | 0 | 1600 | 86.0 | A1 |
| R134a | 1360 | 0 | 101.1 | A1 | R290 | 0 | 20 | 96.7 | A3 |
| R32 | 704 | 0 | 78.1 | A2L | R600a | 0 | 20 | 134.6 | A3 |
| R152a | 0 | 148 | 113.3 | A2 | R1270 | 0 | 20 | 91.1 | A3 |
| R1234yf | 0 | <1 | 94.7 | A2L | R717 | 0 | <1 | 132.3 | B2L |
| R1234ze | 0 | <1 | 109.4 | A2L | R744 ($CO_2$) | 0 | 1 | 31.1 | A1 |

[a]https://ccri.in/pdf/acbccu18-pre-workshop-Report.pdf

countries like northern Europe, Canada, some part of the USA, etc. especially, for the heating application.

However, in warm climatic conditions, like in the Middle East, including India, where the majority of the applications are for cooling, which has its own challenges. Therefore, the present focus of the research is on warm climatic conditions. Many modifications and upgradations on the system have already been proposed to make this system useful in the warm climatic conditions. The applications of $CO_2$ system are mainly on commercial refrigeration like display cabinet, standalone unit, bottle cooler water cooler, visi cooler, ice cream cabinet, etc. In addition to that, it is also suitable for mobile air conditioning (car, bus, truck, train, etc.).

2. Technical challenges and opportunities.

Despite the most favourable thermophysical properties and its eco-friendly nature, there are many technical challenges associated with the $CO_2$ system. The main drawback of the $CO_2$ is its lower critical temperature (31.1 °C), because of that, while operating the system in warm climatic conditions, it has to be operated in the supercritical region. The whole cycle operates between the subcritical and supercritical region; therefore, the cycle is called a trans-critical cycle. The basic cycle of the $CO_2$ refrigeration system is shown in Fig. 1. Subsequently, Fig. 2. depicts the variation of COP with different gas cooler outlet temperature (which is directly related to climatic conditions) at various high side operating pressure.

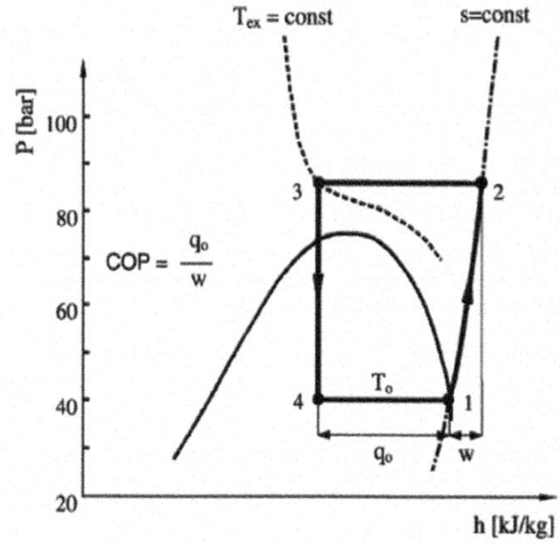


**Fig. 1** Trans-critical cycle for $CO_2$

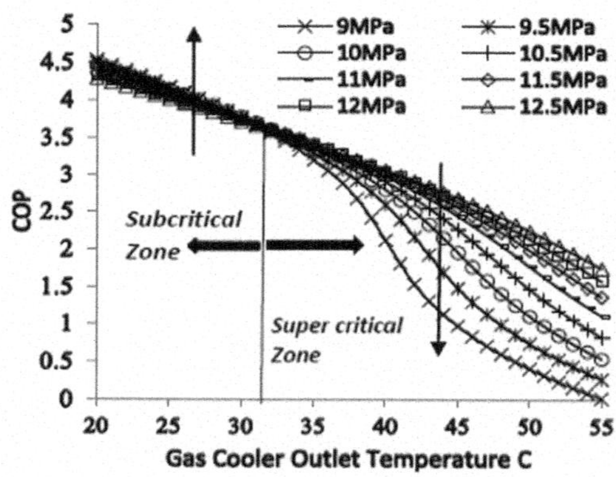


**Fig. 2** COP of the basic system with GC temperature

It was observed that the trends of COP are same as the conventional system in subcritical operation, whereas in trans-critical operation it follows the reverse trend. The reverse trend is more sensitive with higher temperature ambient conditions. This is one of the major challenges while designing the component, i.e., gas cooler for the system, when it is operating at the trans-critical cycle, especially at high-temperature climatic conditions. Therefore, the gas cooler is the most sensitive component of the system. The technological development in the present scenario for making the system possible to operate in trans-critical range, the component has been designed based on the requirement. However, using this system has still lower performance compared to the existing system, especially in warm climatic conditions. Therefore, systems need to be tuned based on the application wise. Subsequently, various modified cycles have been proposed. In the following section, the development of the $CO_2$ refrigeration system has been discussed for supermarket and cold chain applications.

## 3 Development of the $CO_2$ Systems

There are several modifications for the performance improvement of the $CO_2$ system including the different possible technologies of its supermarket application have been investigated by various researchers. Some of the developments of these systems are summarized below; Sawalha [16] analysed the performance of two different $CO_2$ supermarket systems and found 5–22% improvement using two-stage compression, operating in 10–40 °C climate. Subsequently, the performance was compared with R404A system [17].

Ge et al. [2] described that optimum pressure does not significantly depend on intermediate pressure, for LT and MT evaporator and it's superheating for booster system. However, it mainly depends on the effectiveness of suction line heat exchanger, ambient conditions, and isentropic efficiency of HP compressor. The performance of the $CO_2$ booster system and with R404A systems was compared and found that $CO_2$ booster system performance is better at lower ambient conditions.

Minetto et al. [11] studied the $CO_2$ plants with over-feeding of the evaporator in the parallel system, about 13% of the energy saving was reported at 16 °C ambient temperature and 0 °C MT. The performance and control policy for booster system was examined by Karampour and Sawalha [8]. The most applicable methods, the saving of 24 MWh/year was reported by increasing the evaporator temperatures and lowering approach temperature of the gas cooler. Subsequently, a study on integrated booster system with heat recovery, parallel compression and air conditioning was reported [9]. The COP improved by 14% was reported for summer operation.

A comparative analysis of various $CO_2$ cascade/secondary loop and trans-critical refrigeration system for supermarket applications and also performance comparison of selected $CO_2$ refrigeration systems with baseline R404A multiplex direct expansion system has been done by Sharma et al. [18] for eight climate zones of USA. It was found that among all the investigated systems, TBS-BC consumes less energy, at below 8 °C ambient and R404A direct expansion (DX) system consumes less energy for higher ambient temperature.

Similarly, Gullo et al. [4] compared eight different possible configurations of the $CO_2$ booster system supermarket system for warm climates. A comparative study reported on conventional booster system, and parallel compression with a multi-ejector unit and R404A DX system. About 15.6–27.3% of energy saving reported with multi-ejector system over R404A [5].

Fricke et al. [1] presented the experimental study of the booster system for 15.6 °C to 32.2 °C ambient temperature. The booster system was proposed as a promising alternative to the existing multiplex D-X systems, used in commercial applications.

Hafner et al. [6] reported up to 30% energy saving with the multi-ejector system over parallel compression using field data in Spiazzo. Later, Tsamos et al. [19] showed that the 5% performance improvement with parallel compressor over the basic booster system. Recently, Purohit et al. [14] proposed 'all-natural' $NH_3/CO_2$ cascaded booster system and compares the performance based on Life Cycle Climate

Performance and annual combined COP of the systems with R404A direct expansion system, flooded evaporator booster system and multi-ejector system for selected cities of India and the Middle East. Total 12.23% energy saving was reported along with the 11.2% of reduction in the total emission. However, in low temperature, the climate performance of multi-ejector system is better. Most recently, Ramgopal [15] discussed the benefits of using $CO_2$ as secondary fluids for cold storage application. Thermodynamic modelling shows that the performance of the systems can be improved by using $CO_2$ as secondary fluid compared to other systems.

## 4 Results and Discussions

The development of the $CO_2$ refrigeration systems has been discussed for supermarket and cold chain applications and the crux of the important results; i.e., energy performance and the high side optimum pressure have been summarized. The results included are based on the operating conditions and parameter as shown in Table. 3. The different modified, improved and proposed systems have been abbreviated in Table. 4.

Figure 3 indicates the best possible COP at different climatic temperature for various systems for supermarket application. The COP of the total 12 different systems has been compared for given operating parameters Table. 3 with the standard booster system.

It is found that the maximum COP achieved among all system with CCSC system followed by BSBC and BCPCFEWE systems. The results included the comparison of COP at five climatic temperatures, i.e., 25, 30, 35, 40 and 45 °C. The COP decreases as the temperature increases; however, the improvement rate with modified system

**Table 3** Operating parameters of the investigated solutions[a] 13

| Operating parameters | Ranges |
|---|---|
| MT load | 120–130 kW |
| LT load | 30–65 kW |
| Approach temperature for gas cooler/condenser | 2–5 °C |
| Condensing temperature minimum | 9 °C |
| Evaporating temperature (MT) | −12 to −6 °C |
| Evaporating temperature (LT) | −35 to −28 °C |
| Superheating | 5 °C |
| Circulation ratio for LT flooded evaporator (CR) | 2.5 |
| Intermediate vessel pressure for BC1 | 3 to 6 MPa |
| Maximum allowable gas cooler pressure | 10.6–13 MPa |

[a]https://ccri.in/pdf/acbccu18-pre-workshop-Report.pdf

**Table 4** Summary of various $CO_2$-based refrigeration systems

| S. No. | Abbreviations | Systems | Sources |
|---|---|---|---|
| 1 | SBC | Standard booster configuration | Purohit et al. [13], Sharma et al. [18] Ge et al. [3] |
| 2 | BCPC | Booster configuration with parallel compressor | Purohit et al. [13] |
| 3 | BCFE | Booster configuration with flooded evaporator | Purohit et al. [13] |
| 4 | BCWE | Booster configuration with work recovery expander | Purohit et al. [13] |
| 5 | BCPCFEWE | Booster configuration with parallel compressor, flooded evaporator and work recovery expander | Purohit et al. [13] |
| 6 | CCSC | Combined $CO_2$ secondary cascade system | Sharma et al. [18] |
| 7 | CGSCC | Combined glycol secondary/$CO_2$ cascade system | Sharma et al. [18] |
| 8 | BS-UX | Booster system with upstream expansion | Sharma et al. [18] |
| 9 | BSBC | Booster system with bypass compressor | Sharma et al. [18] |
| 10 | CBC | Cascaded booster configuration $NH_3/CO_2$ | Purohit et al. [14] |
| 11 | BCE | Booster configuration system with ejector | Huang et al. [7] |
| 12 | FBRE | Modified booster configuration with ejector | Huang et al. [7] |
| 13 | CS-CC | $NH_3/CO_2$ cascade system cold chain applications | Ramgopal [15] |

increases with the higher temperature. Therefore, the different modified configuration is summarized. Figure 3 gives a glimpse of the possible modification and also the magnitude of improvement with these configurations. Subsequently, Fig. 4 shows the optimum high side pressure of the system at which maximum COP achieved for individual system configurations and corresponding operating conditions.

Figure 5 depicts the application of the COP system for cold storage (cold chain) application. The comparison has been presented for the secondary loop system, working with $NH_3$, $NH_3/CO_2$ (CS-CC) and R1234yf. It was found that the performance is best with $NH_3$ among all three refrigerant, whereas the results are better with $NH_3/CO_2$ (CS-CC) than R1234yf. This could be a possible area to explore in cold chain application, especially for ammonia charge reduction in cold storage.

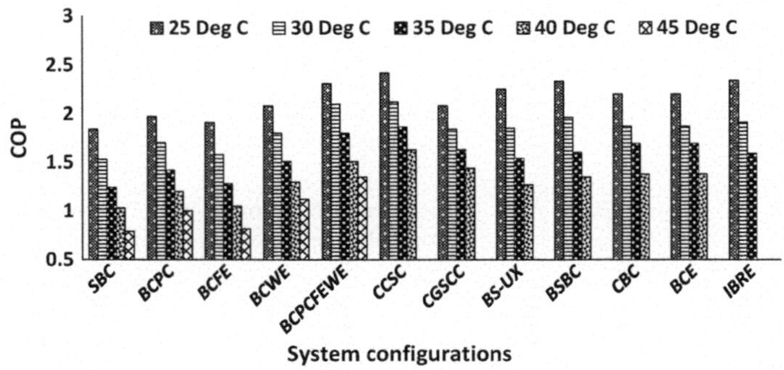


**Fig. 3** Best possible COP at different climatic temperature for various systems for supermarket application

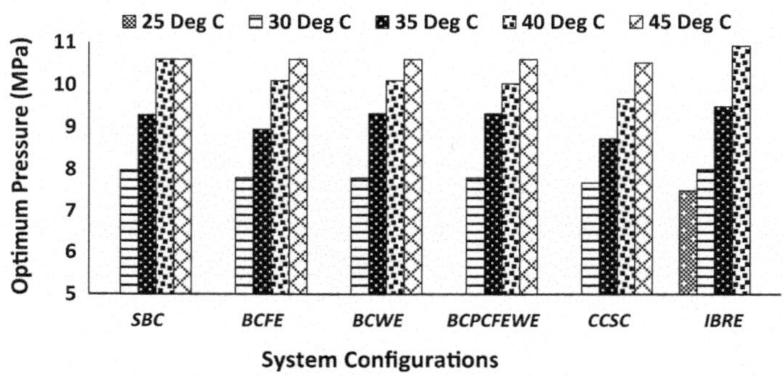


**Fig. 4** Optimum pressure for best possible COP at different climatic temperature for various systems for supermarket application

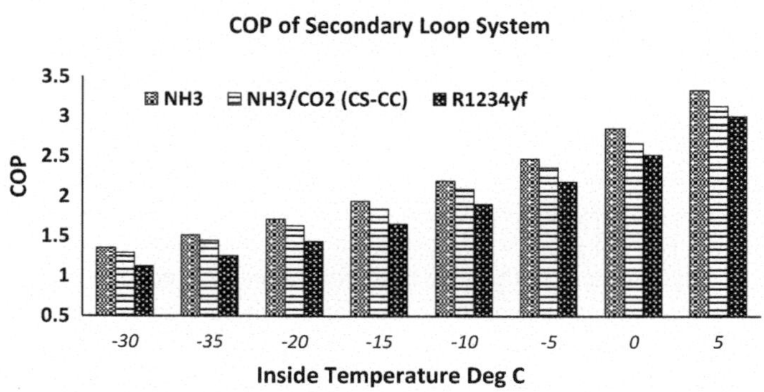


**Fig. 5** Comparison of COP in secondary loop system for cold chain application

## 5 Conclusions and Outlook

The application $CO_2$ as a refrigerant for supermarket and cold chain application has been discussed. The recent development for using the $CO_2$ system for the above application has been summarized. It is concluded that $CO_2$ may be considered as an ecologically safe and natural refrigerant for the next generation of HVAC industries, where the captured $CO_2$ from the different sources can be utilized. There are different challenges and opportunities associated with the $CO_2$ refrigeration system; therefore, many research groups are actively working on these systems to make it feasible. It has been found that the design and selection of $CO_2$ systems are based on its application such as supermarket, dairy, and cold storage. Moreover, many more applications of $CO_2$ refrigerant are to be explored in the future for Indian HVAC industries.

## References

1. Fricke B, Zha S, Sharma S, Newel J (2016) Laboratory evaluation of a commercial CO2 booster refrigeration system. International Refrigeration and Air Conditioning Conference. Paper 1691
2. Ge YT, Tassou SA (2011) Performance evaluation and optimal design of supermarket refrigeration systems with supermarket model “SuperSim”. Part II: Model applications. Int J Refrig 34:540–549
3. Ge YT, Tassou SA (2011) Thermodynamic analysis of transcritical CO2 booster refrigeration systems in supermarket. Energ Conver Manage 52:1868–1875
4. Gullo P, Elmegaard B, Cortella G (2016) Energy and environmental performance assessment of R744 booster supermarket refrigeration systems operating in warm climates. Int J Refrig 64:61–79
5. Gullo P, Hafner A, Cortella G (2017) Multi-ejector R744 booster refrigerating plant and air conditioning system integration – A theoretical evaluation of energy benefits for supermarket applications. Int J Refrig 75:164–176
6. Hafner A, Banasiak K, Herdlitschka T, Fredslund K, Girotto S, Haida et al (2016) R744 ejector system, case: Italian Supermarket, Apiazzo, Presented at: the 12th IIR Gustav Lorentzen Conference on Natural Refrigerants, Edinburgh, Scotland
7. Huang Z, Zhao H, Yu Z, Han J (2018) Simulation and optimization of a R744 two- temperature supermarket refrigeration system with an ejector. Int J Refrig
8. Karampour M, Sawalha S (2014) Performance and control strategies analysis of a CO2 transcritical booster system. In: 3rd IIR International Conference on Sustainability and the Cold Chain, London, UK
9. Karampour M, Sawalha S (2015) Theoretical analysis of CO2 trans-critical system with parallel compression for heat recovery and air conditioning in supermarkets. In: 24th IIR Refrigeration Congress of Refrigeration. IIF/IIR, Yokohama, Japan
10. Kim MH, Pettersen J, Bullard CW (2004) Fundamental process and system design issues in CO2 vapor compression systems. Prog Energ Combust Sci 30(2):119–174
11. Minetto S, Brignoli R, Zilio C, Marinetti S (2014) Experimental analysis of a new method for overfeeding multiple evaporators in refrigeration systems. Int J Refrig 38:1–9
12. National Centre for Cold-chain Development (NCCD) Report “All India Cold-chain Infrastructure Capacity (Assessment of Status & Gap)” 2014 prepared by NABARD Building, New Delhi
13. Purohit N, Gupta DK, Dasgupta MS (2017) Energetic and economic analysis of trans-critical CO2 booster system for refrigeration in warm climatic condition. Int J Refrig 80:182–196

14. Purohit N, Sharma V, Sawalha S, Fricke B, Llopis R, Dasgupta MS (2018) Integrated supermarket refrigeration for very high ambient temperature
15. Ramgopal M (2018) CO2 based Secondary Loops for Cold Storage Applications. Air Cond Refrig 21(5)
16. Sawalha S (2008) Theoretical evaluation of trans-critical CO2 systems in supermarket refrigeration. Part I: Modeling, simulation and optimization of two system solution. Int J Refrig 31:516–524
17. Sawalha S (2008) Theoretical evaluation of trans-critical CO2 systems in supermarket refrigeration. Part II: System modifications and comparisons of different solutions. Int J Refrig 31:525–534
18. Sharma V, Fricke B, Bansal P (2014) Comparative analysis of various CO2 configurations in supermarket refrigeration systems. Int J Refrig 46:86–99
19. Tsamos KM, Ge YT, Santosa I, Tassou SA, Bianchi G, Mylona Z (2017) Energy analysis of alternative CO2 refrigeration system configuration for retail food applications in moderate and warm climates. Energ Convers Manage 150:822–829

# Research & Development Pathways/Challenges in Direct Air Capture of $CO_2$

**Gautam Sen**

**Abstract** Economic growth, due to increased consumerism, extravagance and human greed, necessitates higher energy demand. Fossil fuels burning, which generates carbon dioxide, a greenhouse gas, provides this energy resulting in a quantum jump in carbon dioxide emission and accumulation. Carbon cycle and other natural processes like mineralization of carbon dioxide through weathering of rocks rich in magnesium or calcium are inadequate to remove excess greenhouse gases. This has caused global climate change, threatening the existence of living beings including Homo sapiens. Removing carbon dioxide from atmosphere, either at the sites of carbon dioxide generation like power plants, called point capture of carbon dioxide (PCC) or in ambient air popularly known as direct air capture (DAC), therefore has become a necessity. In this chapter, DAC techniques like biosequestration, absorption and adsorption are discussed along with the challenges in their implementation. The DAC machines offer a greater flexibility of usage, as they are not tethered to the generation plants and can be placed at the site of subsequent usage of carbon dioxide or sequestration. Challenges in DAC are requirement of high energy, high water consumption and economic viability. Few companies have incorporated these techniques in their startups, and these have been deliberated along with the financial costs. It is concluded that initiating direct air capture as research and development projects on larger scale will help in developing viable options for carbon dioxide reduction as a future strategy.

**Keywords** Direct air capture · Biosequestration · Absorption · Adsorption · R&D challenges

G. Sen (✉)
Former Senior Vice-President Reliance Industries Ltd. (E and P), Former Executive Director, ONGC, Res: C-498 Defence Colony, New Delhi 110024, India
e-mail: gautamsenindia@yahoo.com

 
M. Goel et al. (eds.), *Climate Change and Green Chemistry of $CO_2$ Sequestration*, Green Energy and Technology, https://doi.org/10.1007/978-981-16-0029-6_12

## Abbreviations

| | |
|---|---|
| PCC | Point Capture of Carbon Dioxide |
| DAC | Direct Air Capture |
| NOAA | National Oceanic and Atmospheric Administration |
| GDP | Gross Domestic Product |
| NCR | National Capital Region |
| TSA | Temperature Swing Adsorption |
| CE | Carbon Engineering |
| CC | Carbon Capture |
| EASAC | European Academies Science Advisory Council |
| NRL | Naval Research Laboratory |
| NASEM | National Academies of Science - Engineering – Medicine |

## 1 Introduction

Carbon dioxide ($CO_2$), concentration in the air, has crossed 410 parts per million. Carbon dioxide being a greenhouse gas causes global warming, defined as long-term warming of the planet's overall temperature. Though this warming trend has been going on for a long time, its pace has significantly increased in the last one hundred years due to the burning of fossil fuels and more so in last twenty years. Burning fossil fuels, which includes coal, oil, and natural gas, release carbon dioxide which causes "enhanced greenhouse effect," in the Earth's atmosphere resulting in global warming. Catastrophic effects of increase in Earth's temperature is well known with ice caps melting, increase in sea level, inundation of coastal areas across the world, ocean turning acidic and frequent extreme weather conditions. These are bound to destroy the ecology, the flora and the fauna including Homo sapiens. As per National Oceanic and Atmospheric Administration (NOAA), USA, the average Earth temperature has increased by 1.0 °C since 1880, with the steepest rise in the last four decades [1].

As the Earth has been in an interglacial period for around 11,700 years, the next glacial period is estimated to begin at least 50,000 years from now. Since these changes, called Milankovitch cycles are long term, their effects do not ring a bell. Natural Solar cycles with 11-year periods are also well known, but humans really do not have any control on this and other natural changes. The anthropogenic changes are within human control and therefore concerns mankind. There has been an average global sea level rise of 1.8 mm per year in recent decades, mainly due to global warming but its effects are not the same everywhere. Unfortunately, sea level rise in Indian coasts is higher, putting at risk Indian coastal cities. According to the Ministry of Earth Sciences, Government of India [2], sea level rise is 5.16 mm per year at Diamond harbour, 3.18 mm per year at Kandla, 2.89 per year at Haldia, 2.2 mm per year at Port Blair during the same period. Rising sea levels can exacerbate the impact of coastal hazards, such as Tsunami, cyclones, storms besides loss of coastal land.

Global energy demand, on an average, is rising at a rate of around 2% per year and is linked with global GDP growth. Currently, fossil fuels constitute almost 80% of the energy basket and the best estimates by global energy experts predict that even with incentives for renewable and nuclear energies, the share of fossil fuels is expected to remain around 67% in 2050. In absolute terms, the amount of fossil fuels burnt will not go down. It may therefore be prudent to increase the share of natural gas and reduce share of coal and even oil. India's predicament is still worse since India is endowed mainly with coal reserves. BP's outlook predicts that India's share of coal in the energy basket may reduce only marginally, from 56% in 2017 to 48% by 2040. Even, electric vehicles need electricity and if fossil fuels are used to generate that electricity, reduction in emission will not be a game changer. Summarizing, it can be reasonably assumed that the world will not stop use of fossil fuels in short and medium terms in spite of the cherished dream of doing away with fossil fuels by 2100.

A carbon capture and sequestration (CCS) system extracts $CO_2$ from the flue gases from the fossil fuel-based plants by various gas separation methods. Several options of post-combustion capture, pre-combustion capture are in advance stages of development using absorption, adsorption, cryogenic cooling, and membrane separation among other novel techniques. The carbon capture system involves integration with a power plant technology or industrial process plant. The captured $CO_2$ gas is liquefied under pressure and is transported to an appropriate site for its storage underground or utilization into value-added products. Post-combustion carbon capture (PCC) has been used successfully in the oil and gas industry by injecting liquefied $CO_2$ for enhanced oil recovery. A number of demonstration facilities using PCC exist around the world.

Terrestrial sequestration of $CO_2$ by photosynthesis leads many environmental benefits such as protection of biodiversity, afforestation and water management, etc. in addition to $CO_2$ removal from the atmosphere. But in a recent report published in Nature [3] in March 2020, scientists from University of Leeds claimed that the ability of the World's tropical forests to act as carbon sink is diminishing based on tracking 300,000 trees over 30 years. The new analysis of tree growth and death in the last three decades from 565 forests have found that the overall uptake of carbon into Earth's tropical forest peaked in 1990s and by 2010 the average ability dropped by one-third largely driven by carbon losses from trees dying. Croplands have the potential to store an additional 1.85 gigatons carbon each year,as much as the global transportation sector emits annually and soils could continue to sequester carbon for 20 to 40 years before they become saturated as per a study conducted in 2017. Natural biosequestration can't keep pace up with the emissions which are released on daily basis in the environment. Mineralization of carbon dioxide also occurs naturally during weathering of silicates and rocks rich in calcium or magnesium. These rocks are present in upper mantle and basaltic lava. However, natural processes occur at a very slow rate and the rise in the carbon dioxide concentration continues unabated.

Direct air capture (DAC) of carbon dioxide can become an acceptable strategy to achieve stabilization of carbon dioxide by augmenting the existing natural processes. Hitherto, most of the research to remove carbon dioxide from atmosphere has focused on point source capture of carbon dioxide (PCC), at the sites of generation. Ambient air near such power plants have higher concentration of carbon dioxide and therefore energy and costs requirement to capture may not be as high as when attempts are made to capture carbon dioxide in ambient air at places away from such generation sites. Yet, DAC methods offer a higher degree of flexibility and the machines can be installed at the sites where they could be reused or sequestered. It is because of this flexibility that it is becoming the preferred choice.

## 2 Direct Air Capture

How can we capture carbon dioxide ($CO_2$) from air, reuse it and/or store it safely? Capturing 400 molecules of carbon dioxide in millions of molecules of nitrogen and oxygen (i.e., air: $CO_2$ = 2500:1) is not easy, but today carbon engineering has been able to make some progress. The $CO_2$ removal strategies like direct air capture (DAC) at first might allow the world to pursue a strict stabilization target for the $CO_2$ concentration, by overshooting the target and then later approaching the target from above through net negative global emissions [4]. DAC works in addition to the natural process of carbon cycle which, by itself, is far too slow to take remedial action against anthropogenic changes. There are numerous ways that carbon dioxide can be captured directly from air. DAC would enhance the process if it can be applied cost-effectively to mitigate climate change. However, DAC is currently an expensive approach to mitigating climate change.

Natural biosequestration is a process of removing carbon dioxide from air directly, but with deforestation on one hand and increase of the emissions on the other, and it is insufficient by itself to reduce the concentration of carbon dioxide. DAC machines, placed anywhere, even away from the source of emissions, are capable of sucking massive quantities of carbon dioxide out of the air. $CO_2$ gets distributed evenly within the atmosphere and therefore there is no necessity of installing DAC machines at the source of carbon dioxide emissions. The PCC can take $CO_2$ directly from smoke stacks and the pump it into the ground for permanent sequestration, but that technology is also expensive and required massive industrial plants tethered to $CO_2$ emission point sources, making the technology inflexible. DAC [5] is a carbon capture technique which produces a concentrated carbon dioxide gas stream from the ambient air. The process involves ambient air flows through a filter where either through adsorption, absorption or mineralization, $CO_2$ is separated from the air. This concentrated $CO_2$ stream may be utilized commercially or sequestered or reused for enhanced oil recovery to produce low-carbon footprints.

The various methods of direct air capture are described below.

## 2.1 Biosequestration Through Forests and Soil-Terrestrial Biological CDR

Biosequestration can be considered one of the most economical and natural methods of direct air capture. This can be done either by more dense plantings or by adding ecologically inert biocarbon (e.g., charcoal) to the soil. Living biome and soil is the biological scrubber for capturing the anthropogenic carbon. Terrestrial plants need carbon dioxide for their growth and preparation of their food. Photosynthesis removes carbon dioxide naturally, and trees are especially good at storing carbon from the atmosphere. Expanding forests, managing existing forests and introducing urban forests to encourage more carbon uptake can leverage the power of photosynthesis to convert carbon dioxide in the air into carbon stored in wood and soils. A single urban tree has a potential to biosequester equivalent to three to five forest trees [6] due to the higher concentration of carbon dioxide in the air.

Soils naturally store carbon and in turn carbon increases soil health and crop yields. Soil contains thrice the amount of carbon stored in living plants and animals [7]. Different soils have different capacities. Soil carbon pool and carbon dioxide emission are also highly influenced by the type of vegetation and environmental factors. Planting cover crops especially with deeper roots when fields are otherwise bare can extend photosynthesis throughout the year, sequestering about half a metric ton of $CO_2$ per acre per year. Using compost can improve yields while storing the compost's carbon content in the soil and soils could continue to sequester carbon for 20 to 40 years before they become saturated as per a study conducted in 2017.

Availability of land can, however, be a challenge for desired results, and therefore, vertical growth of forests is a viable option using a frame or other support systems. The technique is being seen as an instrument to fight the increasing air pollution in urban areas. Stefano Boeri Architect, known for its innovative residential projects that blend greenery into the exterior of buildings, has a number of new projects under development in major Chinese cities (https://www.scmp.com/business/article/). Noida in National Capital Region (NCR), India, has set up vertical gardens to boost the aesthetics of the city and to help curb pollution [8]. Pillars under the metro rail corridors, elevated roads, flyovers among other places in Noida and parts of Delhi are being targeted for plantation.

## 2.2 Absorption

Absorption is a common method for direct air capture as has been used for $CO_2$ capture from point sources. In DAC, $CO_2$ is captured using liquid solvents like water or alkaline solution such as sodium or potassium hydroxide, or amine-based solvents. A fan blows air into a filter that is lined with a hydroxide-based liquid solvent absorbent. There are techniques to increase the liquid gas contact area, and that helps in increasing the efficiency of the absorption process. Absorption with a caustic

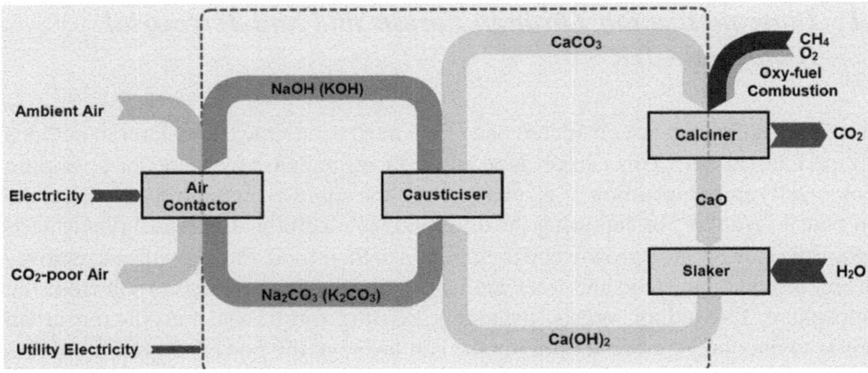


**Fig. 1** $CO_2$ direct air capture based on aqueous solution of NaOH or KOH as an alternative [9]

solvent is an acid–base reaction in which $CO_2$ reacts with sodium hydroxide and precipitates a stable sodium carbonate. In the second cycle, known as regeneration, sodium carbonate is mixed with calcium hydroxide in the Causticizer unit, where calcium carbonate is formed and sodium hydroxide is regenerated and is sent back to the contactor to start another absorption cycle. Meanwhile, in the most energy intensive step, calcium carbonate is heated up to around 900 °C in the Calciner unit to release $CO_2$. The overall heat demand is in the range of 1420–2250 kWh per ton of $CO_2$. The outputs of this reaction are calcium oxide and a pure stream of $CO_2$. $CO_2$ is collected, and calcium oxide is mixed with water in the Slaker unit for regeneration of calcium hydroxide. The process is shown schematically in Fig. 1.

Mono-ethanolamine is another promising liquid absorbent to capture carbon dioxide from the atmosphere. Costs are currently high, but the potential for savings is thought to be greater, due to the modular design that could be made on an industrial production line, and with lower temperatures needed to release $CO_2$.

## 2.3 *Adsorption*

Adsorption uses solid solvents in contrast to liquid solvents. For direct air capture, like PCC, primary amine-based solids are used to remove carbon dioxide from the air by chemisorption, and the processed $CO_2$ binds to a gas–solid contact. Chemisorption is a process which involves a chemical reaction between the surface and the absorbate. The strong interaction between the adsorbate and the substrate surface creates new types of bonds, and the absorbate is held by these chemical bonds. Through heat and vacuum, the $CO_2$ is then desorbed from the solid sorbent. This results in a gaseous $CO_2$ stream that can be cooled and cycled back into the system for reuse or storage. The filter of the DAC technologies using solid adsorbents is made of the ion-exchange resin. There is a variety of filter shapes because of the flexibility of the resin. Air

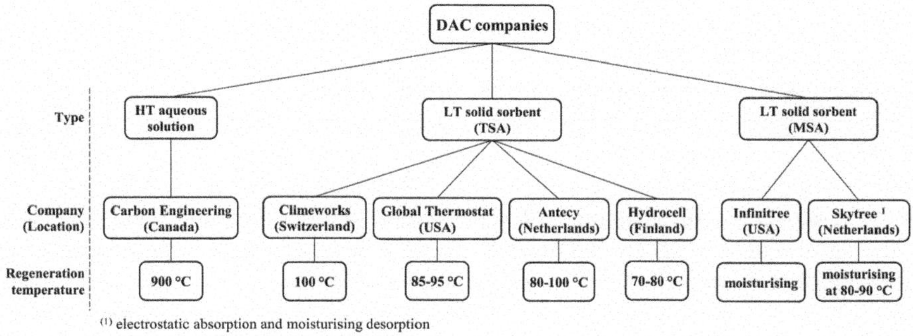


**Fig. 2** Companies active in the field of $CO_2$ DAC

flow into the filter is produced by a fan or blower. Natural air flow could also be used. Saturated filters are regenerated by putting it in moisture circumstance or heating it to release captured $CO_2$. Though not very efficient, this method requires little water and has a smaller footprint. Temperature swing adsorption (TSA) is the main DAC method in this category [9]. Unlike typical aqueous solution-based systems, the regeneration in solid sorbent DAC happens at relatively lower temperatures (80–100 °C), which is cheaper to produce or could be available as waste heat from some industrial plants. Moisture swing adsorption (MSA) is the other method in this category in which the regeneration happens by moisturising of $CO_2$-rich sorbent. [10] examined the possibility of MSA $CO_2$ capture on amine-based ion-exchange resin at low temperatures at around 45 °C. Figure 2 depicts the methodology adopted by various companies and the regeneration temperature for DAC by absorption and adsorption.

Abbreviations used: high temperature—*HT*, low temperature—*LT*, moisture swing adsorption—*MSA*, temperature swing adsorption—*TSA* [9].

## 2.4 *Mineralization*

Mineralization occurs naturally during weathering of silicates and rocks rich in calcium or magnesium. These rocks are present in upper mantle and basaltic lava. The calcite and magnetite have diversified applications, like they are used in cosmetics, pharmaceuticals, inks, resins, paints, rubbers, explosives, construction industries, detergents; calcium carbonate is used for the surface treatment of plastics. During ex situ carbon mineralization, solid reactants are reacted with fluid or gas rich in carbon dioxide, at the site of carbon dioxide capture. Alkaline industrial wastes can also be used. In in situ carbon mineralization, carbon-dioxide-rich fluids are circulated within basalts as inter-trappean at depths for a permanent storage. Mineralization methods emulate natural process of carbon dioxide uptake, while direct methods require additional technologies and energy input. Current technology requires 1.6

to 3.7 tonnes of rock to fix one tonne of $CO_2$. Since the reaction is exothermic, it produces energy that can be recycled back to source the reactor. The kinetics of this process is slower than adsorption and absorption processes.

The surface area of minerals/industrial waste comprising of olivine and serpentine or others is needed to be increased by finely grounding and spreading them on the land exposed to water and the atmosphere. This enhances the rate of $CO_2$ capture. For it to be successful grinding to PM10 or lower may be required, which is cost and energy intensive. There is currently no large-scale implementation of mineral carbonation. One impediment to implementation of mineralization is the large costs involved in the preparation of the solid reactant and its additives. More research would be needed to develop improved techniques before enhanced weathering could be considered a viable option for $CO_2$ removal on human timescales.

## 3 Direct Air Capture Projects—Commercial Outlook

There are currently seven DAC projects worldwide taken up by different startups. The stages of each project vary from, in planning, operational or in completed stages [9]. The processes adopted in them are discussed below.

**Antecy**. The Dutch company has developed non-amine technology for both DAC ($CO_2$-From-Air) and PCC ($CO_2$-From-Point Source). Antecy founded in 2010 has a "first of its kind" DAC pilot plant. Its unique design allows operation at large scale considering minimal use of energy, materials and footprint while maximizing capacity. As opposed to amine sorbents, Antecy's inorganic solid sorbent provides high stability, robustness and long life. The materials used are non-toxic, disposable and fully recyclable. The $CO_2$ produced does not contain any toxic contaminants, which makes this DAC technology environmentally friendly. Based on current achievements, the estimated cost of $CO_2$ captured from air at large scale is expected to be 50–80 USD/t. The cost for $CO_2$ captured from point sources with higher $CO_2$ concentrations is lower. Another company Climeworks has now joined Antecy (www.climeworks.com).

**Carbon Engineering (CE)**. The CE is one of the commercial companies which is engaged direct air capture. It is located in Calgary (Canada) and was founded in 2009 by David Keith [11] and became operational in 2011. The plant is currently capturing 1 ton of $CO_2$ a day and produces 1 barrel of fuels using its AIR to FUELS technology. A direct air capture pilot project in British Columbia (Canada) has converted a portion of its concentrated $CO_2$ into synthetic fuels using enhanced oil recovery technique in 2017 and by 2021, the project is expected to implement a full-scale direct air capture to liquid fuels plant. As per the proposed scale of the project, it would be 2000 barrels a day of fuels that would be distributed to areas with low-carbon fuels standards. They have partnered with Greyrock.

**Climeworks**. The Climeworks located in Zurich is capable of capturing 900 tonnes of $CO_2$/year. A $CO_2$ filter which is a sort of solid adsorbent is developed. A $23 million plant of Climeworks with 18 capturing units traps enough $CO_2$ to grow vegetables in a nearby greenhouse. The approximate cost to capture one ton of $CO_2$ from the air is about $600. In collaboration with Union Engineering (a Danish Company), a plant which provides beverage grade $CO_2$ [12] has been developed. In collaboration with Reykjavik energy, it is proposed to design a pilot plant using direct air capture and Carbfix technology. This large-scale plant is expected to remove $CO_2$ from air and sequester it in basalt rock in Hellisheidi, Iceland. The direct air capture project would run alongside a geothermal power plant that currently injects and mineralizes $CO_2$ into basalt bedrock. Clime works is aiming to be a first carbon-negative power plant [12].

**Global Thermostat**. The Global Thermostat located in Manhattan, New York, was founded in 2010. It is a privately funded carbon capture (CC) company and uses amine-based chemical sorbents bonded to porous honeycomb ceramic monoliths [12] to remove $CO_2$ from the atmosphere. The company has a range of projects—from small scale: 40 tons/year to large scale: 50,000 ton/year. The company has claimed to produce 98% pure stream of $CO_2$ from direct air capture. The company has no post-capture process and sells its $CO_2$ transported from its remote capture locations.

**Center for Negative Emission**. The Center for Negative Emission of Arizona State University is currently working on a DAC process using ion-exchange resin. The equipment is stored in a shipping container [13]. The estimated cost of capture is expected to be $30–200 USD/t-$CO_2$. Water is used to desorb the carbon dioxide.

**Carbon Sink**. Carbon sink uses amine-based ion-exchange capture technology and focuses on the technology to supply carbon dioxide for greenhouses. The cost of capturing was less expensive than through fossil fuels combustion point sources [12].

**Coaway**. The Coaway, an American company first to start in 2008, uses alkaline solution to capture atmospheric $CO_2$. The resulting material was regenerated using waste heat from adjacent oil refineries or existing power plant cooling towers that already move large quantities of air to create the necessary throughput for the carbon capture process [12]. The resulting material is "regenerated" in a thermal process that also releases the captured $CO_2$ as a concentrated stream ready for commercial use or sequestration. The company however has stopped working (Table 1).

## 4 Challenges in Direct Air Capture (DAC)

Direct air capture requires 1.8 to 3.6 times more energy than PCC from a flue gas because the separation occurs from a highly dilute $CO_2$ stream. Consequently, direct air capture machines are energy intensive and absorption technology, especially, needs either huge water, or caustic chemicals to capture carbon dioxide. One estimate is that amine-based technology absorption needs 300 cubic KL of water to capture 3.3 giga tons of carbon dioxide a year. It is estimated that scrubbing 1 gigaton of carbon dioxide from the air would require around 7% of all projected US energy production in 2050 and unless it is powered by low- or zero-carbon energy sources, result will not be net carbon removal. In a recent paper (realmonte) on DAC machines

**Table 1** Cost/energy estimate of DAC for different startups

| S. No. | Name of the company | Type of DAC used | Direct air capture cost/energy used | References |
|---|---|---|---|---|
| 1 | Antecy | Non-amine technology-inorganic solid sorbent | Capture is expected to cost 50–80 USD/t-$CO_2$ | www.climeworks.com |
| 2 | Carbon Engineering | Combination of KOH and $Ca(OH)_2$ | The cost is 136USD/t-$CO_2$ under 4USD/GJ of heat and carbon free electricity of 7 cent/kWh | Keith [11]; Ishimoto et al. [12] |
| 3 | Climeworks Jointly with Union engineering | Solid adsorbent | Energy required for regeneration—heat 1800-2500 kWh/t-$CO_2$ (100 °C) and electricity 350—450 kWh/t-$CO_2$ | Ishimoto et al. [12] |
| 4 | Global thermostat | Amine-based chemical sorbents bonded to porous honeycomb ceramic monoliths | At 1 million tons per year, cost is estimated to be 10–35 USD/t-CO2<br>Energy required for capture is 4.4 GJ/t-$CO_2$ (95 °C steam) | Ishimoto et al. [12] |
| 5 | Center for negative emission of Arizona state University | Ion-exchange resin | Capture cost is 30 ~ 200 USD/t-$CO_2$<br>Energy required for regeneration is 50 MJ/mol | Klaus [10], Arizona state university |
| 6 | Carbon sink | Amine-based ion exchange | For application to greenhouses, the cost is less than the supplied by Cylinders | Ishimoto et al. [12] |
| 7 | Coaway | Alkaline solution Capture | Capture cost is less than 20 USD/t-$CO_2$ in case power plant waste heat can be utilized | Ishimoto et al. [12] |

*Source*—compiled by the author from various sources

that can capture carbon dioxide from the atmosphere, an estimate was made on the energy requirement for negative emission by 2100, and it was found that the energy requirement will be astounding with 300 exajoules each year. This is more than half of overall global energy demand of today, from all sources, and despite rising demand in this century, it would still be a quarter of expected demand in 2100. Capturing 30 Gt $CO_2$ /year which is around the current global emissions would mean building some 30,000 large-scale DAC factories, while coal-fired power stations in the world are less than 10,000 today. If DAC were to be carried out using small modular systems, then as many as 30 million machines will be needed by 2100.

## 5 Financial Costs

A DAC plant designed to capture 1 mega ton of carbon dioxide a year is estimated to cost $2.2 billion [14]. Point source carbon dioxide levels are 8–14%, while in ambient air it is 0.003%. The concentrations are therefore 300–400 times more. There is still a wide variety in costs for PCC depending on the methods used, so a cost comparison is difficult between PCC and DAC methods. Using the same costing methodology and also keeping many other features of the two systems the same (such as the final pressure to which captured $CO_2$ is compressed: ~100 atmospheres pressure), the ratio of the DAC and PCC cost estimates was made. The comparison is expected to be more robust than the cost estimates for either PCC or DAC alone as shown by POPA, APS Panel on Public Affairs in the Direct Air Capture of $CO_2$ with Chemicals (DACCC) report [14]. In comparison relative to the benchmark DAC system, the normalized capital cost for the reference PCC system is estimated to be $40/t $CO_2$, which is five times smaller. While the operating costs of PCC is $40/t of $CO_2$, more than four times smaller. At present, capacity to estimate the cost for DAC is limited. With better understanding, it would help to set a limit for climate change mitigation and adaptation targets [15]. The total avoided cost for PCC has been estimated to be about $80/t $CO_2$, which is about eight times smaller than DAC.

The estimated capture cost of this DAC system is seven to nine times higher than the estimated cost of post-combustion capture system, with respective cost estimates of DAC and PCC are $430 to $550 and $60 per ton of $CO_2$ captured. If we take in a renewable source of power to supply the energy to DAC, then for each ten $CO_2$ molecules removed from the atmosphere by the DAC system, three $CO_2$ molecules are emitted to the atmosphere by the energy source, so that total capture costs are spread over seven-tenths as much captured $CO_2$ as would have been the case if the electricity had been produced without $CO_2$ emissions [14]. Hence, the power requirement for DAC is quite high. DAC costs have come down considerably since then. A comparative study carried out by Yuki et al. [12] is reproduced below graphically, and the gist is reproduced below (Fig. 3).

$CO_2$ capture cost for hydroxide-based solvents is around $150/ton of $CO_2$.

Current liquid amine-based separation is $10–35/ton of $CO_2$.

Adsorption-based $CO_2$ capture costs are between $30–200/ton of $CO_2$.

New startups are joining to develop DAC demonstrating facilities [16]. Solitaire Power commissioned a demonstration facility using technology developed by the VTT Technical Research Centre of Finland. It combines DAC, an electrolyte for hydrogen production, and a synthesis reactor for hydrocarbon production. In 2018, the pilot plant operated 300 h, producing 6.2 kg of oil and wax per day. In 2019, the Swiss Synhelion started testing a small refinery in Zürich. The demonstration project using solar energy to extract $CO_2$ and water from ambient air the demonstration project produced syngas mixture at 1,500 °C comprising of hydrogen and carbon monoxide. The syngas mixture can afterward be processed into hydrocarbons such as kerosene or methanol. According to Synhelion, by using a photovoltaic system with an area of 1 km$^2$ it could produce 20,000 L of kerosene a day.

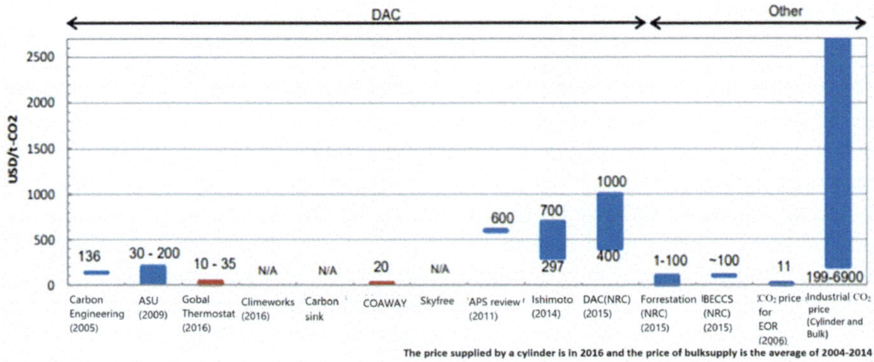


**Fig. 3** Direct capture costs of carbon dioxide through different methods [12]

# 6 Direct $CO_2$ Capture from Seawater

Extracting $CO_2$ from seawater capture is akin to direct air capture, except $CO_2$ is extracted from seawater instead of air. By reducing $CO_2$ concentration in the ocean, the water draws in more carbon from air to regain balance. Seawater is a more concentrated solution of $CO_2$ than the ambient air, which means less work is required to separate it out. But seawater is also considerably heavier than air, which means more work is required to move it through the system. Addition of CaO enhances the seawater absorption of $CO_2$ and its solubility in seawater with 0.4% CaO increased by 79.25% to improve the $CO_2$ absorption ability of seawater (https://doi.org/10.1016/j.energy.2018.09.066) and promotes $CO_2$ fixation in carbonate precipitation. It shows that the addition of CaO can enhance greatly seawater capture of $CO_2$. In seawater, there is a potential to capture $CO_2$, but it has to deal with the added complexities of technology deployment in harsh maritime environments.

Because of its strategic uses, the U.S. Naval Research Laboratory (NRL) has developed a prototype seawater device for $CO_2$ capture (navy.mil/submit/display.asp?story_id = 102,700) using Fischer–Tropsch technology. Because $CO_2$ can be converted to fuels by adding energy (and some Navy vessels tend to have spare nuclear reactors on board), such a technology could allow vessels to create their own fuels and avoid having to stop to refuel. Although if the captured carbon is converted to fuels and combusted, it just returns to the atmosphere, but future applications of this kind of technology could provide long-term storage for captured carbon.

Another alternative was proposed to use phytoplankton to capture $CO_2$. When plants die, they go to bottom of sea thus sequestering $CO_2$. Phytoplankton growth can be enhanced by seeding it with iron filings. The ocean may be seeded with iron at selective locations, encouraging phytoplankton growth, as a means of accelerating $CO_2$ capture. There are several challenges in the efficacy of iron seeding. When phytoplankton sinks, the bacteria that decompose them may affect distribution of nitrogen and oxygen in water. In place of iron filings, use of biogenic iron dust nanoparticles has been suggested for carbon dioxide removal. More research would be needed to know whether it offers significant $CO_2$ removal potential.

## 7 Conclusions

Climate scientists predict that countries will need to drop $CO_2$ emissions to near zero by midcentury and then remove more $CO_2$ than they emit, if the planet is to avoid a catastrophic 2 °C warming. In order to avoid dangerous levels of global warming, capturing and utilizing or storing carbon already in the air must be part of climate strategy around the world. It's time to begin investing across the portfolio of carbon-removal approaches. This endeavor needs to gather steam. Pulling carbon dioxide ($CO_2$) from the air and using it to make synthetic fuels seem like the ultimate solution to climate change: Instead of adding ever more $CO_2$ to the air from fossil fuels, we can simply recycle the same $CO_2$ molecules over and over. But such technology is expensive—about $600 per ton of $CO_2$, by one recent estimate. Now, in a new study, scientists say future chemical plants could drop that cost below $100 per ton—which could make synthetic fuels a reality in places such as California that incentivize low-carbon fuels.

**Acknowledgements** The author is thankful to Dr. Malti Goel, President, Climate Change Research Institute & Former Senior Adviser, Ministry of Science & Technology, Government of India, for the support.

## References

1. NOAA, National Oceanic and Atmospheric Administration (NOAA) (2019) USA along with its National centre for environmental information
2. Global climate report, Government of India, ministry of earth science report, 2019; Sea level Rise in India
3. Phillips O, Bitariho R, Taylor D, Sheil D (2020) Hubau et al nature 2020 carbon sink saturation in African and Amazonian forests. Nature 579:80–87. https://doi.org/10.1038/s41586-020-2035-0
4. NASEM, National Academies of Science-Engineering-Medicines (2019) Negative emission
5. NASEM, National Academies of Science -Engineering-Medicines (2018) Direct air capture and mineral carbonation approaches for carbon dioxide removal and reliable sequestration, Washington D.C. USA<Date of Conference>
6. Puri A, Satyanarayana T (2010) Enzyme and microbe mediated carbon sequestration; $CO_2$ sequestration technologies for clean energy. In: Qasim SZ, Goel M (eds) Daya Publishing House
7. Thokchom A, Yadava PS (2014) Soil $CO_2$ flux in different ecosystem of North-East India. Curr Sci 107(1):99–105
8. PTI (2018) Vertical gardens: noida's big plan to cut down pollution, enhance aesthetics Press Trust of India, August 12, 2018

9. Fasihi M (2019) Technoeconomic assessment of $CO_2$ direct air capture plants. J Cleaner Prod 224:957–980
10. Lackner K (2009) Director of center for negative carbon emissions at Arizona state university, capture of carbon dioxide from ambient air. Eur Phys J-Spec Top 176(1):93–106
11. Keith DW, Ha-Duong M, Stolaroff JK (2006) Climate strategy with $CO_2$ capture from the air. Climatic Change 74(1):17–45
12. Ishimoto Y, Sugiyama M, Kato E, Moriyama R, Tsuzuki K, Kurosawa A (2017) Putting costs of direct air capture in context. FCEA working paper series: 002. Washington DC: forum for climate engineering assessment
13. European Academies Science Advisory Council (2018) What role in paris agreement targets; Negative emission technologies
14. POPA, APS Panel on Public Affairs (2011) Direct air capture of $CO_2$ with chemicals (DACCC)
15. Pielke RA Jr (2009) An idealized assessment of the economics of air capture of carbon dioxide in mitigation policy. Environ Sci Policy 12:216–225
16. Cressy D (2015) Commercial boost for firms that suck carbon from air. Nature 526:306–307

# Climate Change Mitigation Through Bio-route and Assimilation of $CO_2$ Fluxes

# Mitigating Global Warming Through Carbonic Anhydrase-Mediated Carbon Sequestration

**Himadri Bose and T. Satyanarayana**

**Abstract** Industrial Revolution has led to an unprecedented rise in carbon dioxide concentrations in the atmosphere. Among various methods available for carbon capture from the industrial emissions such as flue gas, carbonic anhydrase (CA)-based carbon capture techniques have been evolved and gained immense attention in the recent years. Carbonic anhydrase (CA) is a zinc metalloenzyme, which is an essential biocatalyst for all living beings. It plays a role in accelerating the hydration and dehydration of carbon dioxide. This enzyme can be utilized in vitro for capturing carbon from industrial emissions. Thermo-alkali stable CAs from prokaryotes are the most promising candidates for biomimetic carbon sequestration owing to the high temperature of flue gas and alkaline condition needed for precipitation of calcium carbonate formed in the reaction. These CAs can be essentially immobilized on various solid supports and matrices for developing bioreactors and their continuous operation. This chapter reviews developments in utilizing CAs of prokaryotes in carbon capture technologies.

**Keywords** Carbonic anhydrase · Carbon dioxide · Carbon capture · Thermo-alkali stable CAs · Biomimetic carbon sequestration

## 1 Introduction

Period between the late eighteenth century and early nineteenth century saw the emergence of new manufacturing processes in Europe and the USA, popularly known as Industrial Revolution. This included going from hand production methods to machines, development of new chemical and iron production technologies, increase in the use of steam power, development of heavy machine tools and the rise of

H. Bose
Department of Biotechnology, Indian Institute of Technology Kharagpur, Kharagpur, India

T. Satyanarayana (✉)
Division of Biological Sciences and Engineering, Netaji Subhas University of Technology, Sector 3, Dwarka, New Delhi, India
e-mail: tsnarayana@gmail.com



M. Goel et al. (eds.), *Climate Change and Green Chemistry of $CO_2$ Sequestration*, Green Energy and Technology, https://doi.org/10.1007/978-981-16-0029-6_13

the factory system. The production of goods and products in every sector (textile, iron, etc.) increased dramatically [55]. Industrial Revolution almost influenced each and every part of common man's life. There was an unprecedented rise in average per capita income and population began to attain an unprecedented sustained growth. Some experts have noted that the standard of living for the general population began to increase significantly. Some eminent economic historians are of the view that it is the most important event in the history of humanity after domestication of animals [92]. One might say that the Industrial Revolution triggered the rise of modern high-throughput technologies which we see today [42, 90, 142]. There have been many serious repercussions of this revolution which we are facing today, and the coming generations will have to face even worse. The humanity has paid heavy price due to one of the most important events during its evolution. Population is on the rise ever since, and this is not showing any signs of slow down. Today, the world population stands at 7 billion, and the numbers are increasing day by day. In order to produce goods and products in large quantities to meet the demand, the industries started large scale usage of fossil fuels. Even today fossil fuels (coal and petroleum) are the primary source of energy in the world. Their large-scale use led to a phenomenon known as 'global warming'. There has been alarming rise in Earth's average temperature in the post-industrial era (0.8 °C in the previous century) [121]. This increase has led to various very detrimental impacts and changes in global climate, which includes decrease in cold temperature extremes and increase in warm temperature extremes. Loss of ice masses at Arctic and Antarctic in the past few decades has increased the water levels in oceans. The oceans have absorbed about 30% of the emitted anthropogenic $CO_2$ (26% more than the pre-industrial era), which has led to the acidification of oceans. Dissolved oxygen levels have decreased in coastal waters. The precipitation pattern around the world has affected to such an extent that has caused an impact in hydrological systems around the world, both in quality and quantity. Regions with high salt concentrations have become more saline, and the regions with high precipitation have become much fresher, thereby disturbing the overall water cycle. Global warming has even affected the global carbon cycle [61]. In fact, a report published by 'Reuters', a leading international news agency, on February 7, 2019 stated that average global surface temperatures have reached to 1.0 °C above pre-industrial times in 2018 [165]. This rise in earth's average temperature is causing weather and climate imbalances all over the world leading to weather extremities in recent times (wildfires in California and Greece, drought in South Africa, floods in Kerala, bone chilling cold in USA and scorching heat in Australia). According to Gavin Schmidt (Director, NASA's Goddard Institute for Space Studies), 'The impacts of long-term global warming are already being felt—in coastal flooding, heat waves, intense precipitation and ecosystem change'. This climate change has severely impacted activities, abundance and interaction pattern of many marine, terrestrial and freshwater species. There has been an abundance and distribution shifts of many phytoplanktons, marine fishes and invertebrates toward cooler waters, pole ward [56]. Many species are either extinct or on the verge of extinction. These detrimental impacts are severely affecting poor peoples' lives directly and indirectly through impact on livelihood, reductions in crop yields, destruction of homes by natural calamities, increase in food prices and food insecurity [2], hence, gradually ‘global warming’ is pushing the humanity toward its peril. The ultimate cause of global warming is the rise in the concentrations of greenhouse gases (GHGs) particularly $CO_2$ in the post-industrial era due to large-scale use of fossil fuels. The level of atmospheric $CO_2$ approximately 290 ppm before Industrial Revolution stands at 410 ppm at the end of January 2019 [102]. Anthropogenic $CO_2$ emissions are mainly caused by thermal power plants and industries. Industry and energy sectors contribute to about 70% of $CO_2$ emissions in India followed by agriculture (18%), Industrial Processes and Products Use (IPPU) [8%] and from waste sector (3%) [19, 48]. The quality of flue gas depends on the amount and quality of coal burnt [131]. Even our modern and lavish lifestyles have been contributing to carbon emissions, which can be measured in terms of carbon footprints [156]. Along with $CO_2$, other gases like nitrogen and its oxides and oxides of sulfur have also played a prominent part in global warming. Main contributor of $CO_2$ is the flue gas which is emitted from coal-based thermal power plants. Our main concern is, therefore, to reduce our carbon emissions. Many measures have been undertaken by the governments around the world to reduce their carbon emissions. In order to combat global warming, almost 200 governments around the world signed the Paris Climate Agreement in 2015 to phase out the use of fossil fuels and limit the rise in temperatures to 2 °C above that of pre-industrial times, while ‘pursuing efforts’ for 1.5 °C. Carbon taxation, emissions and trading schemes are some of the measures taken up by governments all over the world, for mitigating global warming. The reduction of carbon emissions is, however, quite difficult as more than 70% of the world population still depends on fossil fuels for their sustenance. There are other ways (burning of wood as fuels, etc.) apart from usage of fossils through which carbon dioxide is released into the atmosphere. High cost of harvesting alternative renewable sources of energies (solar, wind, biomass, etc.) is another matter of concern. Scientific consensus on global warming and the fear of nonlinear climate transitions has brought the world together in order to develop new technologies and carefully manage the existing ones [109]. An effective solution to this problem is to capture and store the emitted carbon dioxide, popularly known as carbon capture and storage (CCS) or carbon sequestration. The progress made in developing carbon capture technologies has been reviewed from time to time [16, 19, 20, 46]. IPCC has also published a wholesome review on different CCS technologies providing a precious input for policymakers and researchers in developing processes for reducing GHG emissions [60–62]. This chapter will majorly focus on the progress achieved in carbonic anhydrase-based CCS technologies.

## 2 Carbon Capture and Storage

Carbon capture and storage (CCS) can be defined as the mitigation of anthropogenic $CO_2$ entering into the atmosphere by capturing $CO_2$ from large point-source emitters such as industries and storing it in subsurface reservoirs, where it would remain for sequestration [54]. This term has been extended to Carbon Capture Utilization and

Storage (CCUS) that includes utilization of the stored $CO_2$ in various physical, chemical and biological processes [35, 87]. CCS can also be used to describe the scrubbing of $CO_2$ from ambient air as a climate engineering technique and storing it as mineral carbonates [19]. There are basically two broad methods for CCS, viz. physical and biological.

## 2.1 Physical Methods of Carbon Capture

Physical or geological method of carbon capture can be termed as 'capture of $CO_2$ from flue gas which is emitted from large point sources (power plants etc.) and storing deep underground the earth in a leak proof environment'. The stored $CO_2$ can be even utilized for various applications such as Enhanced Oil Recovery (EOR) [52]. Underground unused mines are one of the most potential sources of carbon sequestration. This method involves three stages, namely $CO_2$ capture (pre-combustion and post-combustion), $CO_2$ transport (transported using large gas pipelines to facilities for storage or reutilisation) and $CO_2$ storage (rock pores ~1 km below the surface or abandoned deep saline aquifers and oil or gas reservoirs) [104]. $CO_2$ capture method involves various absorption (liquid sorbents such as monoethanolamine (MEA), adsorption (zeolite and activated carbon), chemical looping combustion (metal oxide), membrane or hydrate-based separation and cryogenic separation techniques [1, 15, 76] for efficient separation of carbon from flue gas. $CO_2$ transport utilizes network of robust pipelines made of carbon steel for transporting $CO_2$ in supercritical (72 atm pressure) condition [12, 140]. $CO_2$ can be transported in high pressures into nearly depleted oil/gas wells in order to extract the residual oil and gases, while the injected $CO_2$ remains stored. One of the best sites for $CO_2$ storage is highly saline deep aquifers 700–1000 m below the surface or it may be injected into deep coal bed to release methane which is trapped in the porous structure of coal seams ($CO_2$—enhanced coal bed methane) [51].

Deep underground injection in the oceans is another method for $CO_2$ storage as it can absorb about two billion metric tons of $CO_2$ each year. Carbon dioxide is denser than sea water at greater depths, and it can be fixed by phytoplanktons and the fixed carbon can be further utilized by sea animals [53]. Oceans can be enriched by phytoplanktons using natural and artificial deposition of iron-rich dust into oceans [66, 98, 145]. Despite the efficiency, physical methods suffer from several disadvantages as elucidated below.

1. The cost of the overall process is quite high which may lead to increase in power tariff all over [148].
2. It increases energy consumption by 25% [51].
3. Use of monoethanolamine leads to corrosion, evaporative losses, yielding toxic degradation products and may require significant energy to remove $CO_2$ during sorbent regeneration [29, 101, 115].

4. A lot of water is lost during post-combustion $CO_2$ capture, and this method may even release toxic byproducts [53].
5. Direct injection of $CO_2$ acidifies the ocean [33] which can be a grave for marine ecosystem. By the year 2100, dissolved inorganic carbon (DIC) in seawater is expected to increase by 6% and in freshwater by 0.6% with a concomitant decrease in pH by 0.279 and 0.298 units, respectively (Stumm and Morgan 1996), [57].
6. The CCS technology is still in pilot scale, and a large-scale commercialization is yet to be undertaken.

That's why geological storage of $CO_2$ in deep saline aquifers and unused mines is an environmentally friendly approach for CCS [120]. Mineralization of $CO_2$ as mineral carbonates using limestone rocks such as wollastonite ($CaSiO_3$), serpentine ($Mg_3Si_2O_5(OH)_4$) and olivine ($Mg_2SiO_4$) [58, 119] and its deep underground storage can be cost-effective and environmentally benign solution for mitigation of global warming [41, 88, 97]. Other advantages include the availability of vast amount of raw materials in the form of minerals and metal oxides (generated as hazardous waste in the form of fly ash). Utilization of fly ash in mineral carbonation reduces the $CO_2$ concentration from 13.0 to 9.6% and $SO_2$ concentration from 107.8 ppm to 15.1 ppm within 2 min [114, 133]. **Mineralization** comprises a set of reactions as follows [40]:

Gaseous $CO_2$ dissolves quickly in water and produces a loosely hydrated aqueous form (1).

$$CO_2(g) \rightarrow CO_2(aq) \qquad (1)$$

Then, carbonic acid is formed when aqueous $CO_2$ reacts with water (2).

$$CO_2(aq) + OH^- \rightarrow H_2CO_3$$

In the next step, carbonic acid dissociates into bicarbonate and carbonate ions (3, 4)

$$H_2CO_3 \rightarrow HCO_3^- + H^+ \qquad (3)$$

$$HCO_3^- + OH^- \rightarrow CO_3^{2-} + H_2O \qquad (4)$$

Ultimately, the carbonate ions get precipitated as mineral carbonates in the presence of divalent metal cations such as

$$Ca^{2+},\ Mg^{2+} \text{and } Fe^{2+} \qquad (5)$$

$$CO_3^{2-} + Ca^{2+} \rightarrow CaCO_3 \downarrow \quad \text{(calcite)}$$

$$CO_3^{2-} + Mg^{2+} \rightarrow MgCO_3 \downarrow \quad \text{(magnesite)}$$

$$CO_3^{2-} + Ca^{2+} + Mg^{2+} \rightarrow CaMg(CO_3)_2 \downarrow \quad \text{(dolomite)}$$

$$CO_3^{2-} + Fe^{2+} \rightarrow FeCO_3 \downarrow \quad \text{(siderite)}$$

It is a pH-dependent process and occurs mainly at alkaline pH (between pH 8 and 10, both reaction (2 and 3) occur and above pH 10 reaction 2 occurs). Due to abundant supply of $OH^-$ at alkaline pH, mainly $HCO_3^-$ (bicarbonate) and $CO_3^-$ formation takes place leading to $CaCO_3$ precipitation. This $CaCO_3$ can be utilized for producing several usable products such as amino acids, limestone, etc.

Mineral carbonation is being studied extensively for its utility in direct capture and mineralization of $CO_2$ of flue gas. Some pilot-scale studies have been undertaken to demonstrate the viability of the process [114]. Mineral carbonates can produce useful byproducts. But this process is quite slow as the reaction 2, i.e., $CO_2$ hydration, which is the rate limiting step is quite slow. The CAs can intensify the whole process by speeding up the process of hydration of dissolved $CO_2$ into bicarbonate, i.e., the reaction 2. Utility of CAs in mitigating global warming will be discussed below.

### 2.2 *Biological Methods of Carbon Capture and Relevance of Carbonic Anhydrase*

Biological sequestration can be defined as net removal of $CO_2$ from the atmosphere by plants and micro-organisms and its storage in vegetative biomass and in soils. Biological or natural way of $CO_2$ removal is the best possible solution to mitigate global warming. $CO_2$ fixation by green plants and incorporation of $CO_2$ into plant biomass is the most natural way of sequestering the atmospheric carbon [75]. Efforts are underway to engineer $C_4$ plants to improve the catalytic efficiency of the enzymes involved in the $CO_2$ fixation pathway [25]. More and more trees and plants should be planted and grown for a long time so that carbon is stored in their biomass for long time. Modified agricultural practices (use of cover crops) and growth of bioenergy crops can be another way of sequestering the $CO_2$ [35]. Algae such as *Lomentaria and Chlorella* (they can grow much better in the presence of excess $CO_2$) besides being more efficient in photosynthesis, they are one of the main the choices of research for solving global warming problem [77]. The biomass thus produced could be used as biofuel for various heating and power purposes. Photosynthetic cyanobacteria have been tested for their application in $CO_2$ sequestration and can be utilized as microbial cell factories [164]. Anaerobic and non photosynthetic bacteria such as methanogens, which produce methane via Wood—Ljungdahl pathway for acetate synthesis, can be used for commercial methane production from atmospheric $CO_2$ [77]. These methanogens can sustain a wide range of temperatures (9–110 °C and above), hence may be useful in developing cost-effective CCS technologies.

Microorganisms are ubiquitous in nature, and their role in global carbon cycling is well established [50]. Initial studies by some researchers on heterotrophic microorganisms such as *Propionibacterium* revealed that $CO_2$ gets reduced during glycerol fermentation, and further, it produces oxaloacetate by reacting with pyruvate [157]. After that, there has been a growing interest in microorganisms in the last few decades to use them as efficient carbon sequestering agents. Another enzyme carbonic anhydrase (CA) can be utilized in this pathway (catalyzes the conversion of $CO_2$ into bicarbonate) can be used for amino acid production lysine [103]. Microbial contribution to carbon sequestration is governed by the interactions between the amount of microbial biomass, microbial community structure and their byproducts. Much work has been going on since the last decade to explore the use of CA as a potential biocatalyst for mitigating global warming, this aspect will be discussed in the following sections. This efficient and fast hydration property of this enzyme has attracted attention in sequestering $CO_2$ by a promising biomimetic approach. CA can surprisingly expedite the process of $CO_2$ hydration by catalyzing the rate-determining step (step 2a) in the conversion of $CO_2$ to $CaCO_3$ [97].

## *2.3 Carbonic Anhydrase*

It is a ubiquitous zinc metalloenzyme (presence of zinc at its active site) (EC No. 4.2.1.1) which catalyzes the interconversion of carbon dioxide into bicarbonate. It was the first zinc metalloenzyme to be discovered in living systems [127]. This enzyme is either found intracellularly in cytoplasm or secreted outside (extracellular) associated with the periplasm and an essential component for survival of nearly all life forms. It helps in concentrating $CO_2$ in the living cells. In most of the organisms, CAs are required for transport, removal of $CO_2$ from lungs and for synthesis of eye secretions. CAs maintain optimum levels of $CO_2$ and $HCO_3^-$ in the body as this is utilized as substrate for many enzymatic reactions. It plays a vital role in respiration, ion transport and maintenance of acid–base balance in mammals. Carbonic anhydrase (hCAII) is present in relatively high concentrations in red blood cells [14]. CA has many important functions, including gluconeogenesis, lipogenesis, tumorigenicity, and signal transduction in humans [86]. Importance of CAs can be attributed to the fact that mutations in CA genes may result in osteoporosis and mental retardation. It has a high $k_{cat}$ value in the order of $10^6\ s^{-1}$, and hence, its rate of reaction is one of the fastest among known enzyme with catalytic efficiency in the order of $10^8\ M^{-1}\ s^{-1}$. Carbonic anhydrase was discovered in bovine erythrocytes in 1933 [93, 138]. Seven evolutionary distinct classes ($\alpha$, $\beta$, $\gamma$, $\delta$, $\zeta$, $\eta$ and $\theta$) have been discovered till date widespread in living beings, and their structure–function and mechanism of action have been widely discussed by several researchers [9–11, 19, 44, 74, 139]. No specific amino acid sequence similarity has been observed in these classes. This enzyme is a classic case of convergent evolution. All seven types of CAs have a similar metal binding active site, and most of them are inhibited by various sulfonamide inhibitors and their derivatives such as acetazolamide, benezenesulfonamide, methazolamide, 3-fluorosulfanilamide, 3-chlorosulfanilamide, sulfamic acid, sulfananilamide, phenolics and others [138, 139]. These inhibitors have a significant role in developing drugs for many pathogens such as *Helicobacter pylori*, *Mycobacterium tuberculosis*, *Brucella spp.*, *Streptococcus pneumoniae*, *Salmonella enterica*, and *Haemophilus influenzae* in which there is an upregulation of CAs during pathogenesis [138]. The distinct features of all the classes are summarized in Table 1.

## 2.4 Catalytic Mechanism

It is a ping-pong reaction that catalyzes the reversible hydration/dehydration of $CO_2$ into bicarbonate and a proton [125, 126]. It consists of following steps:

(i) **Concentration of $CO_2$**: The hydrophobic residues like Val, Leu and Trp present at the active site get associated with $CO_2$ and concentrate it within the active site cleft. A histidine residue present near the water molecule accepts a proton ($H^+$) released from the water molecule, leaving only the zinc-bound hydroxide ion.

(ii) **Nucleophilic attack**: The zinc-bound hydroxide ion causes nucleophilic attack on the carbonyl group to form zinc-bound bicarbonate ion (Eq. 1). Furthermore, one of the oxygen atoms of $CO_2$ binds to the $Zn^{2+}$ making it penta-coordinated metal ion.

(iii) **Water-bicarbonate exchange**: A water molecule is exchanged for the zinc-bound bicarbonate ion (Eq. 2).

$$\mathrm{EZnOH^- + CO_2 \leftrightarrow ZnHCO_3^-} \quad (1)$$

$$\mathrm{ZnHCO_3^- \leftrightarrow EZnH_2O + HCO_3^-} \quad (2)$$

$$\mathrm{EZnH_2O + PSR \leftrightarrow EZnOH^- + PSR - H^+} \quad (3a)$$

$$\mathrm{PSR - H^+ + B \leftrightarrow PSR + B - H^+} \quad (3b)$$

(iv) **Proton transfer**: This step involves intramolecular $H^+$ transfer (Eq. 3a) from zinc-bound water to ionizable proton shuttle residues (PSRs) like His, Glu, Tyr, Asp or Glu (based on the type of CA) followed by intermolecular $H^+$ transfer (Eq. 3b) to exogenous buffer or solvent as the ultimate acceptor. The enzyme is then regenerated, and the bicarbonate ion is released. The proton transfer step is the rate-limiting step [147].

**Table 1** Characteristic features of CA families

| Characteristics | α CA | β CA | γ CA | δ CA | ζ CA | η CA | θ CA |
|---|---|---|---|---|---|---|---|
| Occurrence | Predominantly in animal kingdom also found in protozoa, algae, green plants, and in some archaea, bacteria and fungi | Reported in all three domains of life | Eubacteria and archaea *Methanosarcina thermophila* | *Thalassiosira weissfogii* | *Thalassiosira weissfogii* (marine diatom) | *Plasmodium falciparum* | *Phaeodactylum tricornutum* (marine diatom) |
| Molecular weight | Monomeric (Mw: 26–37 kDa) | Dimer can form various oligomeric (trimer to octamers) structures 45–200 kDa | Homotrimer 60 kDa (native) | Monomer 27 kDa | Monomer 69 kDa | Monomer 45 kDa | 85 kDa |
| Active site amino acids | Three histidine residues | Two cysteine and one histidine residues | Three histidine residues | Three histidine residues | Two cysteine and one histidine residue | Two histidine residues and one Glutamine residue | Cysteine, Glycine, Histidine |
| Function | Maintenance of physiological pH, homeostasis, secretion of $HCO_3^-$, ion exchange | Cyanate degradation (*E.coli*), $CO_2$ fixation (cyanobacteria), overexpressed during pathogenesis of some pathogens (*Salmonella*) | Acetate metabolism and conversion of acetate to methane (methanogens) | $CO_2$ fixation by marine diatoms | $CO_2$ fixation by marine diatoms | Overexpressed during pathogenesis | $CO_2$ fixation |

(continued)

**Table 1** (continued)

| Characteristics | α CA | β CA | γ CA | δ CA | ζ CA | η CA | θ CA |
|---|---|---|---|---|---|---|---|
| Subclasses and isoforms and salient structural features | 15 isoforms (CAI—CAXV) found in mammals divided into 4 subclasses<br>Majorly dominated by β sheets surrounded by α-helices | Two structural classes<br>Type 1: the active site is open with zinc ion is located on the fundamental dimer interface and is ligated to Cys160, His220, and Cys223<br>Type 2: the active site zinc ion is closed structure 4 protein ligands: Cys42, Asp44, His98, and Cys101<br>Contains α/β fold, being composed of an N-terminal arm, and a C-terminal domain dominated by α-helices | Two homologs based on crystal structures Cam (*M. thermophila* and CamH (*P. horikoshii*) adopts a left-handed parallel β-helix fold | The structure consists of seven α helices, three $3_{10}$-helixes and nine β-strands, which are organized in three β-sheets | Seven α-helices, three $3_{10}$-helices and nine β-strands, which are organized in three β-sheets | Metal ion coordination pattern different from other classes of CAs | rich conserved domain of 110 amino acids |
| Activity | Esterase activity, $CO_2$ hydration activity | $CO_2$ hydration activity | $CO_2$ hydration activity | $CO_2$ hydration activity | $CO_2$ hydration activity | $CO_2$ hydration activity | Esterase activity, $CO_2$ hydration activity |
| Activators | NA | Histamine, dopamine, serotonin, 2-pyridyl-methylamine and 4-(2-aminoethyl)-morpholine | NA | Tyrosine | NA | Glutamic acid, Aspartic acid | NA |

(continued)

**Table 1** (continued)

| Characteristics | α CA | β CA | γ CA | δ CA | ζ CA | η CA | θ CA |
|---|---|---|---|---|---|---|---|
| Other features | N—terminal signal peptide for periplasmic localization or extracellular secretion<br>Cobalt may substitute zinc at its active site in certain conditions | Basic component of carbon concentrating mechanism (CCM) | Oldest form of CA. Fe(II) may present at active sites in anaerobic conditions | – | Cambialistic | Aids during the pathogenesis of malarial parasite *Plasmodium falciparum* | Basic component of Carbon Concentrating Mechanism (CCM)<br>In diatoms |

## 3 Carbonic Anhydrases of Prokaryotes

Bacteria encode enzymes belonging to α-, β-, and/or γ-CA families. The first α-CA characterized from a prokaryotic genus was that of *Neisseria gonorrhoeae* [152]. The next type (β CA) was discovered initially in plants. Later it was also found in the bacterial as well as archaeal genera. This type of CA is the most widely distributed among all the classes [128]. Gamma CA was discovered in hyperthermophilic archaeon *Methanosarcina* [5]. This CA has been found to be associated mainly with archaeal domain and was also discovered in archaeon Pyrococcus horikoshii. Recently, γ-CA was also discovered in several other bacteria too. The γ class CA (CcmM) from the aerobic cyanobacterium *Thermosynechococcus elongates* strain BP-1 had been characterized [105] whose crystal structure was similar to γ-CA of *Methanosarcina.* β and γ CAs have been mostly characterized from plants and prokaryotes. *Synechococcus* harbors multiple CA genes (ccaA, isfA and others) which have a close resemblance to cynT of *E. coli* [81]. The gene encoding γ-CA was also cloned and expressed heterologously, but it was not active. δ, ζ and η CAs have been recently isolated from cyanobacteria, marine diatoms and some dinoflagellates [158]. In cyanobacteria, microalgae and *Plasmodium*, CAs are involved in the $CO_2$ concentrating mechanism (CCM), which helps the cells in photosynthesis in the absence of inorganic carbon [13]. *Rhodospirillum rubrum* harbors a CA with a molecular mass of 28 kDa [49]. Acetogens such as *Acetobacterium woodii and Clostridium thermoaceticum* also possess CA activity. Among them Acetobacterium woodii has the highest CA activity [128]. *Propionibacterium* CA helps in $CO_2$ reduction during glycerol fermentation pathway ultimately leading to the formation of oxaloacetate [157]. Extracellular CA in alkaliphilic cyanobacteria aids in their survival in high alkaline conditions in alkaline soda lakes [80, 81, 130, 132]. In *Microcoleus* and *Rhabdoderma,* CAs do not allow $CO_2$ leakage from cell by converting it into bicarbonate, thus preserving the intracellular $C_i$ pool for photoautotrophic assimilation. In *Rhabdoderma,* out of the two CAs present, one is bound to photosystem II (PSII) of thylakoid membranes. It participates in the light photosynthetic reactions, regulating operation of the water oxidizing complex via its protection against excess of protons, similar to luminal CA of microalgae and higher plants [124], ultimately helping in the formation of oxaloacetic acid from pyruvate and phosphoenolpyruvate (PEP) catalyzed by pyruvate carboxylase and PEP carboxylase by providing bicarbonate [103]. This oxaloacetate is used for the synthesis of aspartate family of amino acids. Lysine production increases in elevated $CO_2$ conditions owing to the action of CA, and pyruvate and PEP carboxylases [111]. *Halothiobacillus neapolitanus,* a chemolithoautotrophic cyanobacterium (CsoS3), contains a CA in its carboxysomes which supply the active sites of RuBisCO with high concentrations of $CO_2$ necessary for its activity and efficient carbon fixation [130]. *E. coli* β-CA present in *E. coli* is a major player in the cyanate degradation pathway of the organism. β-CA plays a vital role in pathogenesis of some bacteria such as *S. typhimurium, H. pylori,* etc. [138, 151]. In methanogens, CA plays an active role in acetate metabolism by converting the excess carbon dioxide into bicarbonates. It also helps in the conversion of acetate to methane. Indeed, a β class CA (Cab) has been characterized from *Methanothermobacter thermoautotrophicus* [129] which reduces $CO_2$ to methane. Methanosarcina thermophila contains two subclasses of γ CAs, Cam and CamH, postulated to facilitate transport of acetate into the cell and $CO_2$ out via an acetate/bicarbonate anion exchanger [136]. γ—class homologs of the CamH subclass are found in mitochondria, where it might have a role in the carbon transport system to increase the efficiency of photosynthetic carbon fixation [146]. Thermophiles such as *Sulfurihydrogenibium yellowstonense and Sulfurihydrogenibium azorense* have been reported to harbor extremely thermo-alkalistable α—CAs [99]. Characterization of the α class CAs from these microorganisms has revealed that they are the fastest among the known CAs with a $k_{cat}$ of $4.40 \times 10^6$ $s^{-1}$ and a $k_{cat}/Km$ of $3.5 \times 10^8$ $M^{-1}$ $s^{-1}$. These species, isolated from terrestrial hot springs, grow chemolithotrophically with $CO_2$ as the carbon source and $H_2$ as the energy source suggesting a potential role for CA in acquisition of $CO_2$ for cell carbon. *Geobacillus kaustophilus* strain HTA426, a heterotrophic thermophile isolated from deep sea sediment at 10,897 m below sea level, contains a γ CA [143]. Deletion of β-CA gene from *Ralstonia eutrophica* resulted in high $CO_2$ requirement for the growth of the mutant strain [82]. Some of the prokaryotes such as *Symbiobacterium, Buchnera and Rickettsia* are deficient in CA [149]. Comparisons between genomes of CA-retaining and CA-deficient prokaryotes indicated that CA was lost in some species during the course of evolution as a consequence of dispensability. On the other hand, pathogenic prokaryotes seem to be adapted to high $CO_2$ concentrations in their host environment, as they usually require 5–10% (vol/vol) $CO_2$ for growth [78]. Table 2 highlights the characteristics and purification strategies employed for CAs of prokaryotes.

## 3.1 *Carbonic Anhydrases Obtained from Prokaryotes Using Molecular Approaches*

Industrial utilization of CAs requires the enzyme to be produced in high titers so that it can be used for carbon sequestration. Utilization of conventional approaches (as discussed above) to produce the enzyme in high titers is labor intensive which adds to the cost as well as time. Cloning and overexpression of CA encoding genes in prokaryotic (*E. coli, Bacillus subtilis*) as well as eukaryotic (*Pichia pastoris*) hosts increases the enzyme production significantly [141]. In the last 50 years, we have discovered many life forms (mostly prokaryotes) which survive and thrive in extreme conditions (temperature, pH, pressure, etc.) known as extremophiles [95]. These organisms prove to be useful in bioprospecting [8, 30, 79] and harbor wide variety of products (enzymes, antibiotics, biofuels, antioxidants) which can retain their activities under extreme conditions [22]. Although we have successfully grown some of them in the laboratory, yet most of them remain unculturable. Developments of metagenomics and new computational tools have given us an opportunity

**Table 2** Purification strategies and characteristics of wild type microbial carbonic anhydrases. Modified from [111]

| Source | Size (kDa) | Purification strategy | pH stability | Thermostability (°C) | pI | $k_{cat}$ ($S^{-1}$) | $k_{cat}/k_m$ ($M^{-1}$ $s^{-1}$) | Reference (s) |
|---|---|---|---|---|---|---|---|---|
| *N. sicca* (α-CA) | 28.6 | Anion-exchange chromatography (DEAE, CM and SE-Sephadex) and gel filtration chromatography (Sephadex G-75) | 9.0–10.0 | 37–45 | | | | Adler et al. [3] |
| *N. gonorrhoeae* (α-CA) | 28 | Affinity chromatography and cation-exchange chromatography (Mono-S) | 8.0–9.0 | 37 | | $1.7 \times 10^6$ | $2.6 \times 10^6$ | Chirica et al. [25] Elleby et al. [34] |
| *R. rubrum* | | Ammonium sulfate precipitation, gel filtration chromatography (Sephacryl S-200), and chromate focussinggel filtration (Sephacryl S-200), and chromatofocussing | 7.5 | ND | 6.2 | ND | ND | Gill et al. [49] |

(continued)

**Table 2** (continued)

| Source | Size (kDa) | Purification strategy | pH stability | Thermostability (°C) | pI | $k_{cat}$ ($S^{-1}$) | $k_{cat}/k_m$ ($M^{-1}$ $s^{-1}$) | Reference (s) |
|---|---|---|---|---|---|---|---|---|
| *H. influenzae* (β-CA) | 99 | Anion-exchange (Q-Sepharose) chromatography Hydrophobic interaction chromatography (butyl-Sepharose), and gel filtration chromatography (Superdex 200) | 7.5–9.0 | – | – | 64 | 5.6 | Rowlett et al. [116] |
| *M. thermophila* (γ-CA) | 84 | Anion-exchange (Q-Sepharose) and HIC (phenyl-Sepharose) | – | 55 | 4.0 | $11.6 \times 10^4$ | $7.5 \times 10^6$ | Alber and Ferry [5] |
| *M. thermoautotrophicum* β-CA | 90 | Thermal precipitation and Anion exchange | 8.5 | 60–75 | – | $1.7 \times 10^4$ | $5.9 \times 10^6$ | Smith et al. [128] |
| *C. freundii* | – | Ammonium sulfate precipitation, Gel filtration (SephadexG-75), anion exchange (DEAE cellulose) | ND | ND | ND | ND | ND | Ramanan et al. [112] |
| *B. subtilis* (γ-CA) | | Gel filtration, anion-exchange chromatography (DEAE cellulose) and affinity chromatography | 8.3 | 37 | 6.1 | | | Puri [111] |

(continued)

**Table 2** (continued)

| Source | Size (kDa) | Purification strategy | pH stability | Thermostability (°C) | pI | $k_{cat}$ ($S^{-1}$) | $k_{cat}/k_m$ ($M^{-1}$ $s^{-1}$) | Reference (s) |
|---|---|---|---|---|---|---|---|---|
| *P. fragii* | 94 | Ammonium sulfate precipitation, and affinity (p-aminomethylbenzene-sulfonamide) chromatography | 8.0 | 35 | ND | ND | ND | Sharma and Bhattacharya [123] |
| *A. woodii* | | Ammonium sulfate precipitation, Hydrophobic interaction chromatography (ethylamino-Sepharose and phenyl-Sepharose), and anion-exchange chromatography, (DEAE cellulose) | | | | | | Braus-Stromeyer et al. [21] |
| *B. halodurans* (α-CA) | 76 | Anion-exchange chromatography (Hi-captoQ), Affinity (paminomethyl-benzene sulfonamide linked to CNBr activated agarose) chromatography | 6.0–11 | 60–80 | ND | ND | ND | Faridi and Satyanarayana [36] |
| *A. Pallidus* (γ-CA) | 96 | Anion-exchange chromatography (Hicapto anion exchanger) hydrophobic interaction chromatography (phenyl sepharose) | 4.0–11 | 60–80 | ND | ND | ND | Bose and Satyanarayana [17] |

to directly study the organisms and their functions without culturing the life forms [89]. We can even directly clone and maintain the functional genes in different clone libraries directly from environmental samples. These clone libraries harboring the functional genes can be studied at will for their bioprospecting abilities [43, 84]. Cloning of the functional genes also enables us to study their structure and function, and we can even generate novel biomolecules by protein engineering. Researchers have modified Human Carbonic Anhydrase II (hCAII) by altering some of its surface amino acid residues to make it thermostable. Six surface amino acid residues Leu100, Leu224, Leu240, Tyr7, Asn62 and Asn67 were replaced by His, Ser, Pro, Phe, Leu and Gln, respectively, by mutagenesis. These mutations on the surface amino acids increased the number of H-bonds and also enhanced its hydrophilicity, besides increasing thermostability by 6 °C [45]. A novel carbonic anhydrase was developed from *Desulfovibrio vulgaris* (β-CA) via directed evolution technology and achieved 10-million-fold increase in thermostability (40% residual activity at 82 °C) as well as alkalistability. Even it was stable in presence of high (4.2 M) concentrations of N-methyl-diethanolamine. High-throughput screening has also led to the selection of highly stable CA variant which can even retain its activity at 107 °C [7]. But development of these biomolecules adds to the production cost. Hence, it is always better to obtain important enzymes such as CAs directly from extremophilic hosts.

There have been several successful attempts for cloning and overexpression of CA genes from prokaryotes. CA from *Neisseria gonorrhoeae* having a molecular mass of 28 kDa was heterologously expressed in *E.coli*. It was found to encode a protein of 252 amino acid residues and also had a signal peptide [24, 69]. Reference [68] cloned the *N. gonorrhoeae*-CA (NCA) gene without the signal peptide, which enabled the enzyme to be produced in high quantities (106 mg $L^{-1}$ yield with a specific activity of 3090 U $mg^{-1}$). All the CA-encoding genes (*can*, *cynT*, *caiE*, *pay* and *yrdA*) from E. coli have been cloned and overexpressed, and they have been studied extensively for their structure as well as function [28, 47, 94, 137]. CA genes from some of the prokaryotes such as *Bacillus, Serratia* and several others have been cloned and expressed in different prokaryotic as well as eukaryotic hosts [38, 134]. The first thermostable CA to be cloned and overexpressed in *E. coli* is a β-CA from archaeon *Methanobacterium thermoautotrophicum* known as CabCA [128]. Since then there have been multiple studies of cloning and overexpression of CA genes from prokaryotes (mesophiles and extremophiles). Carbonic anhydrases from *Sulfurihydrogenibium, Thermovibrio ammonificans* and others have been cloned and heterologously expressed in *E. coli* Table 3. CAs from thermophilic bacteria were found to be extremely thermostable, and they were found to be useful in CCS [24]. Table 3 highlights some of the major CAs cloned and expressed from thermophilic prokaryotes.

Cloning and overexpression of genes from prokaryotes suffer from their own disadvantages. The most common and important is the getting the enzyme in active form. At times, enzymes/biomolecules could get expressed in very high amounts, but they are devoid of any activity. For example, in *Anabaena*, the gene-encoding alpha CA *ecaA* was cloned in *E. coli*, but it did not show any activity [132]. Many a time, the proteins get aggregated into non-native confirmations and have to be

**Table 3** Reports of cloning, expression and characteristic features of carbonic anhydrases from extremophiles

| Sl. No | Organism | Type of CAs | Molecular weight | Other features | References |
|---|---|---|---|---|---|
| 1 | *Methanobacterium thermoautotrophicum* | β (CabCA) | 21 kDa | $k_{cat} = 1.7^3 \times 10^4$ $s^{-1}$<br>$k_m = 2.9$ mM at pH 8.5<br>Optimal $CO_2$ hydration activity at 75 °C,Tetramer | Smith et al. [128] |
| 2 | *Sulfurihydrogenibium azorense* | α | 26 kDa | $k_{cat} = 4.40 \times 10^6$<br>$k_{cat}/K_m = 3.5 \times 10^8$ $M^{-1}$ $s^{-1}$<br>stable at 90–100 °C for several hours, alkalistable (pH 9.6) | De Luca et al. [31] |
| 3 | *Sulfurihydrogenibium yellowstonense YO3AOP1* | α | 26.1 kDa | $K_m = 26$ mM<br>$V_{max} = 7100$ mM (using $CO_2$) as substrate<br>Stable at 100 °C for 2 h, alkalistable (pH 9.6) | Capasso et al. [23] |
| 4 | *Thermovibrio ammonificans* | α | 23 kDa | $k_{cat}/K_m = 1.6 \times 10^8$<br>Tetramer,<br>Highly thermostable, half life of 77 days at 60 °C | Jo et al. [69] |
| 5 | *Persephonella marina* | α | 23 kDa | $k_{cat}/K_m = 3.0 \times 10^7$<br>Dimer,<br>half life at 100 °C was 88 min ($T_m$ = 84.5 °C) | Jo et al. [69]<br>Kanth et al. [72] |
| 6 | *Methanosarcina thermophila* | γ | 84 kDa (native molecular mass) | Optimal activity at 55 °C<br>Stable at 75 °C for 15 min, activity doubles on replacing zinc with cobalt | Albert and Ferry [6] |

(continued)

**Table 3** (continued)

| Sl. No | Organism | Type of CAs | Molecular weight | Other features | References |
|---|---|---|---|---|---|
| 7 | *Methanosarcina thermophila* | γ (Mt Cam) | 62 kDa (native molecular mass) | $K_m = 3.0$ mM $k_{cat}/K_m = 1.7 \times 10^{-6}$ $M^{-1}$ $s^{-1}$ | MacAuley et al. [91] |
| 8 | *Pyrococcus horikoshii* | γ PhCamH | NA | NA | [65] |
| 9 | *Hahella chejuensis* | α | 25 kDa | Alkalistable, halophilic enzyme also retains its activity at broad pH ranges (pH 6.0–10.0) | Ki et al. [73] |
| 10 | *Bacillus halodurans* | α | 76 kDa (native molecular mass) | Moderately thermostable, alkalistable (pH 6.0–11) and its activity increases in the presence of $SO_4^{2-}$ | Faridi and Satyanarayana [36–38]; Faridi et al. [39] |

extracted/solubilized from the inclusion bodies using several techniques such as dialysis, drop dilution, use of chemical denaturants [150, 162]. There are methods for refolding of inactive protein into an active confirmation such as reversed micelle systems [118], size-exclusion chromatography [85], natural GroEL– GroES chaperone system [165], zeolite absorbing systems [100], and the use of laminar flow in microfluidic chips [64]. All these methods although efficient but is quite time-consuming. Absence of post-translational modifications and codon bias are some of the other problems associated with heterologous protein expression in *E. coli* and *Pichia pastoris*. Optimal codons help in achieving faster translation rates and high accuracy. For example, a codon-optimized α-carbonic anhydrase from *Dunaliella* species was successfully expressed in *E. coli* and was used for $CO_2$ sequestration applications [71]. In another study, synthetic gene based on *Escherichia coli* codon usage of a new α-type CA (HC-aCA) of *Hahella chejuensis*, a marine microorganism, was highly expressed in *E. coli* [73]. The codon-optimized carbonic anhydrase gene of thermophilic *Persephonella marina* EX-H1 (PMCA) was expressed and characterized. The gene was cloned after removal of signal peptide, 5 times more activity was observed than that of the intact gene [72].

## 4 Thermo-alkali Stable CAs and Their Utility in Global Warming Mitigation

Utilization of CAs to mitigate global warming was reported initially by [97]; way back in 2007 using human carbonic anhydrase (hCAII) [23]. Since then there have been several successful attempts to utilize CAs for biomineralization of carbon dioxide. For example, the $CO_2$ Solution Company has proposed large-scale CA-based reactor for capturing $CO_2$ from coal-fired power plant, oil sands, and other $CO_2$-intensive industries [26]. Other mesophilic counterparts from prokaryotes [106, 112] showed good $CO_2$ hydration activity. In order to use this enzyme for biomimetic carbon sequestration, the enzyme must have two key features, i.e., thermostability and alkalistability. As already discussed above, the precipitation of calcium carbonate is favored at alkaline pH,hence, the enzyme needs to be thermostable and the temperature of flue gas is around 140 °C, which is cooled to about 60 °C for post-combustion $CO_2$ capture. Hence, the enzyme is needed which is stable for a long time at 60 °C and above [40]. Extremophilic microorganisms have distinctive structural and physiological characteristics which enable them to thrive in extreme habitats. The presence of specific biomolecules (DNA, proteins/enzymes, lipids) help them to thrive in the extreme environments. These biomolecules can even serve as novel bioproducts for biotechnological applications, for example, thermo-alkali stable carbonic anhydrases of extremophilic prokaryotes can be efficient biomolecules for biomimetic carbon sequestration from flue gas. There have been several reports of alkalistable, thermostable and thermoalkalistable CAs from prokaryotes. The enzyme must be also stable in the presence of different divalent metal ions and other chemicals present in flue gas. *Microcoleus chthonoplastes*, a photosynthetic cyanobacterium, harbors an alkalistable CA (pH 9) having the molecular mass of 34 kDa [80]. The nucleotide sequence of the gene is similar to that of β-CA from *Synechococcus* sp. PCC 7942 (IcfA) and *Synechocystis* sp. PCC 6803 (CcaA) [81]. A putative α-CA localized in the glycocalyx was also characterized from *M. chthonoplastes*. This CA has a molecular mass of about 34 kDa and a pI of 3.5. It displays two peaks of activity at pH 10.0 and 7.5 with the highest at the former [81]. CA from *E. coli* has also been shown to display activity at alkaline pH 9.1. A highly thermo-alkali-stable α-CAs (SazCA and SspCA) have been discovered from thermophilic archaea *Sulfurihydrogenibium azorense* and *Sulfurihydrogenibium yellowstonense* YO3AOP1 ($T_{opt.}$ 110 °C). The SspCA has a molecular mass of 26 kDa. This enzyme retains activity even after incubation for 3 h at 70 °C. It also retains activity even after prolonged incubation at 100 °C [4, 23]. More stability was achieved after immobilization of the enzyme in polyurethane foam (50 h at 100 °C), thermoactivity analysis revealed an optimum working temperature of 95 °C [31]. It retains carbon dioxide hydration activity even after incubation at 80 and 90 °C for several hours. Half life values of SazCA at 40 and 70 °C were 53 and 8 days, respectively [117]. The inhibition constants are in the range of 0.58–0.86 mM for the anions $NO^{2-}$, $NO^{3-}$ and $SO_4{}^{2-}$. Both of the enzymes retain activity at pH 9.6 (De Luca et al. 2012), [153]. X-ray crystallography of both the enzymes revealed high structural compactness, together with an increased number of charged residues on the protein surface and a greater number of ionic networks appeared to be the key factors involved in the high thermostability of the enzymes with respect to their mesophilic homologues [32]. The complete structure determination of these extremostable CAs enables us to design and engineer CAs in such a way so that they can be easily utilized for biomimetic sequestration strategies. *Dunaliella salina*, a unicellular green alga that thrives in a wide array of natural environments ranging from fresh water to hyper-saline lakes, shows the presence of carbonic anhydrase localized on its surface. This CA even retains activity at 4.0 M NaCl and at low salinities [108]. An alkalistable α–CA was characterized from *Serratia sp.*, which retains 75% of activity even at pH 12.0. [134]. Two CAs have been isolated from poly extremophilic bacteria *Aeribacillus pallidus* (ApCA) and *Bacillus halodurans* (BhCA), and their role in global warming mitigation was successfully demonstrated by our group. Utility of these two enzymes in biomimetic carbon sequestration has been described separately in the next section.

## 5 Biomimetic Carbon Sequestration Using BhCA and ApCA

A thermostable and alkalistable ($T_{opt}$ 60 °C and pH 8–11) γ-CA (ApCA) was reported by [17], from extremophilic bacterium *Aeribacillus pallidus* TSHB1 isolated from Choti Anhoni hot spring of Pipariya (Madhya Pradesh, India). Molecular mass of the purified protein determined by SDS PAGE is 32 ± 2 kDa and native mol. wt. is 96 kDa. Since, initially the enzyme production was quite low, it was optimized using various statistical approaches and about 3.7-fold increase in production was achieved [17]. The enzyme is stable in the pH range between 8.0 and 11.0 and thus alkali-stable and moderately thermostable with $T_{1/2}$ values of 40 ± 1, 15 ± 1, and 8 ± 0.5 min at 60, 70, and 80 °C, respectively. Activation energy for irreversible inactivation $E_d$ of carbonic anhydrase is 67.119 kJ $mol^{-1}$. The enzyme is stable in the presence of various flue gas contaminants such as $SO_3{}^{2-}$, $SO_4{}^{2-}$, and $NO_3{}^{-}$ and cations $Mg^{2+}$, $Mn^{2+}$, $Ca^{2+}$, and $Ba^{2+}$. ApCA was successfully used in biomineralization of $CO_2$ from flue gas (emitted from motorcycle exhaust) (Fig. 1b, c). Replacement of active site $Zn^{2+}$ with $Mn^{2+}$ enabled ApCA to function as a virtual peroxidase (like in the case of BhCA) which exhibited alkali-stability and moderate thermostability like the native enzyme [18]. ApCA can be efficiently utilized for biomimetic carbon sequestration.

A moderately thermostable and alkalistable α-CA (BhCA) was produced from *Bacillus halodurans* TSLV1, a moderately thermophilic, alkaliphilic, and halotolerant bacterium. The enzyme showed complete inhibition in the presence of α-CA-specific sulphonamide inhibitors. The enzyme was purified, and it was found to be a homodimeric enzyme with a subunit molecular mass of ~37 kDa and with a native molecular mass of ~ 75 kDa. BhCA was stable in a broad pH range (6.0–11), with a $T_{1/2}$ of 24.0 ± 1.0 min at 60 °C (moderately thermostable). This has a unique property of stimulation by $SO_4{}^{2-}$, and it remained unaffected by $SO_3{}^{2-}$, NOx, and

**Fig. 1** **a, b** Exhaust fumes from the petrol-driven car and motorcycle were collected. **c** Mineralization of exhaust gas $CO_2$ using BhCA and ApCA in the presence of $Ca^{2+}$. **d** Scanning electron microscopic picture of calcium carbonate crystals formed in the presence of enzyme. (Adapted from [15, 33, 34])

other flue gas contaminants. BhCA is highly efficient in accelerating mineralization of $CO_2$ and was successfully utilized in the sequestration of $CO_2$ from the exhaust of petrol driven car (Fig. 1a, c) (which has a similar chemical composition to that of flue gas) [36]. Site directed mutagenesis revealed the presence of four histidine residues (H137, H139, H156 and H110) at the active site of the enzyme. When $Zn^{2+}$ was substituted with $Mn^{2+}$, rBhCA displayed peroxidase activity. This virtual peroxidase retained the native properties of original enzyme, i.e., rBhCA. The industrial application of this enzyme requires its production in high amount. The CA production from the wild type was quite low even after production optimization; hence, the gene that encodes CA (BhCA) has been cloned and expressed in *Escherichia coli.* A 31.4-fold enhancement in rCA production was achieved due to cloning and expression in *E. coli.* When the cloned BhCA (BhCA-pET22b) was cultivated in a medium with EDTA and lysozyme about 50% of the CA (rBhCA) produced was secreted into the medium due to the presence of pelB leader sequence [37]. Both the native and rBhCA produced in *E. coli* are intracellular, thus require cell disruption to release the enzyme, which is energy intensive and increases the production cost. In order to bring down the cost of production of BhCA, extracellular production of the enzyme is desirable; hence, the gene encoding BhCA was expressed in active form in Pichia pastoris under methanol inducible (AOX1) as well as constitutive (GAP) promoters and complete extracellular production had been achieved [38]. A marked improvement in rBhCA production was attained. The enzyme expressed in *Pichia pastoris* is a glycosylated protein with a native molecular weight of 79.5 kDa. Glycosylation resulted in a slight improvement in thermostability of the enzyme.

Calcium carbonate ($CaCO_3$) exists in different crystal phases [calcite (rhombic), aragonite (needle-like) and vaterite (spherical)]. Calcite phase is thermodynamically the most stable phase, while the vaterite is the metastable phase [41]. Scanning electron microscopy of the calcium carbonate crystals formed after the mineral carbonation reaction of $CO_2$ (present in vehicle exhausts) using both ApCA and BhCA revealed the formation of rhomboid calcite crystals, i.e., the carbon dioxide is converted into the thermodynamically most stable phase of calcium carbonate, which can be stored for long time, and it may be also utilized for several applications Fig. 1d.

## *5.1 Immobilization of BhCA on Silanized Magenetite Iron Nanoparticles (Si-MNPs)*

Although BhCA is quite efficient and suits well in sequestration of $CO_2$ from flue gas, there are a number of practical problems in using the free enzyme in solution which includes low stability and limited reusability [35]. Immobilization of the enzyme onto a solid support [chitosan beads magnetite iron nanoparticles (MNPs), etc.] is the most successful method which overcomes the above problems. Immobilization offers easier separation of products from the enzymes and continuous reuse of the enzyme, besides the possibility of improving the stability and performance of enzymes under harsh operational conditions. Nanoparticle-based immobilization techniques have been established as an efficient method to improve storage stability and performance of immobilized enzymes [158]. Magnetic iron nanoparticles (MNPs) offer huge advantage due to their high surface-to-volume ratio and easy separation of product by applying a magnetic field [83]. These MNPs can be silanized to improve their stability in acidic conditions. Recombinant BhCA was immobilized on the surface of modified (silanized) magnetic iron oxide nanoparticles (Si-MNPs) resulting in improved thermo-alkalistability as compared to that of free BhCA. The immobilized enzyme (rBhCA-Si-MNPs) showed high reusability, and it even retained 50% of its initial activity even after 22 cycles of reuse. The immobilization did not affect the amount of $CaCO_3$ precipitation by the enzyme [39]. rBhCA-Si-MNPs is quite suitable for use in harsh operational conditions and could be efficiently utilized to develop bioreactors for sequestration of $CO_2$ from flue gas.

## 6 Immobilized CAs and Development of CA-Based Bioreactors

As already stated above in order to recycle and reuse the enzymes multiple times immobilization techniques had been utilized. Immobilization enhances the stability of the enzymes greatly. Immobilized carbonic anhydrases have been found to be more efficient than free enzymes. Mammalian CAs and CAs from prokaryotes (*P. Fragi* and *B. pumilus*) have displayed improved $CO_2$ hydration activity after their immobilization in polyurethane foam [155]. Thermostability enhancement is another usefulness of immobilization. Immobilized CAs have been showed to retain 60% of their initial activity after 90 days at 50 °C as compared to that of free enzyme (30%). It also shields the enzyme from various inhibitors [70, 107]. Even carbonate precipitation rate doubled in the case of immobilized CA (in chitosan beads) of *Pseudomonas fragi* [155, 162]. Several other matrices (ordered mesoporous aluminosilicate), nanoparticles [octa (aminophenyl) silsesquioxane-functionalized $Fe_3O_4/SiO_2$ nanoparticles, silica nanoparticles] have been successfully attempted [113, 123] with improved stability and $CO_2$ sequestration efficiency [113, 160]. Ssp CA was immobilized on solid matrix made of silica particles (silanized Sipernat® 22 particles) and polyurethane foam which resulted in exceptional thermostability for very long duration even at 70 °C. Its $CO_2$ absorption efficiency of PU-SspCA was verified in a three-phase trickled bed bioreactor which mimics the post-combustion processes in a thermal power plant. SspCA showed good $CO_2$ capture performance when the PU-SspCA-shredded foam was used in the bioreactor [96]. Immobilization also involves unwanted release of enzyme on the surface of the bioreactor. This problem was overcome by using novel immobilization techniques involving the use of nanoassemblies. For this purpose, γ–CA from *Pyrococcus horikoshii* and *Methanosarcina* thermophila were used. The immobilization techniques allowed the development of γ-CA nanoassemblies [19]. Yoshimoto and Walde [163], have recently published a comprehensive review which deals with immobilization of CAs and their applications. Apart from the utilization of immobilized CAs for capturing and sequestration of carbon dioxide, they can be even used for CA-supported bioelectrochemical conversion of $CO_2$ into organic molecules. Furthermore, it can be even used in biomimetic material synthesis (for bone regeneration based on calcium carbonate formation from urea involving urease and CA). Figure 2 contains schematic diagram of bioreactor-based systems which use immobilized CA for biomimetic carbon sequestration.

Gas–liquid membrane contractors are one of the bioreactors for using CAs for $CO_2$ capture. Iliuta and Iluita (2017), demonstrated a novel, cost-effective and environmentally friendly approach for $CO_2$ capture by immobilizing CA in a hollow-fiber membrane bioreactor (HFMB) in multiscale steady-state model, under partially

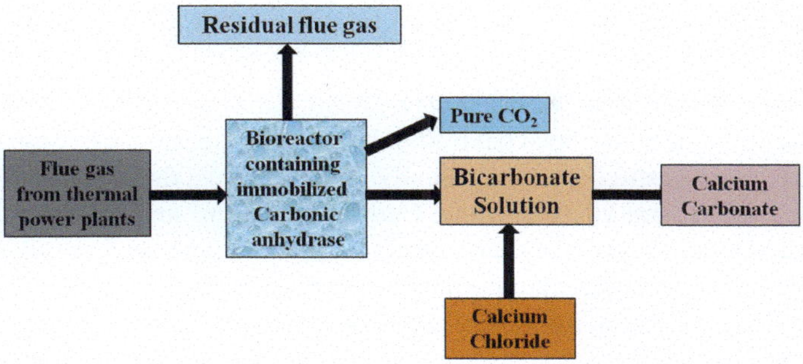


**Fig. 2** Schematic diagram of CA-based bioreactor for biomimetic $CO_2$ sequestration from flue gas

liquid-filled and gas-filled membrane pore conditions. NASA and Carbozyme Inc. have developed several reactor systems for CA-based carbon capture. In NASA, the system comprises a liquid membrane whose core contains 330 μm substantial layer of enzyme solution in an appropriate buffer system squeezed between two micro-porous hydrophobic polypropylene membranes. Two thin metal grids protect the membrane. $CO_2$ diffuses across the membrane and evaporates out from the other side into vacuum or carrier gas [27]. Carbozyme's system comprises hollow micro-porous polypropylene microfiber on which the enzyme (γ–CA) is immobilized and water vapor has been used as sweeping gas [144]. This technology was found highly efficient in carbon capture from flue gas in a broad range of temperatures (15–85 °C). A 3-phase monolith bioreactor based system was developed by Iliuta and Larachi [60], they used immobilized HCA II enzyme. Modification in $CO_2$ hydration kinetics due to gas–liquid and liquid–solid mass transfer exchange was demonstrated in this 3-phase reactor system. Several other reactor systems have been developed with new features for continuous $CO_2$ capture and reduced solvent evaporation. This is achieved by trapping the CA between membrane pores, packed with an ionic liquid. Another gas-to-liquid membrane $CO_2$ sequestration system combines mineralization module supplied with $Ca^{2+}$ for precipitation of bicarbonate to form solid $CaCO_3$. Alvizo et al. [7] developed a technique which involves the flue gas to be transferred to an absorber column where $CO_2$ chemisorbs into an amine solvent (MDEA) and then on this column CA catalyzes the formation of a proton and a bicarbonate ion. Residual flue gas is released in the air and CA containing the bicarbonate-loaded amine solvent transferred to a second column where $CO_2$ is removed at elevated temperatures (>87 °C), and solvent is regenerated. The pure $CO_2$ stream can be compressed and stored in depositories or used in other industrial applications. The regenerated solvent is returned to the absorber column to repeat the process. A highly thermostable variant of DvCA developed using directed evolution technique was used in this process. Although there have been several studies and reports on CA-based bioreactors, most of them are still in pilot stages. Hence, in the present scenario these processes need to be scaled up in order to develop industrial level bioreactors for large-scale $CO_2$ capture from flue gas.

## 7 Conclusions

Among all the problems that we face today, global warming needs immediate attention. Several strategies have been devised in the last few decades for combating this menace. The world is still dependent heavily on fossil fuels; therefore, this problem will continue for several years to come. Carbon emissions are not going to significantly decline to safer levels anytime soon. Natural methods such as afforestation and enzymatic carbon capture are promising approaches for $CO_2$ removal from the atmosphere. In this context, biomineraliztion of carbon dioxide using ‘nature’s own catalyst’ CA is emerging as a significant strategy. The advent of new technologies (computational and non-computational) has led to thermo-alkalistable CAs which are efficient candidates for enzymatic carbon capture generating useful calcium carbonate. The immobilization of CAs permitted the development of CA-based bioreactors for $CO_2$ capture from flue gas and industrial emissions. We are still in the early stage of developing the efficient bioreactors. There is an urgent need to speed up the process of developing reactors for industrial scale operations.

## References

1. Aaron D, Tsouris C (2005) Separation of $CO_2$ from flue gas: a review. Sep Sci Technol 40:321–348
2. Adams MR, Hurd HB, Lenhart S et al (1998) Effects of global climate change on agriculture: an interpretative review. Clim Res 11:19–30
3. Adler L, Brundell J, Falkbring SO, Nyman PO (1972) Carbonic anhydrase from *Neisseria sicca*, strain 6021 I. Bacterial growth and purification of the enzyme. Biochim Biophys Acta Enzym 284:298–310
4. Akdemir A, Vullo D, De Luca V et al (2013) The extremo-α-carbonic anhydrase (CA) from *Sulfurihydrogenibium azorense*, the fastest CA known, is highly activated by amino acids and amines. Bioorg Med Chem Lett 23:1087–1090
5. Alber BE, Ferry JG (1994) A carbonic anhydrase from the archaeon Methanosarcina thermophila. Proc Nat Acad Sci U.S.A. 91:6909–6913
6. Alber BE, Ferry JG (1996) Characterization of heterologously produced carbonic Anhydrase from *Methanosarcina thermophila* J Bacteriol 178:3270–3274
7. Alvizo O, Nguyen LJ, Savile CK et al (2014) Directed evolution of an ultrastable carbonic anhydrase for highly efficient carbon capture from flue gas. Proc Nat Acad Sci USA 111:16436–16441
8. Anand A, Kumar V, Satyanarayana T (2013) Characteristics of thermostable endoxylanase and β-xylosidase of the extremely thermophilic bacterium *Geobacillus thermodenitrificans* TSAA1 and its applicability in generating xylooligosaccharides and xylose from agro-residues. Extremophiles 17: 357–366
9. Angeli A, Del Prete S, Alasmary FA et al (2018) The first activation studies of the η-carbonic anhydrase from the malaria parasite *Plasmodium falciparum* with amines and amino acids. Bioorg Chem 80:94–98
10. Angeli A, Alasmary FA, Del Prete S et al (2018) The first activation study of a δ-carbonic anhydrase: TweCAδ from the diatom *Thalassiosira weissflogii* is effectively activated by amines and amino acids. J Enzyme Inhib Med Chem 33:680–685

11. Angeli A, Donald WA, Parkkila S, Supuran CT (2018) Activation studies with amines and amino acids of the β-carbonic anhydrase from the pathogenic protozoan *Leishmania donovani chagasi*. Bioorg Chem 78:406–410
12. Aspelund A, Molnvik MJ, DeKoeijer G (2006) Ship transport of $CO_2$, technical solutions and analysis of costs, energy utilization, energy efficiency and $CO_2$ emissions. Chem Eng Res Des 84:847–855
13. Badger MR, Price GD (2003) $CO_2$ concentrating mechanisms in Cyanobacteria: Molecular components, their diversity and evolution. J Exp Bot 54:609–622
14. Berg JM, Tymoczko JL, Stryer L (2002) Biochemistry, 5th edn. W H Freeman, New York
15. Bhown AS, Freeman BC (2011) Analysis and status of post-combustion carbon dioxide capture technologies. Environ Sci Technol 45:8624–8632
16. Boone CD, Gill S, Habibzadegan A et al (2013) Carbonic anhydrase: an efficient enzyme with possible global implications. Int J Chem Eng 1–7
17. Bose H, Satyanarayana T (2016) Suitability of the alkalistable carbonic anhydrase from a polyextremophilic bacterium *Aeribacillus pallidus* TSHB1 in biomimetic carbon sequestration. Bioprocess Biosyst Eng 39:1515–1525
18. Bose H, Satyanarayana T (2017a) Utility of thermo-alkali-stable γ-CA from polyextremophilic bacterium *Aeribacillus pallidus* TSHB1 in biomimetic sequestration of $CO_2$ and as a virtual peroxidase. Environ Sci Pollut R 24:10869–10884
19. Bose H, Satyanarayana T (2017b) Microbial carbonic anhydrases in biomimetic carbon sequestration for mitigating global warming: prospects and perspectives. Front Microbiol 8:1615
20. Bose H, Satyanarayana T (2018) Carbonic anhydrases of extremophilic microbes and their applicability in mitigating global warming through carbon sequestration. In: Durvasula VR, Subba Rao DV (eds) Extremophiles. CRC Press, pp 249–276
21. Braus-Stromeyer SA, Schnappauf G, Braus GH et al (1997) Carbonic anhydrase in *Acetobacterium woodii* and other acetogenic bacteria. J Bacteriol 179:7197–7200
22. Canganella F, Wiegel J (2011) Extremophiles: from abyssal to terrestrial ecosystems and possibly beyond. Naturwissenschaften 98:253–279
23. Capasso C, De Luca V, Carginalea V et al (2012) Characterization and properties of a new thermoactive and thermostable carbonic anhydrase. Chem Eng 27:271–276
24. Chirică LC, Elleby B, Jonsson BH et al (1997) The complete sequence, expression in *Escherichia coli*, purification and some properties of carbonic anhydrase from *Neisseria gonorrhoeae*. Eur J Biochem 24:55–760
25. Christer J, Wullschleger D, Stan KC et al (2010) Phytosequestration: carbon biosequestration by plants and the prospects of genetic engineering. Bioscience 60:685–696
26. $CO_2$ solutions. www.co2solutions.com/en/the-process. Accessed on 27th March 2019
27. Cowan RM, Ge JJ, Qin YJ et al (2003) $CO_2$ capture by means of an enzyme-based reactor. Ann N Y Acad Sci 984:453–469
28. Cronk JD, Endrizzi JA, Cronk MR et al (2001) Crystal structure of *E. coli* β–carbonic anhydrase, an enzyme with an unusual pH–dependent activity. Prot Sci 10:911–922
29. Da Silva EF, Booth AM (2013) Emissions from post-combustion $CO_2$ capture plants. Environ Sci Technol 47:659–660
30. Dahlberg L, Hoist O, Kristjansson KJ (1993) Thermostable xylanolytic enzymes from *Rhodothermus marinus* grown on xylan. Appl Microbiol Biotechnol 40:63–68
31. De Luca V, Vullo D, Scozzafava A et al (2013) An α-carbonic anhydrase from the thermophilic bacterium *Sulphurihydrogenibium azorense* is the fastest enzyme known for the $CO_2$ hydration reaction. Bioorg Med Chem 21:1465–1469
32. Di Fiore A, Capasso C, De Luca V et al (2013) X-ray structure of the firstextremo-α-carbonic anhydrase', a dimeric enzyme from the thermophilic bacterium *Sulfurihydrogenibium yellowstonense* YO3AOP1. Acta Crystallogr D 69:1150–1159
33. Dutreuil S, Bopp L, Tagliabue A (2009) Impact of enhanced vertical mixing on marine biogeochemistry: lessons for geo-engineering and natural variability. Biogeosciences 6:901–912

34. Elleby B, Chirica LC, Tu C et al (2001) Characterization of carbonic anhydrase from *Neisseria gonorrhoeae*. Eur J Biochem 268:1613–1619
35. Faridi S, Satyanarayana T (2015) Bioconversion of industrial $CO_2$ emissions into utilizable products. In: Chandra R (ed) Industrial waste management. CRC Press, New York, pp 111–156
36. Faridi S, Satyanarayana T (2016a) Novel alkalistable α-carbonic anhydrase from the polyextremophilic bacterium *Bacillus halodurans*: Characteristics and applicability in flue gas $CO_2$ sequestration. Environ Sci Pollut R 15:15236–15249
37. Faridi S, Satyanarayana T (2016b) Characteristics of recombinant α-carbonic anhydrase of polyextremophilic bacterium *Bacillus halodurans* TSLV1. Int J Biol Macromol 89:659–668
38. Faridi S, Satyanarayana T (2018) Thermo-alkali-stable α-carbonic anhydrase of *Bacillus halodurans*: heterologous expression in *Pichia pastoris* and applicability in carbon sequestration. Environ Sci Pollut R 25:6838–6849
39. Faridi S, Bose H, Satyanarayana T (2017) Utility of immobilized recombinant carbonic anhydrase of *Bacillus halodurans* TSLV1 on the surface of modified iron magnetic nanoparticles in Carbon Sequestration. Energ Fuel 31:3002–3009
40. Farrell A (2011) Carbon dioxide storage in stable carbonate minerals. Basalt laboratory studies of interest to carbon capture and storage. Advisor MN Evans University of Maryland Geology 1–24
41. Favre N, Christ ML, Pierre AC (2009) Biocatalytic capture of $CO_2$ with carbonic anhydrase and its transformation to solid carbonate. J Mol Catal B Enzym 60:163–170
42. Feinstein CH (1998) Pessimism perpetuated: real wages and the standard of living in Britain during and after the industrial revolution. J Econ Hist 58:625–658
43. Ferrer M, Martínez-Martínez M, Bargiela R et al (2016) Estimating the success of enzyme bioprospecting through metagenomics: current status and future trends. Microb Biotechnol 9:22–34
44. Ferry JG (2010) The gamma class of carbonic anhydrases. Biochim Biophys Acta 1804:374–381
45. Fisher Z, Boone CD, Biswas SM, Venkatakrishnan B, Aggarwal M, Tu C, Agbandje-McKenna M, Silverman D, McKenna R (2012) Kinetic and structural characterization of thermostabilized mutants of human carbonic anhydrase II. Protein Eng Des Sel 25:347–355
46. Frost SC, McKenna R (eds) (2013) Carbonic anhydrase: mechanism, regulation, links to disease, and industrial applications. Springer Science & Business Media
47. Fujita N, Mori H, Yura T et al (1994) Systematic sequencing of the *Escherichia coli* genome: analysis of the 2.4–4.1 min (110,917–193,643 bp) region. Nucleic Acids Res 22:1637–1639
48. Garg A, Shukla PR, Kankal B et al (2017) $CO_2$ emission in India: trends and management at sectoral, sub-regional and plant levels. Carbon Manag 8:111–123
49. Gill SR, Fedorka-Cray PJ et al (1984) Purification and properties of the carbonic anhydrase of Rhodospirillum rubrum. Arch Microbiol 138:113–118
50. Gougoulias C, Clark JM, Shaw LJ (2014) The role of soil microbes in the global carbon cycle: tracking the below-ground microbial processing of plant-derived carbon for manipulating carbon dynamics in agricultural systems. J Sci Food Agric 94:2362–2371
51. Hart A, Gnanendran N (2009) Cryogenic $CO_2$ capture in natural gas. Energy Proc 1:697–706
52. Haszeldine SR (2009) Carbon capture and storage: how green can black be. Science 325:1647–1652
53. Herzog HJ (1998) Ocean sequestration of $CO_2$—an overview. In: fourth international conference on greenhouse gas control technologies August 30–September 2 Interlaken, Switzerland 1–5
54. Hicks N, Vik U, Taylor P et al (2017) Using prokaryotes for carbon capture storage. Trends Biotechnol 35:22–32
55. Horn J, Rosenband LN and Smith MR (eds) (2010) Reconceptualizing the industrial revolution. MIT press
56. Houghton J (2005) Global Warm Rep Progress Phys 68:1343–1403
57. Hu H, Gao K (2008) Impacts of $CO_2$ enrichment on growth and photosynthesis in freshwater and marine diatoms. Chin J Oceanol Limnol 26:407–414

58. Huijgen WJ, Comans RN, Witkamp GJ (2007) Cost evaluation of $CO_2$ sequestration by aqueous mineral carbonation. Energy Convers Manag 48:1923–1935
59. Iliuta I, Iliuta MC (2017) Investigation of $CO_2$ Removal by immobilized carbonic anhydrase enzyme in a hollow-fiber membrane bioreactor. AIChE J 63:2996–3007. https://doi.org/10.1002/aic.15646
60. Iliuta I, Larachi F (2012) New scrubber concept for catalytic $CO_2$ hydration by immobilized carbonic anhydrase II & in-situ inhibitor removal in three-phase monolith slurry reactor. Sep Purif Technol 86:199–214
61. IPCC (Inter-governmental Panel on Climate Change) (2014) Climate change 2014 synthesis report. Cambridge University Press, Cambridge
62. IPCC (2013) Fourth assessment report: climate change (AR5) 2013 from IPCC website https://ipcc.ch/publications_and_data/publications_and_data_reports.shtml/. Accessed on 26th March 20179
63. IPCC (2000) Special report on emissions scenarios: a special report of working group III of the intergovernmental panel on climate change. Cambridge University Press, New York
64. Jahn A, Vreeland WN, DeVoe DL et al (2007) Microfluidic directed formation of liposomes of controlled size. Langmuir 23:6289–6293
65. Jeyakanthan J, Rangarajan S, Mridula P et al (2008) Observation of a calcium-binding site in the $\gamma$-class carbonic anhydrase from *Pyrococcus horikoshii.* Acta Crystallogr. D 64:1012–1019
66. Jin X, Gruber N, Frenzel H et al (2008) The impact on atmospheric $CO_2$ of iron fertilization induced changes in the ocean's biological pump. Biogeosciences 5:385–406
67. Jo BH, Kang DG, Kim CS, Choi YS, Cha HJ (2012) Biomineralization-based conversion of carbon dioxide to calcium carbonate using recombinant carbonic anhydrase. Chemosphere 87:1091–1096
68. Jo BH, Kim IG, Seo JH et al (2013) Engineered *Escherichia coli* with periplasmic carbonic anhydrase as a biocatalyst for $CO_2$ sequestration. Appl Environ Microbiol 79:6697–6705
69. Jo BH, Seo JH, Cha JC (2014) Bacterial extremo- $\alpha$-carbonic anhydrases from deep-sea hydrothermal vents as potential biocatalysts for $CO_2$ sequestration. J Mol Catal B Enzym 109:31–39
70. Kanbar B, Ozdemir E (2010) Thermal stability of carbonic anhydrase immobilized within polyurethane foam. Biotechnol Prog 26:1474−1480
71. Kanth BK, Min K, Kumari S et al (2012) Expression and characterization of codon-optimized carbonic anhydrase from *Dunaliella* species for $CO_2$ sequestration application. Appl Biochem Biotechnol 167:2341–2356
72. Kanth BK, Jun SY, Kumari S (2014) Highly thermostable carbonic anhydrase from *Persephonella marina* EX-H1: its expression and characterization for $CO_2$ sequestration applications. Process Biochem 49:2114–2121
73. Ki MR, Min K, Kanth BK et al (2013) Expression, reconstruction and characterization of codon-optimized carbonic anhydrase from *Hahella chejuensis* for $CO_2$ sequestration application. Bioprocess Biosyst Eng 36:375–381
74. Kikutani S, Nakajima K, Nagasato C et al (2016) Thylakoid luminal $\theta$-carbonic anhydrase critical for growth and photosynthesis in the marine diatom *Phaeodactylum tricornutum.* Proc Nat Acad Sci USA 113:9828–9833
75. Kindermann G, Obersteiner M, Sohngen B et al (2008) Global cost estimates of reducing carbon emissions through avoided deforestation. Proc Nat Acad Sci USA 105:10302–10307
76. Knudsen JN, Jensen JN, Vilhelmsen PJ, Biede O (2009) Experience with $CO_2$ capture from coal flue gas in pilot-scale: testing of different amine solvents. Energy Proc 1:783–790
77. Kubler JE, Johnston AM, Raven JA (1999) The effects of reduced and elevated $CO_2$ and $O_2$ on the seaweed *Lomentaria articulata.* Plant Cell Environ 22:1303–1310
78. Kumar RSS, Ferry JG (2014) Prokaryotic carbonic anhydrases of Earth's environment. In: Frost S, McKenna R (eds) Carbonic anhydrase: mechanism, regulation, links to disease, and industrial applications. Springer, Dordrecht, pp 77–87

79. Kumar V, Satyanarayana T (2011) Applicability of thermo-alkali-stable and cellulase-free xylanase from a novel thermo-halo-alkaliphilic *Bacillus halodurans* in producing xylooligosaccharides. Biotechnol Lett 33: 2279–2285
80. Kupriyanova EV, Villarejo A, Markelova AG et al (2007) Extracellular carbonic anhydrases of the stromatolite-forming cyanobacterium *Microcoleus chthonoplastes.* Microbiology 153:1149–1156
81. Kupriyanova EV, Sinetova MA, Markelova AG et al (2011) Extracellular β_class carbonic anhydrase of the alka-liphilic cyanobacterium *Microcoleus chthonoplastes.* J Photochem Photobiol B 103:78–86
82. Kusian B, Sultemeyer D, Bowien B (2002) Carbonic anhydrase is essential for growth of *Ralstonia eutropha* at ambient $CO_2$ concentrations. J Bacteriol 184:5018–5026
83. Laurent S, Forge D, Port M et al (2008) Magnetic iron oxide nanoparticles: synthesis, stabilization, vectorization, physicochemical characterizations, and biological applications. Chem Rev 108:2064–2110
84. Lee MH, Lee SW (2013) Bioprospecting potential of the soil metagenome: novel enzymes and bioactivities. Genome Inform 11:114
85. Li M, Su ZG, Janson, JC (2004) In vitro protein refolding by chromatographic procedures. Protein Expr Purif 33:1–10
86. Li S, An L, Duan Q et al (2018) Determining the rate of carbonic anhydrase reaction in the human brain. Sci Rep 8:2328
87. Lim X (2015) How to make most of carbon dioxide. Nature 526:629–630
88. Liu N, Bond GM, Abel TA et al (2005) Biomimetic sequestration of $CO_2$ in carbonate form: role of produced waters and other brines. Fuel Process Technol 86:1615–1625
89. Lorenz P, Eck J (2005) Metagenomics and industrial applications. Nat Rev Microbiol 3:510
90. Lucas RE (ed) (2002) Lectures on economic growth. Harvard University Press
91. MacAuley SR, Zimmerman SA, Apolinario EE et al (2009) The archetype γ-class carbonic anhydrase (cam) contains iron when synthesized in vivo. Biochemistry 48:817–819
92. McCloskey D (2004) Review of the Cambridge economic history of modern Britain. In: Floud R, Paul J (eds) Times higher education supplement, 15 Jan 2004
93. Meldrum NU, Roughton FJ (1933) Carbonic anhydrase. Its preparation and properties. J Physiol 80:113
94. Merlin C, Masters M, McAteer S (2003) Why is carbonic anhydrase essential to *Escherichia coli.* J Bacteriol 185:6415–6424
95. Mesbah NM, Wiegel J (2012) Life under multiple extreme conditions: diversity and physiology of the halophilic alkalithermophiles. Appl Environ Microbiol 78:4074–4082
96. Migliardini F, De Luca V, Carginale V et al (2014) Biomimetic $CO_2$ capture using a highly thermostable bacterial α-carbonic anhydrase immobilized on a polyurethane foam. J Enzyme Inhib Med Chem 29:146–150
97. Mirjafari P, Asghari K, Mahinpey N (2007) Investigating the application of enzyme carbonic anhydrase for $CO_2$ sequestration purposes. Ind Eng Chem Res 46:921–926
98. Monastersky R (1995) Iron versus the greenhouse: Oceanographers cautiously explore a global warming therapy. Sci News 148:220
99. Nakagawa S, Shtaih Z, Banta A, Beveridge TJ et al (2005) *Sulfurihydrogenibium yellowstonense* sp. nov., an extremely thermophilic, facultatively heterotrophic, sulfur-oxidizing bacterium from Yellowstone National Park, and emended descriptions of the genus *Sulfurihydrogenibium, Sulfurihydrogenibium subterraneum* and *Sulfurihydrogenibium azorense.* Int J Syst Evol Microbiol 55:2263–2268
100. Nara TY, Togashi H, Sekikawa C et al (2009) Use of zeolite to refold a disulfide-bonded protein. Colloids Surf B 68:68–73
101. Nielsen CJ, Herrmann H, Weller C (2012) Atmospheric chemistry and environmental impact of the use of amines in carbon capture and storage (CCS). Chem Soc Rev 41:6684–6704
102. NOAA (2019) National oceanic and atmospheric administration. https://www.esrl.noaa.gov/gmd/ccgg/trends/. Accessed on 26th March 2019

103. Norici A, Dalsass A, Giordano M (2002) Role of phosphoenolpyruvate carboxylase in anaplerosis in the green microalga *Dunaliella salina* cultured under different nitrogen regimes. Physiol Plant 116:186–191
104. Olajire AA (2010) $CO_2$ capture and separation technologies for end-of-pipe applications–a review. Energy 35:2610–2628
105. Peña KL, Castel SE, de Araujo C et al (2010) Structural basis of the oxidative activation of the carboxysomal γ-carbonic anhydrase, CcmM. Proc Nat Acad Sci USA 107:2455–2460
106. Prabhu C, Valechha A, Wanjari S et al (2011) Carbon composite beads for immobilization of carbonic anhydrase. J Mol Catal B Enzym 71:71–78
107. Prabhu C, Wanjari S, Gawande S et al (2009) Immobilization of carbonic anhydrase enriched microorganism on biopolymer based materials. J Mol Catal B Enzym 60:13–21
108. Premkumar L, Bageshwar UK, Gokhman I et al (2003) An unusual halotolerant alpha-type carbonic anhydrase from the alga *Dunaliella salina* functionally expressed in *Escherichia coli*. Protein Expr Purif 28:151–157
109. Princiotta FT (2007) The role of power generation technology in mitigating global climate change. In: Cen K, Chi Y, Wang F (eds) Challenges of power engineering and environment. Springer, Berlin, pp 3–13
110. Puri AK (2012) Carbon sequestration using heterotrophic bacteria
111. Puri AK, Satyanarayana T (2010) Carbon sequestration for mitigating disastrous effects of global warming. In: Goel M (ed) Natural and Manmade Disasters. MD Publications, New Delhi, pp 229−252
112. Ramanan R, Kannan K, Sivanesan SD et al (2009) Bio-sequestration of carbon dioxide using carbonic anhydrase enzyme purified from *Citrobacter freundii*. World J Microbiol Biotechnol 25:981–987
113. Rayalu S, Yadav R, Wanjari S et al (2012) Nanobiocatalysts for carbon capture, sequestration and valorisation. Top Catal 55:1217–1230
114. Reddy KJ, Weber H, Bhattacharya P et al (2010) Instantaneous capture and mineralization of flue gas carbon dioxide: pilot scale study. Nat Preceed 1–11
115. Reynolds AJ, Verheyen TV, Adeloju SB et al (2012) Towards commercial scale post combustion capture of $CO_2$ with monoethanolamine solvent: key considerations for solvent management and environmental impacts. Environ Sci Technol 46:3643–3654
116. Rowlett RS, Hoffmann KM, Failing H et al (2010) Evidence for a bicarbonate "escort" site in *Haemophilus influenzae* β-carbonic anhydrase. Biochemistry 49:3640–3647
117. Russo ME, Olivieri G, Capasso C et al (2013) Kinetic study of a novel thermo-stable α-carbonic anhydrase for biomimetic $CO_2$ capture. Enzym Microb Technol 53:271–277
118. Sakono M, Kawashima YM, Ichinose H et al (2004) Direct refolding of inclusion bodies using reversed micelles. Biotechnol Prog 201783–1787
119. Santos A, Toledo-Fernandez JA, Mendoza-Serna R et al (2007) Chemically active silica aerogel– wollastonite composites for CO2 fixation by carbonation reactions. Ind Eng Chem Res 46:103–107
120. Shahbazi A, Nasab BR (2016) Carbon capture and storage (CCS) and its impacts on climate change and global warming. J Pet Environ Biotechnol 7:29
121. Shakun JD, Clark PU, He F et al (2012) Global warming preceded by increasing carbon dioxide concentrations during the last deglaciation. Nature 484:49–54
122. Sharma A, Bhattacharya A (2010) Enhanced biomimetic sequestration of $CO_2$ into $CaCO_3$ using purified carbonic anhydrase from indigenous bacterial strains. J Mol Catal B Enzym 67:122–128
123. Sharma A, Bhattacharya A, Shrivastava A (2011) Biomimetic $CO_2$ sequestration using purified carbonic anhydrase from indigenous bacterial strains immobilized on biopolymeric materials. Enzyme Microb Technol 48:416–426
124. Shutova T, Kenneweg H, Buchta J et al (2008) The photosystem II_associated Cah3 in *Chlamydomonas* enhances the $O_2$ evolution rate by proton removal. EMBO J 27:782–791
125. Silverman DN (1982) Carbonic anhydrase: Oxygen-18 exchange catalyzed by an enzyme with rate-contributing Proton-transfer steps. Methods Enzymol 87:732–752

126. Silverman DN, Lindskog S (1988) The catalytic mechanism of carbonic anhydrase: implications of a rate-limiting photolysis of water. Acc Chem Res 21:30–36
127. Smith KS, Ferry JG (2000) Prokaryotic carbonic anhydrases. FEMS Microbiol Rev 24:335–366
128. Smith KS, Cosper NJ, Stalhandske C et al (2000) Structural and kinetic characterization of an archaeal β-class carbonic anhydrase. J Bacteriol 182:6605–6613
129. Smith KS, Jakubzick C, Whittam TS, Ferry JG (1999) Carbonic anhydrase is an ancient enzyme widespread in prokaryotes. Proc Nat Acad Sci 96:15184–15189
130. So AK, Espie GS, Williams EB et al (2004) A novel evolutionary lineage of carbonic anhydrase (ε class) is a component of the carboxysome shell. J Bacteriol 186:623–630
131. Solomon S, Plattner GK, Knutti R et al (2009) Irreversible climate change due to carbon dioxide emissions. Proc Natl Acad Sci U.S.A. 106:17041709
132. Soltes-Rak E, Mulligan ME, Coleman JR (1997) Identification and characterization of a gene encoding a vertebrate-type carbonic anhydrase in cyanobacteria. J Bacteriol 179:769–774
133. Soong Y, Fauth DL, Howard BH (2006) $CO_2$ sequestration with brine solution and fly ashes. Energy Converse Manage 47:1676–1685
134. Srivastava S, Bharti RK, Verma PK et al (2015) Cloning and expression of gamma carbonic anhydrase from *Serratia* sp. ISTD04 for sequestration of carbon dioxide and formation of calcite. Biores Technol 188:209–213
135. Stadie WC, O'Brien H (1933) The catalysis of the hydration of carbon dioxide and dehydration of carbonic acid by an enzyme isolated from red blood cells. J Biol Chem 103:521–529
136. Sterling D, Alvarez BV, Casey JR (2002) The extracellular component of a transport metabolon. Extracellular loop 4 of the human AE1 Cl/$HCO_3$ exchanger binds carbonic anhydrase IV. J Biol Chem 277:25239–25246
137. Sung YC, Fuchs JA (1988) Characterization of the *cyn* Operon in *Escherichia coli* K12. J Biol Chem 263:14769–14775
138. Supuran CT (2011) Carbonic anhydrase inhibitors and activators for novel therapeutic applications. Future Med Chem 3:1165–1180
139. Supuran CT (2016) How many carbonic anhydrase inhibition mechanisms exist? J Enzyme Inhib Med Chem 31:345–360
140. Svensson R, Odenberger M, Johnsson F, Strömberg L (2004) Transportation systems for CO2—application to carbon capture and storage. Energy Convers Manag 45: 2343–2353
141. Swartz JR (2001) Advances in *Escherichia coli* production of therapeutic proteins. Curr Opin Biotechnol 12:195–201
142. Szreter S, Mooney G (1998) Urbanization, mortality, and the standard of living debate: new estimates of the expectation of life at birth in nineteenth—century British cities. Econ Hist Rev 51:84–112
143. Takami H, Nishi S, Lu J, Shimamura S, Takaki Y (2004) Genomic characterization of thermophilic *Geobacillus* species isolated from the deepest sea mud of the Mariana Trench. Extremophiles 8:351–356
144. Trachtenberg MC, Cowan M, Smith DA et al (2009) Membrane-based, enzyme facilitated, efficient carbon dioxide capture. Energy Proc 1:353–360
145. Traufetter G (2009) Cold carbon sink: Slowing global warming with Antarctic iron. Spiegel Online
146. Tripp BC, Ferry JG (2000) A structure-function study of a proton transport pathway in a novel γ-class carbonic anhydrase from *Methanosarcina thermophila*. Biochemistry 39:9232–9240
147. Tripp BC, Smith K, Ferry JG (2001) Carbonic Anhydrase: New insights for an ancient enzyme. J Biol Chem 276:48615–48618
148. Tsouris C, Aaron DS Williams KA (2010) Is carbon capture and storage really needed? Environ Sci Technol 44:4042–4045
149. Ueda K, Nishida H, Beppu T (2012) Dispensabilities of carbonic anhydrase in proteobacteria. Int J Syst Evol Microbiol 2012
150. Umetsu M, Tsumoto K, Hara M et al (2003) How additives influence the refolding of immunoglobulin-folded proteins in a stepwise dialysis system spectroscopic evidence for highly efficient refolding of a single-chain fv fragment. J Biol Chem 278:8979–8987

151. Valdivia RH, Falkow S (1997) Fluorescence-based isolation of bacterial genes expressed within host cells. Science 277:2007–2011
152. Veitch FP, Blankenship LC (1963) Carbonic anhydrases in bacteria Nature 197:76–77
153. Vullo D, De Luca V, Scozzafava A et al (2012) Anion inhibition studies of the fastest carbonic anhydrase (CA) known, the extremo-CA from the bacterium *Sulfurihydrogenibium azorense.* Bioorg Med Chem 22:7142–7145
154. Wanjari S, Prabhu C, Yadav R (2011) Immobilization of carbonic anhydrase on chitosan beads for enhanced carbonation reaction. Process Biochem 46:1010–1018
155. Wanjari S, Prabhu C, Satyanarayana T et al (2012) Immobilization of carbonic anhydrase on mesoporous aluminosilicate for carbonation reaction. Micropor Mesopor Mat 160:151–158
156. What is a carbon footprint—definition. https://timeforchange.org/what-is-a-carbon-footprint-definition. Accessed on 26th March 2019
157. Wood HG, Werkman CH, Hemingway A et al (1941) Heavy carbon as a tracer in heterotrophic carbon dioxide assimilation. J Biol Chem 139:367–375
158. Xu J, Sun J, Wang Y et al (2014) Application of iron magnetic nanoparticles in protein immobilization. Molecules 19:11465–11486
159. Xu Y, Feng L, Jeffrey PD et al (2008) Structure and metal exchange in the cadmium carbonic anhydrase of marine diatoms. Nature 452:56–61
160. Yadav R, Satyanarayanan T, Kotwal S, Rayalu S (2011) Enhanced carbonation reaction using chitosan-based carbonic anhydrase nanoparticles. Curr Sci (Bangalore) 100:520–524
161. Yadav RR, Mudliar SN, Shekh AY et al (2012) Immobilization of carbonic anhydrase in alginate and its influence on transformation of $CO_2$ to calcite. Process Biochem 47:585–590
162. Yamaguchi S, Yamamoto E, Mannen T et al (2013) Protein refolding using chemical refolding additives. Biotechnology J 8:17–31
163. Yoshimoto M, Walde P (2018) Immobilized carbonic anhydrase: preparation, characteristics and biotechnological applications. World J Microbiol Biotechnol 34:151
164. Yue L, Chen W (2005) Isolation and determination of cultural characteristics of a new highly $CO_2$ tolerant fresh water microalga. Energy Converse Manage 46:1868–1876
165. Zhi W, Landry SJ, Gierasch LM et al (1992) Renaturation of citrate synthase: influence of denaturant and folding assistants. Prot Sci 1:522–529
166. 2018 Was the Fourth Warmest Year on Record With Kerala and California at the receiving end, 2019. https://www.news18.com/news/world/2018-was-the-fourth-warmest-year-on-record-with-kerala-and-california-at-receiving-end-2027999.html. Accessed on 26th Mar 2019

# Biomass-Based $CO_2$ Adsorbents for Biogas Upgradation with Pressure Swing Adsorption

**Pragyan P. Garnaik, Ranjan R. Pradhan, Yi Wai Chiang, and Animesh Dutta**

**Abstract** Numerous types of adsorbents have been studied and used in industrial scale during upgradation of biogas by employing pressure swing adsorption. This chapter provides an insight into the working principle, efficiency, and energy consumption of this method in comparison with other processes. The most recent advancement includes the addition of subsequent units, omissions of additional stages, and the use of novel adsorbents in the process that determine the productivity, separation efficiency, and cost of the technology. Based on the literature review, research gaps were identified concerning the biogas upgradation to deliver renewable natural gas to the combined heat and power industries or to the natural gas pipelines as a vehicular fuel. The main setbacks for this sustainable sector are initially due to the digester conditions that lower the biomass conversion to methane, inadequate pre-treatment of biogas for removal of other contaminants, followed by production and selection of appropriate low-cost adsorbent in the final upgradation stage to maximize carbon dioxide elimination. In recent years, biowastes have been found to have the potential of transforming into mesoporous and nanoporous adsorbents upon carbonization and activation techniques which would enhance the adsorption activity. Also, reformation of the simple biochemical and thermochemical methods has intensified the methane yield at digester and reactor levels, respectively. Nonetheless, analyzing the effects of the process parameters for upgradation of biogas and production of adsorbents will lead to further investigation and innovative outcomes. Conclusions are drawn to augment the recent developments and sustainable technologies for broader adoption of renewable natural gas.

**Keywords** Biogas upgradation · Renewable natural gas · Adsorbents · $CO_2/CH_4$ separation · Pressure swing adsorption · Biowaste

P. P. Garnaik
107-1204 Main Street East, Milton L9T9K7, Canada

R. R. Pradhan (✉)
Shrimp Canada, 65 Periwinkle Way, Guelph, ON N1L 1J2, Canada
e-mail: rpradhan@uoguelph.ca

R. R. Pradhan · Y. W. Chiang · A. Dutta
School of Engineering, University of Guelph, Guelph N1G 2W1, Canada



M. Goel et al. (eds.), *Climate Change and Green Chemistry of $CO_2$ Sequestration*, Green Energy and Technology, https://doi.org/10.1007/978-981-16-0029-6_14

## Abbreviations

| | |
|---|---|
| AC | Activated carbon |
| AD | Anaerobic digesters |
| AF | Anaerobic fermentation |
| CCS | Carbon capture and storage |
| CHP | Combined heat and power |
| EDS | Energy-dispersive spectrometer |
| ESA | Electric swing adsorption |
| FTIR | Fourier transform infrared spectroscopy |
| GHG | Greenhouse gas |
| HHV | Higher heating value |
| HPWS | High-pressure water scrubbing |
| HTC | Hydrothermal carbonization |
| LNG | Liquified natural gas |
| MOFs | Metal organic frameworks |
| PSA | Pressure swing adsorption |
| RNG | Renewable natural gas |
| SRT | Solid retention time |
| SSA | Specific surface area |
| TSA | Temperature swing adsorption |
| VOCs | Volatile organic compounds |
| WWTP | Wastewater treatment plant |

# 1 Introduction

Natural gas production has been rising since the economic crisis of 2009 with a compound annual growth rate of 2.8% [57]. Most of the large consumer countries have a vast network of supply distribution pipelines. For example, Canada is the world's fifth producer of natural gas called TransCanada Pipelines (TransCanada Pipelines [124]. With the ongoing technological advancements in the extraction sites, the gas wells could supply one-third of Canada's national requirement for another 300 years [22, 35]. The global over-reliance on such non-renewable resources to produce energy is one of the most significant concerns of environmentalists, economists, and power engineers [125]. The growing interest in developing and using alternative resources is a result of the following reasons: (i) ever-increasing energy demands due to the exponential rise of the global population that may cross 9 billion by 2040 [128], (ii) the spiking prices of gasoline/vehicular grade fuels in the upcoming years, and (iii) the action taken for greenhouse gas (GHG) mitigation and the Paris Agreement enforced in November 2016, to maintain the carbon capture and storage (CCS) cycle in the highest energy consuming third world countries as well as the developed countries [48, 56].

The conversion of biogas to renewable natural gas (RNG) or biomethane with a high content of methane is known as upgradation. It serves as an alternative solution to cope with the depleting reserves of natural gas and other vehicular grade fuels. In Ontario, the on-farm installations comprise 48% of total Canada's anaerobic digesters (ADs), and a few of them are delivering this 'green renewable energy' to natural gas grid supplies and distribution pipelines. In 2013, 37 biogas installations were operating on-farm in Ontario with 51 other projects under development [20, 28]. The most abundant feedstock to these digesters is usually biomass such as discarded parts of energy crops (corn silage, hay, and grasses), other agricultural waste, oils, and greases from urban centers, and beef and potato rejects [1]. However, according to [60], over 30 sites in Canada generate biogas from landfills and are directly burnt off to produce electricity. European nations are leading with the highest number of biogas plants, i.e., over 17,200 that are focusing on the production of vehicular grade fuel (European Biogas Association [34]. As of January 2015, there were 247 livestock anaerobic digesters, 645 operative landfills, and 1500 wastewater treatment plant (WWTP) producing biogas in the USA [113]. However, most of these sources deliver electrical energy rather than vehicular fuel [129]. After the upgradation, biomethane is compressed and delivered as liquified natural gas (LNG) [10], Kelleher [64]. Figure 1 presents an overview of biogas industry that starts from the collection of wastes, generation stage in ADs, followed by several treatment and purification steps to achieve the final product as biomethane.

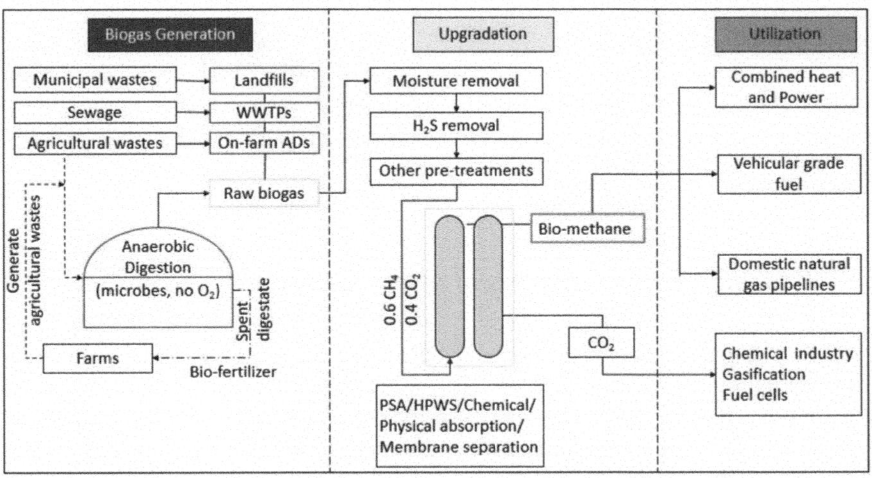


**Fig. 1** Biogas generation–upgradation–utilization flow sheet

## 2 Methods

Writing a review article concerning the biogas upgradation technologies is noteworthy to understand the research gaps and evaluate the need to cite the challenges faced by the current industries. The authors carried out the following steps to complete the review which includes general information gathering about the topic, the purpose of reviewing by summarizing and comparing the obtained articles, sorting the essential components that impact the process as subheadings, and by governing a timeline to support the framework.

### *2.1 Literature Sourcing*

The Elsevier Web site, Springer Journal Finder, Google Scholar search engine, and databases from University of Guelph library are the significant contributors to arrange sufficient literature for this paper. A few papers were asked for purchase to access; thus, the authors were requested to provide us with full text.

### *2.2 Screening of Articles*

At the beginning of the search, the authors listed out several keywords. The paper focuses on the application of biomass-based adsorbents in the biogas upgradation process. The keywords search started with North American renewable energy/biogas upgradation techniques/pressure swing adsorption/adsorbents for carbon dioxide capture.

### *2.3 Prioritizing and Assortment of the Literature*

After reviewing the abstracts for the selected articles, the search was narrowed down to 42 papers which included journal articles, conference proceedings, books, technical reports, and two masters and two doctoral theses. This step assisted the authors to generate the scope of the manuscript. The area of interest was segregated into three major sections: (i) maximizing the removal of contaminants in the biogas using bio-adsorbents, (ii) methane conversion of biomass to biogas at the digester level, and (iii) exploring the sustainable ways to produce bio-adsorbents using bio-waste. The in-depth search for each of the sections provided the researchers with over 150 references in the context to the manuscript.

## 2.4 *Constructive Analysis*

The most crucial step was reading the texts and giving attention to details such as:

- Ongoing research to improve the methane conversion and comparison of outcomes from the selected citations,
- Morphological structure of the adsorbents to suit the adsorption of the contaminants in the biogas,
- Environmental and economic factors that affect the upgradation processes, and
- Providing possible suggestions in accordance with the sustainable use of bio-waste in the industrial PSAs.

# 3 Measures and Challenges in Biogas Generation

The significant challenges cited by various researchers for efficient enrichment of biogas to appropriate grade of natural gas are associated with the conversion of organic matter via different processes. There are three types of anaerobic environments those contribute to GHG emissions, i.e., ADs, WWTPs, and landfills. In Ontario, Canada, the feedstocks are mostly organic waste matter obtained from agricultural farms, food wastes, energy crops, wastewater/sewage, and municipal wastes [20]. These feedstocks are broken down by microorganisms into simple compounds such as $CH_4$ and $CO_2$ in the absence of $O_2$ with a solid retention time (SRT) from 3 to 20 days under optimum operating conditions of pH range within 6.5–7.5 and temperature of 35–38 °C [21, 133].

## 3.1 *Anaerobic Digestion*

The on-farm ADs in Ontario, those run-on feedstocks such as manure do not produce siloxanes (Si–O) compounds, which eliminates the apparent step for removing them [64]. However, in the 1980s these ADs failed due to issues while operating and, thus, were not economically effective [28]. Since countries in the northern hemisphere have a cold climatic condition, a significant challenge to operating the digesters is providing thermal insulation for maintaining the optimum temperature, thus increasing the overall energy consumption.

Holm-Nielsen et al. [53], identified that regulation of the most critical parameters during this biological process such as temperature, organic loading rate, hydraulic and solid retention time, carbon: nitrogen (C/N) ratio, and pH will enhance the biogas yield. The whole process of purifying and enriching biogas to RNG from anaerobic digestion of waste generates additional wastes that are beneficial for plant growth along and minimizes the effect of human-made $CH_4$ emissions into the open atmosphere. Harris et al. [50], listed some of the microbial and chemical factors influencing methanogenesis using low-ranked subsurface coal (which comprises about 60% of the total US reserves) such as hydrolytic depolymerization of organic material, enzymatic activity of microbes, consumption of $H_2/CO_2$ or acetates by methanogens, and the intraspecific competition against homoacetogens. This study is an evident proof that by controlling the parameters mentioned above, the anaerobic digestion of biochar, viz., equivalent to low-ranked subsurface coal, can increase the $CH_4$ production route. Panjičko et al. [94], used brewery spent grain in the two-stage anaerobic digestion process that enhanced hydrolysis and acidogenesis in the solid-state digester and methanogenesis in a granular biomass reactor. The results showed an efficient total solid degradation up to 83%. For commercial digesters, the feed is usually a mixture of various types of wastes,thus, it undergoes the co-digestion process.

### *3.2 Anaerobic Fermentation*

Recent developments in this sector in European nations have helped in finding another source to obtain biogas by anaerobic fermentation (AF) of lignocellulosic biomass integrated to the co-digestion [34]. The hemicellulose and cellulose present in this biomass convert into $CH_4$ and $CO_2$. However, this method requires pre-treatment to convert the compact structure of the lignocellulosic biomass into simpler carbohydrates [18, 27]. Some of the pre-treatment methods are using acidic, alkaline, chemicals, hot water, or steam explosion methods [65]. Pobeheim et al. [99], studied the impact of heavy metals, i.e., nickel and cobalt on AF of maize silage model substrate. They found that with an organic loading rate of 4.3 g of organic dry mass stabilized the fermentation irrespective of increasing the concentration of the metal ions. Wang et al. [132], compared two-stage anaerobic fermentation of rice straw using NaOH (2%) and HCl (1%) treatment as well as studied the effect of hydrolysate concentration to improve biogas production. The digestion of hydrolysate after acidic treatment enriched the $CH_4$ yield by 43.89%. During two-stage AF of 400 g of corncob, Li H. et al. [73] and Li W. et al. [74], combined black soldier fly larvae to the process that produced 86.7 L of biogas and 3.17 g of biodiesel from the digestate.

### *3.3 Catalytic Gasification*

A non-biological, thermochemical method to produce biogas is by catalytic gasification. The biomass, usually wood, is gasified in the presence of air/oxygen and steam at a temperature of 850 °C and converts to producer gas, which primarily consists of $H_2$ and CO. It is further processed to remove tar, sulfur, and chlorides with simultaneous cooling [24]. The cooled producer gas stabilizes and transforms into $CH_4$, $CO_2$, and water upon methanation reaction in the presence of nickel catalyst [13,

85, 126]. However, the formation of nonflammable gases ($N_2$ and $CO_2$) and catalyst poisoning caused by sulfur and carbon depositions are some of the challenges that need to be resolved to achieve maximum conversion. But, the researchers have found that changing and enhancing the Ni-based catalyst with ruthenium can boost this process [102].

## 4 The Need for Upgradation

Many developing countries have been using biogas as a source of cooking fuel [59], but it cannot be used directly as a high-grade vehicular fuel or in combined heat and power (CHP) industries [115]. The primary challenge to use this resource directly is due to the presence of contaminants such as carbon dioxide ($CO_2$), hydrogen sulfide ($H_2S$), water vapor ($H_2O$), volatile organic compounds (VOCs), Si–O compounds, ammonia ($NH_3$), halides, hydrogen ($H_2$), and oxygen ($O_2$) [76]. Table 1 enlists the composition of biogas produced from various methods, and it can be noted that the higher heating value (HHV) of untreated raw biogas should be to nearly double so that it can be used consistently as a substitute for NG.

**Table 1** Typical composition of biogas compared to natural gas grid specifications

| Component | Units | [91] | [91] | [101,135] | TransCanada natural gas pipeline [33, 124] |
|---|---|---|---|---|---|
| $CH_4$ | $m^3$ | 0.53–0.70 | 0.30–0.65 | 0.58–0.70 | 0.9321 |
| $CO_2$ | $m^3$ | 0.30–0.50 | 0.25–0.47 | 0.30–0.43 | 0.0054 |
| $N_2$ | $m^3$ | 0.02–0.06 | <0.01–0.17 | 0.012–0.071 | 0.0101 |
| $O_2$ | $m^3$ | 0–0.05 | <0.01–3 | 0.001–0.02 | 0.004 |
| $H_2$ | $m^3$ | n.a | 0–0.03 | 0–0.0001 | n.a |
| Light hydrocarbons | $m^3$ | n.a | n.a | n.a | 0.0524 |
| $H_2S$ | ppm | 0–2000 | 30–500 | 2.5–3450 | 0.023 (permissible) |
| $NH_3$ | ppm | <100 | 0–5 | <1 | n.a |
| Total chlorides | $mg/m^3$ | <0.25 | 0–225 | <0.1 | n.a |
| Siloxanes | ppm | <0.08–0.5 | <0.3–36 | 0–2500 | n.a |
| $H_2O$ | $m^3$ | 0.05 | 0.06 | 0.0001 | 0.065 ppm (permissible) |
| VOCs | $mg/m^3$ | 5–8 | 46–173 | 13–268 | n.a |
| *Power* | $kWh/m^3$ | 6.388 | 4.444 | 6.722 | 10.5 |

Considering the basis of biogas volume = 1 $m^3$; ppm = parts per million; $mg/m^3$ = milligrams per $m^3$

Contaminants in biogas are harmful to CHP engines due to their corrosive nature. Table 4.2 lists some of the problems associated with each of the undesirable components present in raw biogas stream and the methods to remove them. The temperature of raw biogas generated at the ADs, WWTPs, or landfills is nearly 40 °C which consists of much moisture. When it passes through pipelines at an ambient temperature of a colder climatic region (~0 °C), it tends to lose moisture naturally by condensation. This turns out to be favorable when practiced on a large scale. In addition to natural condensation, installation of a column with hygroscopic salts will help to maximize the removal efficiency of water vapor from biogas. At the farm sites, passing the biogas in columns filled with ferrous oxides is one of the most efficient ways to remove $H_2S$. However, these spent metal oxides are a concern to environment/soil health management due to its disposal on open grounds. The current technologies have been able to eliminate $H_2S$, water vapor, and $NH_3$ during the pre-treatment stages such as chemical absorption, drying, and physical adsorption on activated carbon that results in the change of biogas composition to be satisfactory for the further upgradation processes [100].

Table 2 summarizes that adsorption on activated carbon is most commonly employed methods for contaminant removal for most of the purification steps and, thus, can be a potential solution to deal with the choice of adsorbents which is a substantial dilemma for the industries working with PSA units. PSA and HPWS are industrially operative around the globe to perform the conversion of biogas to achieve the RNG specifications. In British Columbia, Canada, Xebec Inc. has installed PSAs to upgrade biogas from the ADs that have proven to be 96% efficient. With 99.8% $CH_4$ recovery, they have also installed high-performance polymer membrane separators in China and North America [136]. HPWS is a separation technology that falls under the category of physical absorption [92]. In this process, the pressurized biogas (6–12 bar) is pumped in a countercurrent direction to the water which scrubs $CO_2$ out of the system in dissolved form. This stream of dissolved $CO_2$ and water passes into a regenerative column at low pressure that releases the captured $CO_{2,}$ and the water is recycled back into the system [23]. In Hamilton, Ontario, Flotech, Inc. operates a water scrubber with a capacity of about 10,000 $Nm^3/h$ to enrich biogas from WWTPs [46]. Molino et al. [89], developed a method to produce biomethane at low pressure by using a smart grid injection. They used polymeric membrane module and operated it at a pressure ranging from 2 to 15 bar. They found the purity varied from 60 to 93% at the expense of decreasing $CH_4$ recovery from 98 to 70%. Grande et al. [42], studied another $CO_2$ capturing method known as electric swing adsorption for post-combustion flue gas cleaning. It can be inferred that this technique can be used to upgrade biogas. However, it is an energy-intensive process. Since pre-treatment stages consume additional energy, the synergistic approach is studied to overcome the challenges to upgrade biogas economically and remove maximum contaminants in a single step. A laboratory-scale packed bed reactor was designed by Kashikaphaiboon et al. [63], that used monoethanolamine (MEA), calcium, and sodium hydroxide solutions to scrub both $CO_2$ and $H_2S$ efficiently over 90%. Recently developed second-generation upgrading technologies are very efficient (~99%) but are dependent on high energy requirements in the form of steam or heat, pressure, cryogenic cooling,

**Table 2** Problems with each undesirable component and respective methods of removal

| Component | Problem | Method of removal[a] | Refs |
|---|---|---|---|
| $CO_2$ | Lower heating value | PSA, HPWS, membrane separation, cryogenic distillation, ESA, TSA, chemical absorption, physical scrubbing, membrane absorption | [92] |
| $H_2O$ | Corrosion, rust, washes away lubrication | Drying, chillers, compression, silica gel, hygroscopic salts such as CaO and $Na_2SO_4$, molecular sieves, activated carbon | [100] |
| $H_2S$ | Unpleasant odor, leads to formation of $SO_2$ and $H2SO_4$, toxic to adsorbents of PSA | Chemical absorption, activated carbon, Fe-chelate method, reaction with $FeCl_2$, $Fe_2O_3$, biological treatment in digesters | [54, 100] |
| $NH_3$ | Forms nitrogen oxides during combustion in gas engines, foul odor | Acidic solvent-based gas-scrubbing | [1] |
| $N_2$ and $O_2$ | Lowers heating value, corrosion | Adsorption on activated carbon, molecular sieves, membranes | [96] |
| Siloxanes | Abrasion of engines, insulators, and plugs | Cooling, scrubbing with organic solvents, chemical adsorption with activated carbon, silica gel | [68] |
| Halides | Forms corrosive acids (HCl and HF), increases frequency to change engine oil | Adsorption on activated carbon at 200 °C | [68] |

[a]PSA: Pressure swing adsorption; HPWS: high-pressure water scrubbing; ESA: electric swing adsorption; TSA: temperature swing adsorption

and electrical. These aspects add on to the expenses, thus, industries prefer mostly PSAs [105, 119]. Table 3 lists and briefly describes the available technologies to purify biogas and derive RNG. It is important to assess all the technological advancements and research outcomes to upgrade/enrich biogas for contribution to the natural gas pipelines [16].

## 5 Pressure Swing Adsorption

### *5.1 Working Principle*

Pressure swing adsorption (PSA) is a conventional separation technique used to adsorb $CO_2$ on the porous adsorbent at high pressures, operated cyclically and repetitively. This method includes pressurizing the feed as well as the upgraded product streams, followed by high-pressure adsorption of the gases varying from 8 bars, pressure equalization steps, blowdown to depressurize the column to desorb $CO_2$ and simultaneous regeneration of the adsorbents, and low-pressure purging. The sequence is represented in Fig. 2. PSA is an established technology that works on the Skarstrom cycle based on the principle of physisorption [66]. Grande [39], has explained the sequential stages of the working of PSA cycle based on modified Skarstrom cycle. Skarstrom cycle denotes PSA as a two-step process which involves adsorption at high pressure and regeneration/desorption by lowering the pressure,

**Table 3** A brief overview of other competitive processes to upgrade biogas

| Methods | Description | Efficiency (%) | Energy consumption ($kWh/Nm^3$) | Usage | Refs |
|---|---|---|---|---|---|
| Physical absorption (high-pressure water scrubbing) | Solvent, i.e., water scrubs the highly dissolvable $CO_2$ at high pressure, and the solvent is regenerated by depressurizing. $H_2S$ removal is not necessary prior to this upgradation technique. Higher $CH_4$ slips | 88.9–96 | 0.23–0.90 | As HPWS | [93, 103] |
| Organic solvent scrubbing | Polyethyl glycol esters and other commercial organic solvents have higher affinity to dissolve $CO_2$. The used solvent is renewed by heating at 55–80 °C. $H_2S$ removal is necessary before this scrubbing technology | 90–95.5 | 0.2–0.67 | Limited due to high installation costs and the solvents are corrosive | [127] |
| Membrane separation | Selective gas permeation. The semipermeable membrane allows $CO_2$ to pass as off-gas and $CH_4$ stays as retentate. Highly intensified technology with operating pressures of 8–20 bars | 82.4–98 | 0.15–0.43 | Mostly European | [36, 86] |

(continued)

**Table 3** (continued)

| Methods | Description | Efficiency (%) | Energy consumption ($kWh/Nm^3$) | Usage | Refs |
|---|---|---|---|---|---|
| Chemical scrubbing | Potassium arginate, mono-, di-, tri-ethanolamine reacts specifically with $CO_2$. High $CH_4$ recovery. The solvent requires high temperature (120–170 °C) to regenerate | 88.5–97.7 | 0.1–1.5 | As amine scrubbing | [2, 71] |
| Cryogenic separation | Based on the freezing point of $CO_2$ (~ss110 °C). $CO_2$ condensed out from biogas, and 98% pure $CH_4$ is obtained. High-pressure (200 bars) compression of biogas consumes a lot of energy | 84.9–96.7 | 0.2–1.54 | Limited as it is energy-intensive | [ 97, 140] |
| TSA | In PSA, the desorption of gases occurs at low pressure, but in TSA it happens at high temperature (120 °C). Shortens the duration of the cycle | n.a | n.a | Limited as it is energy-intensive | [80, 138] |
| Hybrid membrane techniques | Some examples are: membrane absorption, TSA followed by a membrane, etc. These are process-intensified techniques when compared to the conventional methods to obtain high-purity $CH_4$ | n.a | n.a | Limited as it is in the developing stage | [111, 141] |

(continued)

**Table 3** (continued)

| Methods | Description | Efficiency (%) | Energy consumption (kWh/Nm$^3$) | Usage | Refs |
|---|---|---|---|---|---|
| Biological enrichment | Methanation using chemo-autotrophic and sulfur-reducing bacteria, and creating a most favorable environment for the microorganisms | n.a | n.a | Growing widely | [73, 74] |
| Gas hydration | Using the phase diagram, gas hydration is a combined method of $H_2S$ and $CO_2$ removal that optimizes energy consumption. These are crystalline solids which are formed by van der Walls force bonding to water molecules. High-pressure (40–60 bar) operation and exothermic process with internal temperature maintained at 3 °C. Requires multistage operation for maximum purity | 24.5 | 3.8–3.9% as specific energy consumption | Emerging technology | [17, 88] |

n.a.—not available

but Yan [138], altered this process by adding depressurization co-current blowdown (BD) step and pressure equalization (EQ1 and EQ2) step(s). This changing resulted in the loading of $CH_4$ per cycle to be constant,i.e., $CH_4$ concentration in the column is 0, which is called a cyclic steady state (CSS). It also resulted in better productivity and minimizing the energy consumption to re-pressurize (PR) the feed.

(a) High-pressure adsorption step: In the first stage, the pre-treated biogas (PT-biogas) is pressurized and fed at 6–8 bars of high pressure. The adsorption of $CO_2$ is initiated on the adsorbent, and the $CH_4$-rich stream is passed to another column (biomethane).

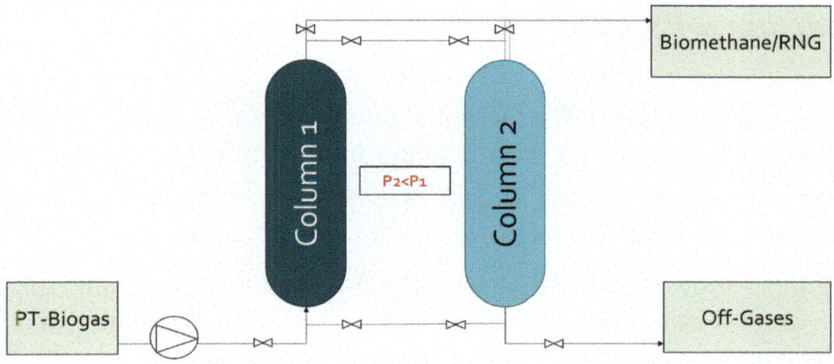


**Fig. 2** Two-column PSA based on Skarstrom cycle

(b) Intermediate pressure equalization step (EQ2): The pressurized gas present in the interstitial spaces at the end of the column partially pressurizes another column. This action increases the efficiency and lowers energy consumption.
(c) Depressurization blowdown (BD) step: It succeeds the EQ2 stage till the pressure goes down to nearly vacuum conditions. This change causes the column temperature to increase.
(d) Secondary equalization step (EQ1): It occurs in the first column which is like the step described by EQ2, thus making the first column partially pressurized and reducing the compressor energy consumption to pressurize the column to 6–8 bars.
(e) Final pressurization purge step (PU): The pressurized PT-biogas enters the first unit of PSA as feed, and with this, the cycle continues initial phase as described in the first stage (a).

Hence, PSA is an attractive method due to its higher separation efficiency and lower operational costs as well as the energy input. Figure 2.1 depicts a simple schematic of PSA which delivers RNG using the feed as PT-biogas, when the pressure in column 2 ($P_2$) is lower than column 1 ($P_1$).

## *5.2 Design Considerations*

The adsorbent is the crux of this process, and its selective adsorption performance to capture $CO_2$ from the pre-treated biogas (PT-biogas) enables conversion to biomethane. The operational expenses include the cost of changing the adsorbents after few years along with the mechanical energy invested in the compression of the feed [1]. There are different types of configurations available based on different types of adsorbents used, a number of cycles, and the design of adsorption columns installed to obtain 99% pure biomethane but the mostly employed is the two columns with packed bed design [38]. The selected adsorbent is filled in the central zone of the columns with a void fraction nearly 78% for maximum $CO_2$ capture [39].

## 5.3 Modeling a PSA Unit

With the help of a dynamic non-isothermal model based on linear driving force (LDF) approximation of mass transfer with a single lumped transport parameter ($k_{LDF,i}$), the kinetics of the adsorption process can be described. The model of a PSA is based on the following assumptions:

i. The gases follow axially dispersed plug flow and ideal gas behavior.
ii. There are no radial concentration and temperature gradients; thus, thermal equilibrium is maintained between solid and gas phases.
iii. Moreover, there is adiabatic interaction with surroundings.

The system is solved according to a set of partial differential equations. Agarwal [4], mathematically modeled and designed a PSA using gPROMS software. The experimental setup by Lestinsky et al. [72], as well as the numerical modeling by Agarwal suggests that high-purity $CH_4$ (98%) and recovery of 80% can be obtained with a single PSA unit. Augelletti et al. [12], modeled the most recent configuration of upgradation process that takes place in four interconnected columns and produces biomethane. They suggested that recovery can be as high as 99% with minimum energy consumption of 1250 kJ/kg of biomethane when compared to conventional methods. It was also confirmed that the $CH_4$ slips/losses are minimum with PSA (<2%). Vemula et al. [130], modeled rapid PSA using COMSOL Multiphysics and MATLAB software for separation of $N_2$/He feed mixture on zeolite-5A. Rapid PSA is the time-based intensification of the conventional PSAs. They considered functional adiabatic adsorber properties and found 99.9% pure helium could be obtained in 5.5 s. This development can be implemented to separate $CH_4/CO_2$ mixtures at a faster rate. Some of the PSAs used for $CH_4/CO_2$ streams are listed in Table 4. The performance of PSA can be calculated by evaluating: (i) purity of $CH_4$ obtained in the product, (ii) recovery of $CH_4$ after the purge step, and (iii) productivity of a unit for the adsorbent used and time is taken to complete one cycle [39].

## 5.4 Adsorbents

Adsorption is an exothermic separation phenomenon where a solid porous medium (adsorbent) adsorbs physically/chemically a component (adsorbate) to be separated [138]. Recently, low-cost adsorbents are primarily of interest for industrial applications [61]. The efficiency of the upgradation and pre-treatment processes is based on the selection of adsorbent. Many researchers have experimented $CO_2$ capture in PSA using carbon-based adsorbents, zeolites, metal organic frameworks (MOFs), and silica in the process, but the most effective one was found to be carbon-based adsorbents [41, 69]. Some of the commonly used carbon-based adsorbents in PSAs are activated carbon (AC) and carbon molecular sieve (CMS-3 K and CMS-5A).

**Table 4** Types of PSAs known to separate a $CO_2$1 /$CH_4$ mixture

| Types of PSA | Feed | Feed pressure (bar) | Feed temp. (°C) | Adsorbent used | Total no. of cols | % $CH_4$ purity | % $CH_4$ recovery | Time (s) | Productivity[a] | Refs |
|---|---|---|---|---|---|---|---|---|---|---|
| 2-bed and 4-steps | 55% $CH_4$ 45% $CO_2$ | 3.2 | 31–35 | CMS-3 K | 1 | 97.1 | 79.4–86.3 | 400 | n.a | [39] |
| 4-bed and 9-steps | 67% $CH_4$ 33% $CO_2$ | 4 | 50 | CMS-3 K | 4 | 97.88–98.3 | 88.5–88.93 | 900 | 5.5 | [26] |
| Duplex PSA | 55% $CH_4$ 45% $CO_2$ | 2–4 | 25 | Zeolite-5A | | 95–99 | n.a | 2–5.5 | 1.3 | [122] |
| Vacuum PSA 5-steps | 55% $CH_4$ 45% $CO_2$ | 6–8 | 33 | CMS-3 K | 1 | 98.1 | 79.7 | 400 | 3.83 | [112] |
| Vacuum PSA 5-steps | 55% $CH_4$ 45% $CO_2$ | 6–8 | 33 | Zeolite-13X | 1 | 96.58 | 48.6 | 400 | 3.20 | [43] |
| Layered VPSA 5-steps | 55% $CH_4$ 45% $CO_2$ | 8 | n.a | CMS-3 K and Zeolite-13X | 2 | 98.0 | 60.1 | 400 | 7.24 | [44] |
| Layered PSA | 30–45% $CO_2$ | 8 | 25 | CMS-3 K and Zeolite-13X | 2 | 98.7 | 74.8 | 400 | 2 | [40] |
| 28-bed kinetic rotary PSA | n.a | n.a | n.a | Zeolite molecular sieves or CMS | 1 | 97–99 | 95–99 | 2 | n.a | [6] |

[a]moles of CH4 per hour per kilogram of adsorbent
n.a.—not available

The International Union of Pure and Applied Chemistry (IUPAC) classifies the adsorbents based on the pore morphology and size, i.e., microporous (>50 nm), mesoporous (1–2 nm), and nanoporous (<2 nm) adsorbents [29]. Also, adsorbents can be broadly classified into two categories based on adsorption mechanism: equilibrium- and kinetic-based [43]. On the surface of kinetic-based adsorbents, the adsorption occurs due to the kinetics of reaction according to the Freundlich or Langmuir models [70]. The equilibrium-based adsorbents work on the physisorption isotherms which get unstable after few cycles [51]. Bagreev et al. [14], have studied the kinetics of both types of adsorption and found that chemisorption is more efficient than physical adsorption, but the regeneration of spent kinetic-based adsorbents consumes a high amount of energy. Use of zeolites and other non-carbon-based commercial adsorbents is a threat to the environment due to their disposal after exhaustion [7].

## 5.5 *Activated Carbon*

It is the most widely used adsorbent for various industrial applications since the time adsorption kinetics were first studied and the expected revenue in the coming five years will reach $3 billion with 10% rise in demand globally [121]. Due to the high specific surface area (>600 $m^2/g$), high porosity (0.2–2 $cm^3/g$), nominal bulk density (~450 $kg/m^3$), abrasion resistance, and known capacities for gas and wastewater purification properties, ACs are considered as adsorbents in pre-treatment steps and PSA for biogas upgradation. Highly developed twisted pores (tortuosity), ability to withstand repetitive uses, and economical (low cost) are some more factors to consider while selecting the adsorbent for prolonged industrial applications. Also, the heat adsorption also plays an essential parameter to choose adsorbent for a specific process [67]. Other carbon-based adsorbents such as CMS, monolithic carbon, carbon nanotubes, and ordered mesoporous carbon are extensively researched. When ACs are compared to zeolites and metallic organic frameworks (MOFs), it is observed that ACs are more hydrophobic which makes them suitable for any wet gas streamed processes [131].

There are three methods to activate the carbon mass to maximize its specific surface area (SSA): (i) physical activation by using $CO_2$/steam at high temperature (>800 °C) which makes the process energy-intensive, (ii) chemical activation using strong acids (nitric acid—$HNO_3$, phosphoric acid—$H_3PO_4$) that act as the dehydrating agents in an inert atmosphere at relatively lower temperature (400–700 °C) or bases (sodium hydroxide—NaOH, potassium hydroxide—KOH) or salts (potassium carbonate—$K_2CO_3$, zinc chloride—$ZnCl_2$, tin chloride—$SnCl_2$), and (iii) physio-chemical activation using combined techniques. With alkaline activation, the presence of the $OH^-$ ions on the surface increases its efficiency to adsorb $CO_2$. Song et al. [116], performed both types of activation experiments on a corn cob and found that KOH chemical activation is better than steam activation when one is aiming for better $CO_2$ adsorption capacity. $CO_2$ adsorption capacity is directly proportional to high specific surface area (SSA) and presence of nitrogen-containing functional groups [107].

After repeated cycles of adsorption and desorption, ACs can be regenerated by thermal treatment and steam regeneration techniques. This step is necessary to address the issue of PSA operations for long run. The most common method to resurrect the spent adsorbents is by thermal treatment. Bagreev et al. [14], studied the impact of thermal treatment of used adsorbent at 300 °C and found 100% regeneration of the adsorbents. Shah et al. [109], reviewed the experimental and technical aspects of steam regeneration method and found that using steam increases the cost of the operation, and thus there is a need to study steam adsorbate reactivity.

## *5.6 Effects of Activation Parameters*

The impregnation ratio (IPR), time of exposure for the activation and carbonization, and activation temperature are correlated factors. With the extreme increase in temperature and time, the overall mass yield decreases. Gratuito et al. [45], chose response surface methodology to optimize the process to produce AC from coconut shell. They found that the optimal IPR of 1.725 ($H_3PO_4$: coconut shell), time of 19.5 min, and temperature of 416 °C resulted in the formation of AC with an average pore diameter of 5 μm. For $H_3PO_4$ activation at a lower temperature, Yorgun and Yıldız [139], concluded that increase in IPR increases the mesoporosity but decreases the pyrolytic decomposition and SSA whereas, at higher temperatures, better adsorbents are produced with higher pore volume. The presence of $C = O$ functional groups (ketones, carboxyl, and lactone) in the AC disappears with an increase in temperature and IPR, which enhances the adsorbent properties. For KOH or NaOH activation, KOH activation occurs at a lower temperature (~400 °C) when compared to NaOH (>570 °C) [15]. This method is chosen to increase aromaticity ($C = C$) index in the biochar as per the required adsorption characteristics. Moupong et al. [90], characterized the KOH-activated AC from tamarind seeds and confirmed the above using the Fourier transform infrared spectroscopy (FTIR) analysis. It was also observed that carbon, oxygen, silicon, and potassium are the only elements detectable by energy-dispersive spectrometer (EDS) on AC's surface, and the carboxylic functional groups decreased with increase in temperature and use of KOH as the activating agent.

# 6 Raw Materials for Activated Carbon Production

## 6.1 Coal

The most conventional method to produce carbon-based adsorbents is sourced from coal. As per the ultimate and proximate analysis of different types of coal, the one with maximum fixed carbon is used for the preparation of activated carbon. Teng et al. [55], used Australian bituminous coal to produce AC via chemical activation using $H_3PO_4$ and further compared with physical activation using $CO_2$. They found that combined activation of the coal produced the best highly mesoporous AC with SSA as 687 $m^2/g$. The limitations of using AC derived from coal when compared to the wood-based ACs are high raw material costs, lower SSA, and unsuitable structure for the gas separation. This implies that not many studies are done to overcome these problems.

## 6.2 Biomass

The current issue with biomass industry is the generation of the massive amount of rejects after the grinding and milling operations of biomass before any thermochemical conversion. This powdered biomass is found to have a potential to become adsorbents (activated carbon) after carbonization [79]. Since biomass has a high content of volatile matter, there is a need to decrease the hydrogen: carbon (H/C) ratio by carbonizing at high temperature and pressure as per Van Krevelen diagram [8]. One of the processes to do so is by using hydrothermal carbonization (HTC) within a temperature range of 180–300 °C and a pressure of 40–50 bars in the absence of $O_2$ with specific biomass to solvent ratio. The hydrochar obtained is chemically or physically activated at high temperature (400–900 °C) for a prolonged period (0.5–2 h), further breaking the complex bonds and increasing the porosity. Hence, it shows remarkable nanoscale adsorptive properties upon activation that can help capture $CO_2$ from $CH_4$-rich streams, since $CH_4$ is not adsorbed if the pore size is over 1 nm [77].

Pyrolysis is another process to decrease the H/C ratio of the biomass for improving characteristics of the biochar that will assist further activation. It is a form of carbonization of solid biomass in an inert environment of $N_2$ at 600–800 °C. Recent studies on many food and agricultural waste show that these are emerging sources to produce adsorbents with ultra-fine pores and higher SSA. It was found that these agricultural wastes upon exposure of different activation techniques can produce ACs of the specific surface area of about 250–2410 $m^2/g$ and pore volumes of 0.022–91.4 $cm^3/g$. A few of them such as wheat, pinecone, rapeseed, pecan nut shells, rice straws, husks, hulls, bagasse, eucalyptus, peach stones, grape seeds, and oat hulls were reviewed by Ioannidou and Zabaniotou [58]. They reviewed the pyrolysis reaction kinetics and segregated the study based on the isothermal and non-isothermal operation. As biomass constitutes a network of lignocellulosic compounds, it was difficult defining the order of the reaction. Hence, devolatilization reaction was categorized as the first-order reaction. In Natalie Ho's thesis [52], the adsorbents were made by pyrolysis of anaerobic digestate by-product at a temperature ranging from 500 to 700 °C to maximize the pore size by breaking the lignocellulosic bonds. The most common raw material used for pyrolysis is coconut shell due to its sturdy structure.

Recent developments in the structural modification such as developing monolithic pores enhance the adsorption of $CO_2$ on the AC. Some non-carbon-based adsorbents such as calcium oxide (CaO), iron oxide ($Fe_2O_3$), zero valent iron ($Fe^o$), iron chloride ($FeCl_2$), silica gel, sodium sulfate ($Na_2SO_4$), and alumina are also used to separate $CO_2$, water vapor, siloxanes, and $H_2S$ from the raw biogas [83]. Table 5 summarizes the current progress of biomass is driven adsorbent industry for gas separation, wastewater treatment, and other applications. Some of the commonly used adsorbents in numerous pre-treatment steps for the purification of biogas such as H2S, H2O, and siloxanes removal, as well as in the $CO_2/CH_4$ adsorption, are listed in Table 6.

Since $CO_2$ intake depends on the features of the ACs used, adsorbents in PSA when compared to those of the super-AC produced after the carbonization/activation of biomass show that super-AC is better adsorbents. Not only would they help minimize the operational costs, but they also help in omitting several stages for separate upgradation techniques for each of the contaminants in the biogas. Hence, the adsorbent selection is an essential step to acquire maximum biomethane and lowers the number of adsorbent replacements. The use of waste-generated activated carbon will help to reduce the dependency on the zeolites and MOFs which are disposed of after exhaustion. It has an additional benefit as it can restore and maintain the carbon content in the soil without disturbing the natural nutrient cycle.

## 7 Conclusion and Recommendations

The biogas from anaerobic digesters has high amounts of impurities that make it unfit for use as a top-ranked fuel. As the biogas specifications differ according to the sources of the feedstock in the digesters and the climatic conditions of every geographical area are different, there is a need to look for the best-suited method to purify this renewable resource of energy with lower energy consumption (as heating is necessary to maintain the optimum temperature).

HPWS is widely used to upgrade the biogas generated from WWTPs in North American countries due to the high availability of water resources. A significant problem associated with this method is high consumption of water as the absorbent/solvent to capture $CO_2$, and this causes exhaustion of an essential natural resource. In addition to this, there is a demand for energy to dry the product gas in the subsequent stages. Hence, PSA is a well-known method to upgrade the biogas, with lower specific energy consumption and no/little extra requirement of materials

**Table 5** Types of biomass and processes to prepare ACs and their uses

| Raw materials | Activation technique(s) | Adsorption study on | Refs |
|---|---|---|---|
| Cherry stones | Physical steam activated<br>$CO_2$ activated | $CO_2/CH_4$ separation | [9] |
| Camphor leaves | HTC (180–300 °C—5 h) followed by KOH activation (800 °C—1 h) | $CO_2$ capture | [47] |
| Switchgrass | Both impregnated and un-impregnated KOH pyrolyses (500 °C—1 h) or HTC (300 °C—1 h) followed by inert activation (0.5–1 h) | Acetic acid | [134] |
| Miscanthus x giganteus straws | Pyrolysis (500–600 °C—70 min) followed by steam activation | $N_2$/77 K | [87] |
| Sunflower seed oil residue | Carbonization (700 °C) followed by microwave-assisted $K_2CO_3$ activation (2.45 GHz) | Methylene blue and acid 15 blue | [37] |
| Cassava and tapioca flour | HTC (250 °C—8 h) followed by KOH activation (800 °C—1 h) | $N_2$/77 K | [95] |
| Jatropha | Pyrolysis (400–600 °C) followed by:<br>KOH (evaporating 80 °C—24 h),<br>$H_3PO_4$ (drying 60 °C—24 h), and<br>$CO_2$ activation (400 °C—30 min) | $N_2$/77 K | [117] |
| Wood-based | Manufactured by Westvaco, $H_3PO_4$ activated | Wet $H_2S$ gas stream | [3] |
| Cellulose | Carbonization followed by physical or chemical activation | Liquid phase | [118] |
| Olive residues | KOH impregnation followed by carbonization (300 °C) followed by activation (840 °C)<br>Carbonization (300 °C) followed by KOH activation (840 °C)<br>$H_3PO_4$ impregnation followed by carbonization (300 °C) followed by activation (700 °C) | Supercapacitors electrodes | [32] |

(continued)

**Table 5** (continued)

| Raw materials | Activation technique(s) | Adsorption study on | Refs |
|---|---|---|---|
| Cotton fibers | Carbonization (500 °C—0.5 h) followed by microwave-induced H3PO$_4$ activation | Methylene blue dye | [31] |
| Apricot shells | $H_3PO_4$ impregnation followed by carbonization (400 °C—1.5 h) and dried (80 °C—24 h) | Tetracycline hydrochloride salt in aqueous solution | [84] |
| Pistachio nut shells | $H_3PO_4$ impregnation followed by carbonization (500 °C—2 h) | Methylene blue, rhodamine B, phenol, *p*-nitrophenol | [11] |
| Acorn shells | Different conc. of $ZnCl_2$ and HCl impregnation—24 h each, followed by $CO_2/H_2O$ physical activation (600 °C—30 min) | Iodine number | [110] |
| Peanut hulls | Simultaneous activation and carbonization (600 °C—1 h) followed by iron nitrate impregnation | Methylene blue | [114] |
| Date stones | Carbonization followed by physical, chemical, or physio-chemical activation | Phenols, dyes, pesticides, and heavy metals | [5] |
| Paulownia wood | $H_3PO_4$ impregnation (85 °C—6 h) then dried (110 °C—12 h) followed by activation (300–600 °C—1 h) and dried (110 °C—12 h) | $N_2$/77 K | [139] |
| Tamarind seeds | Carbonization (500 °C—1 h) followed by KOH activation (500–700 °C) | Iodine number, methylene blue, iron (III) from aqueous solution | [90] |
| Arundo donax Linn and pomelo peel | Impregnation in 85% $H_3PO_4$ (10 h) followed by activation (450 °C—1 h) | Ciprofloxacin hydrochloride | [120] |
| Reedy grass | 60% $H_3PO_4$ impregnation (4 h) followed by activation (400–800 °C—1–4 h) | $N_2$/77 K, iodine number | [137] |
| Cocoa bean husk | Carbonization (1000 °C—0.5 h) followed by thermal treatments (215–900 °C) $CO_2$ activation (880 °C—1–10 h) | $N_2$/77 K | [98] |

(continued)

**Table 5** (continued)

| Raw materials | Activation technique(s) | Adsorption study on | Refs |
|---|---|---|---|
| Gulfweed | Carbonization (600 °C—0.5 h) KOH activation (350 °C—0.5 h; 700–850 °C—1.5–3 h) | $N_2$/77 K | [75] |
| Waste tea | 40% $H_3PO_4$ impregnation and activation (450 °C—1 h) | $N_2$/77 K, oxytetracycline | [62] |
| Used tea *(Camellia sinensis)* | 50% $H_3PO_4$ impregnation (24 h) followed by activation (200 °C—0.5 h; 400–800 °C) | Iodine number, methylene blue | [82] |
| CCA (chromated copper arsenate) treated tree | Pyrolysis (700 °C—2 h) followed by $H_3PO_4$ impregnation, drying, and final activation (800–900 °C—1–4 h) | $CO_2$ capture | [19] |
| Plane tree seeds | Pyrolysis (850 °C—1 h) followed by $CO_2$ activation (650–850 °C—0.3–1 h) | n.a | [30] |
| *Glebionis coronaria* | Pyrolysis (500–600 °C—2 h) followed by H3PO4 impregnation (1.5–2 h), drying (110 °C—24 h), and activation (400–500 °C—1–2 h) | Iodine number, methylene blue, heavy metals (cadmium, cobalt) | [123] |

n.a.—not available

other than the adsorbent. The hybrid approaches like a combination of PSA and HPWS, or PSA and membrane separation give high-purity $CH_4$ and high recovery. These processes not only achieve the required grade of fuel but also give highly pure $CO_2$ that can be used (i) as a source to physically activate the biochar obtained after carbonization process, (ii) in chemical industries to produce other chemicals, and (iii) as a gasifying agent in gasification of biomass. Based on choice of adsorbent for improving efficiency in the PSAs, the use of activated carbon could be recommended for upgrading biogas to eliminate maximum undesirable components because of the following reasons: (i) It has high separation efficiency, (ii) it is regenerated with mild heat treatment at 300 °C, (iii) it is cost effective, (iv) it can be produced from waste biomass using hydrothermal torrefaction (220–260 °C) or pyrolysis (>600 °C) and further activated, and finally, (v) exhausted activated carbon can be dumped into soil after it has lost its regeneration capacity. With extensive reaction kinetics research and adsorption study on biomass-based adsorbents, limited work is done to testify them in the pre-treatment steps and PSAs for $CH_4/CO_2$ separation in biogas industry. Since these activated carbons are super-AC and are produced from zero-cost waste materials, these are preferable to the commercial chemically derived kinetic and equilibrium-based adsorbents.

**Table 6** Types of universally used adsorbents during upgradation process and their physical characteristics

| Different adsorbents | Type of adsorbent | Pore diameter (nm) | Pellet density($kg/m^3$) | Bulk density ($kg/m^3$) | Pellet porosity | Pore volume ($cm^3/g$) | Surface area ($m^2/g$) | Adsorption capacity | Ref |
|---|---|---|---|---|---|---|---|---|---|
| Activated carbon | Equilibrium | n.a | 984.3 | 670.6 | 0.506 | n.a | 845.87 | 0.25 mmol/s | [112] |
| Activated biochar after HTC | Kinetic | 0.8–2.0<br><2.0 | n.a | 390–430<br>n.a | n.a | 1.15–1.35<br>0.62-0.97 | 2250–2850<br>1260–1940 | 2.9–3.5 $CO_2$ mmol/g<br>3.9–4.8 $CO_2$ mmol/g | [79, 108] |
| Alkaline activated biochar | Kinetic | 2.35–2.84 | n.a | n.a | n.a | >2 | 514.98–1633.71 | 0.12–0.80 $CO_2$ mmol/g | [47] |
| Monolithic activated carbon | Kinetic | 5000–15,000 | 1086–1364 | 550–710 | n.a | 0.469–1.047 | 2200 | 6.71–10.18 $CH_4$ mmol/g | [81, 106, 104] |
| Carbon molecular sieve 3 K | Kinetic equilibrium | 0.4–0.9 | 1060 | 715.43–750 | 0.46 | n.a | n.a | ~98% $CO_2$ capture | [39, 40] |
| Zeolite-13X | Equilibrium kinetic | 0.5–1 | 1324 | 645–660 | 0.338 | n.a | n.a | ~98% $CO_2$ capture | [4, 25] |
| Zeolite-5A | Kinetic | 0.5–1 | 1160 | n.a | 0.64 | n.a | n.a | ~ 99% $CO_2$ capture | [12] |
| Activated carbon from pyrolysis (500–700 °C) | Equilibrium kinetic | n.a | n.a | n.a | n.a | 0.05–0.22 | 62.1–317.6 | 24.22 mg/g $H_2S$ | [52] |

(continued)

**Table 6** (continued)

| Different adsorbents | Type of adsorbent | Pore diameter (nm) | Pellet density($kg/m^3$) | Bulk density ($kg/m^3$) | Pellet porosity | Pore volume ($cm^3/g$) | Surface area ($m^2/g$) | Adsorption capacity | Ref |
|---|---|---|---|---|---|---|---|---|---|
| MOFs | n.a | 0.85–1.36 | n.a | n.a | n.a | n.a | ~ 2000 | 0.5–8 $CO_2$ at 4 MPa<br>1.18–14.6 $H_2S$ | [49, 78] |

n.a.—not available

**Acknowledgements** This study was supported by the NSERC Discovery Grant No. 400495 and Shrimp Canada, 67 Watson Rd. S (Unit -2), Guelph, Ontario N1L 1E3, Canada.

## References

1. Abatzoglou N, Steve B (2009) A review of biogas purification. Biofuels Bioprod Biorefining 3:42–71. https://doi.org/10.1002/bbb.117
2. Abdeen FRH, Mel M, Jami MS, Ihsan SI, Ismail AF (2016) A review of chemical absorption of carbon dioxide for biogas upgrading. Chinese J Chem Eng 24:693–702. https://doi.org/10.1016/j.cjche.2016.05.006
3. Adib F, Bagreev A, Bandosz T (1999) Effect of surface characteristics of wood-based activated carbons on adsorption of hydrogen sulfide. J Colloid Interface Sci 214:407–415. https://doi.org/10.1006/jcis.1999.6200
4. Agarwal A (2010) Advanced strategies for optimal design and operation of pressure swing adsorption processes (Doctoral dissertation). Retrieved from https://dynopt.cheme.cmu.edu/content/thesis/anshul_thesis.pdf
5. Ahmed MJ (2016) Preparation of activated carbons from date (Phoenix dactylifera L.) palm stones and application for wastewater treatments: Review. Process Saf Environ Prot 102:168–182. https://doi.org/10.1016/j.psep.2016.03.010
6. Alizadeh-Khiavi S, Alvaji JR, McLean C (2013) Low Energy, high recovery, rapid cycle kinetic PSA for biogas. US20130247762A1
7. Alonso-Vicario A, Ochoa-Gómez JR, Gil-Río S, Gómez-Jiménez-Aberasturi O, Ramírez-López CA, Torrecilla-Soria J, Domínguez A (2010) Microporous and Mesoporous Materials Purification and upgrading of biogas by pressure swing adsorption on synthetic and natural zeolites. Microporous Mesoporous Mater 134:100–107. https://doi.org/10.1016/j.micromeso.2010.05.014
8. Álvarez-Gutiérrez N, García S, Gil MV, Rubiera F, Pevida C (2014) Towards bio-upgrading of biogas: biomass waste-based adsorbents. Energy Proc 63:6527–6533. https://doi.org/10.1016/j.egypro.2014.11.688
9. Álvarez-Gutiérrez N, Gil MV, Rubiera F, Pevida C (2016) Adsorption performance indicators for the $CO_2$/$CH_4$ separation: application to biomass-based activated carbons. Fuel Process Technol 142:361–369. https://doi.org/10.1016/j.fuproc.2015.10.038
10. Askari MB, Fatemeh G, Saman M, Mirhosseiny LM (2015) Advantages and Disadvantages of Biogas Energy. Bull Adv Sci Res 1:132–135
11. Attia AA, Girgis BS, Khedr SA (2003) Capacity of activated carbon derived from pistachio shells by $H_3PO_4$ in the removal of dyes and phenolics. J Chem Technol Biotechnol 78:611–619. https://doi.org/10.1002/jctb.743
12. Augelletti R, Conti M, Annesini MC (2017) Pressure swing adsorption for biogas upgrading: a new process configuration for the separation of biomethane and carbon dioxide. J Clean Prod 140:1390–1398. https://doi.org/10.1016/j.jclepro.2016.10.013
13. Baeyens J, Appels L, Peng L, Dewil R (2016) The production of bio-energy by microbial (biogas through anaerobic digestion) or thermal (pyrolysis) processes. Renew Energy 96:1055. https://doi.org/10.1016/j.renene.2016.06.012
14. Bagreev A, Rahman H, Bandosz TJ (2001) Thermal regeneration of a spent activated carbon previously used as hydrogen sulfide adsorbent. Carbon N Y 39:1319–1326. https://doi.org/10.1016/S0008-6223(00)00266-9
15. Basso D, Castello D, Baratieri M, Fiori L (2013) Hydrothermal carbonization of waste biomass : progress report and prospect. 21st Eur. Biomass Conf. Exhib. Conf. Proc., pp 1478–1487
16. Bauer F, Persson T, Hulteberg C, Tamm D (2013) Biogas upgrading—technology overview, comparison. Biofuels Bioprod Biorefin 7:499–511. https://doi.org/10.1002/bbb.1423

17. Castellani B, Rossi F, Filipponi M, Nicolini A (2014) Hydrate-based removal of carbon dioxide and hydrogen sulphide from biogas mixtures: Experimental investigation and energy evaluations. Biomass Bioenerg 70:330–338
18. Bong CPC, Lim LY, Lee CT, Klemeš JJ, Ho CS, Ho WS (2018) The characterisation and treatment of food waste for improvement of biogas production during anaerobic digestion—a review. J Clean Prod 172:1545–1558. https://doi.org/10.1016/j.jclepro.2017.10.199
19. Botom ML, Poletto P, Junges J, Perondi D, Dettmer A, Godinho M (2017) Preparation and characterization of a metal-rich activated carbon from CCA-treated wood for $CO_2$ capture. Chem Eng J 321:614–621. https://doi.org/10.1016/j.cej.2017.04.004
20. Bradburn K (2014) CanBio report on the status of bioenergy in Canada https://www.fpac.ca/wp-content/uploads/2014_CanBio_Report.pdf. Accessed 26th Nov 2017
21. Busch G (2013) Biogas technology. In: Yang ST, El-Enshasy HA, Thongchul N (eds), Bioprocessing technologies in biorefinery for sustainable production of fuels, chemicals, and polymers. Wiley, pp. 279–291
22. CAPP (2015) The facts on: Canada's natural gas https://www.capp.ca/publications-and-statistics/publications/272337. Accessed 21st Nov 2017
23. Chandra R, Vijay VK, Subbarao PMV (2012) Vehicular quality biomethane production from biogas by using an automated water scrubbing system. ISRN Renew Energy 2012:1–6. https://doi.org/10.5402/2012/904167
24. Chynoweth DP, Owens JM, Legrand R (2001) Renewable methane from anaerobic digestion of biomass. Renew Energy 22:1–8. https://doi.org/10.1016/S0960-1481(00)00019-7
25. da Silva Santos MP (2011) Advanced modelling of PSA processes for biogas upgrading. (Master's thesis, University of Porto, Porto, Portugal). Retrieved from https://repositorio-aberto.up.pt/bitstream/10216/61452/1/000149424.pdf
26. da Silva Santos MP, Grande CA, Rodrigues E (2011) New cycle configuration to enhance performance of kinetic PSA processes. Chem Eng Sci 66:1590–1599. https://doi.org/10.1016/j.ces.2010.12.032
27. Dahunsi SO, Oranusi S, Efeovbokhan VE (2017) Cleaner energy for cleaner production: modeling and optimization of biogas generation from Carica papayas (Pawpaw) fruit peels. J Clean Prod 156:19–29. https://doi.org/10.1016/j.jclepro.2017.04.042
28. DeBruyn J, Hilborn D (2015) Anaerobic digestion basics: factsheet. https://www.omafra.gov.on.ca/english/engineer/facts/07-057.htm. Accessed 21st Sept 2016
29. Do DD (1998) Adsorption analysis: equilibria and kinetics, 1st edn. Imperial College Press, London
30. Dodevski V, Janković B, Stojmenović M, Krstić S, Popović J, Pagnacco MC, Popović M, Pašalić S (2017) Plane tree seed biomass used for preparation of activated carbons (AC) derived from pyrolysis. Modeling the activation process. Colloids Surfaces A Physicochem Eng Asp 522:83–96. https://doi.org/10.1016/j.colsurfa.2017.03.003
31. Duan X, Srinivasakannan C, Wang X, Wang F, Liu X (2017) Synthesis of activated carbon fibers from cotton by microwave induced $H_3PO_4$ activation. J Taiwan Inst Chem Eng 70:374–381. https://doi.org/10.1016/j.jtice.2016.10.036
32. Elmouwahidi A, Bailón-García E, Pérez-Cadenas AF, Maldonado-Hódar FJ, Carrasco-Marín F (2017) Activated carbons from KOH and $H_3PO_4$-activation of olive residues and its application as supercapacitor electrodes. Electrochim Acta 229:219–228. https://doi.org/10.1016/j.electacta.2017.01.152
33. Enbridge Gas Distribution (2018).Gas analysis—January 2018 Ontario https://www.enbridgegas.com/assets/docs/VictoriaSquare Station-Jan. 2018.pdf. Accessed 2nd Feb 2018
34. European Biogas Association (2015) EBA biomethane & biogas report. https://european-biogas.eu/2015/12/16/biogasreport2015/. Accessed 24th July 2017
35. Faramawy S, Zaki T, Sakr AA (2016) Natural gas origin, composition, and processing: a review. J Nat Gas Sci Eng 34:34–54. https://doi.org/10.1016/j.jngse.2016.06.030
36. Favre E, Bounaceur R, Roizard D (2009) Biogas, membranes and carbon dioxide capture. J Memb Sci 328:11–14. https://doi.org/10.1016/j.memsci.2008.12.017

37. Foo KY, Hameed BH (2011) Preparation and characterization of activated carbon from sunflower seed oil residue via microwave assisted $K_2CO_3$ activation. Bioresour Technol 102:9794–9799. https://doi.org/10.1016/j.biortech.2011.08.007
38. Grande CA (2012) Advances in pressure swing adsorption for gas separation. ISRN Chem Eng 2012:1–13. https://doi.org/10.5402/2012/982934
39. Grande CA (2011) Biogas upgrading by pressure swing adsorption. In: Bernardes DMADS (ed), Biofuel's engineering process technology. InTech, Oslo, pp 65–84
40. Grande CA, Cavenati S, Rodrigues AE (2008) Separation column and pressure swing adsorption process for gas purification. WO2008072215A2
41. Grande CA, Cavenati S, Silva FAD (2005) Carbon molecular sieves for hydrocarbon separations by adsorption. Ind Eng Chem Res 44:7218–7227. https://doi.org/10.1021/ie050376r
42. Grande CA, Ribeiro RPL, Oliveira ELG, Rodrigues AE (2009) Electric swing adsorption as emerging $CO_2$ capture technique. Energy Proc 1:1219–1225. https://doi.org/10.1016/j.egypro.2009.01.160
43. Grande CA, Rodrigues AE (2007a) Biogas to fuel by vacuum pressure swing adsorption I. Behavior of equilibrium and kinetic-based adsorbents. Ind Eng Chem Res 46:4595–4605. https://doi.org/10.1021/ie061341+
44. Grande CA, Rodrigues AE (2007b) Layered vacuum pressure-swing adsorption for biogas upgrading. Ind Eng Chem 46:7844–7848. https://doi.org/10.1021/ie070942d
45. Gratuito MKB, Panyathanmaporn T, Chumnanklang RA, Sirinuntawittaya N, Dutta A (2008) Production of activated carbon from coconut shell: Optimization using response surface methodology. Bioresour Technol 99:4887–4895. https://doi.org/10.1016/j.biortech.2007.09.042
46. Greenlane, Flotech, (2011) Flotech/Greenlane Company Overview https://www2.hamilton.ca/nr/rdonlyres/5d6ba30e-ed94-427a-a21c-132ff9947abd/0/flotechgreenlaneoverviewmarch11woodward1.pdf. Accessed 6th Oct 2016
47. Guangzhi Y, Jinyu Y, Yuhua Y, Zhihong T, Dengguang Y (2017) Preparation and $CO_2$ adsorption properties of porous carbon from camphor leaves by hydrothermal carbonization and sequential KOH activation. RSC Adv 7:4152–4160. https://doi.org/10.1039/C6RA25303B
48. Ham J Van (2008) Canada's fossil energy future the way forward on carbon capture and storage. https://www.nrcan.gc.ca/sites/www.nrcan.gc.ca/files/www/pdf/com/resoress/publications/fosfos/fosfos-eng.pdf. Accessed 8th March 2017
49. Hamon L, Serre C, Devic T, Loiseau T, Millange F (2009) Comparative study of hydrogen sulfide adsorption in the MIL-53 (Al, Cr, Fe), MIL-47 (V), MIL-100 (Cr), and MIL-101 (Cr) metal—organic frameworks at room temperature. J Am Chem Soc 131:8775–8777. https://doi.org/10.1021/ja901587t
50. Harris SH, Smith RL, Barker CE (2008) Microbial and chemical factors influencing methane production in laboratory incubations of low-rank subsurface coals. Int J Coal Geol 76:46–51. https://doi.org/10.1016/j.coal.2008.05.019
51. Hicks JC, Drese JH, Fauth DJ, Gray ML, Qi G, Jones CW (2008) Designing adsorbents for $CO_2$ capture from flue gas-hyperbranched aminosilicas capable of capturing $CO_2$ reversibly. J Am Chem Soc 130:2902–2903. https://doi.org/10.1021/ja077795v
52. Ho N (2012) Modeling hydrogen sulfide adsorption by activated carbon made from anaerobic digestion by-product (Master's Thesis, University of Toronto, Toronto, Canada) Retrieved from https://tspace.library.utoronto.ca/handle/1807/32575
53. Holm-nielsen JB, Seadi TA, Oleskowicz-popiel P (2009) The future of anaerobic digestion and biogas utilization. Bioresour Technol 100:5478–5484. https://doi.org/10.1016/j.biortech.2008.12.046
54. Horikawa MS, Rossi M, Gimenes ML, Costa CM, da Silva MG (2004) Chemical absorption of $H_2S$ for biogas purification. Brazilian J Chem Eng 21:415–422
55. Hsisheng T, Yeh T-S, Shu L-Y (2012) Preparation of activated carbon from bituminous coal with phosphoric acid activation. Carbon N. Y. 36, pp 1387–1395. https://doi.org/10.1016/S0008-6223(98)00127-4

56. IEA (2016) Introduction and scope. World energy outlook 2016. International Energy Agency, Paris, pp 31–55
57. IEA (2019) Natural Gas Information, Statistics report—September-2019. https://www.iea.org/reports/natural-gas-information-2019
58. Ioannidou O, Zabaniotou A (2007) Agricultural residues as precursors for activated carbon production-a review. Renew Sustain Energy Rev 11:1966–2005. https://doi.org/10.1016/j.rser.2006.03.013
59. Kadam R, Panwar NL (2017) Recent advancement in biogas enrichment and its applications. Renew Sustain Energy Rev 73:892–903. https://doi.org/10.1016/j.rser.2017.01.167
60. Kajat B (2006) Bioenergy and biofuels: Canadian industry and market opportunities. https://www.compost.org/CCC_Science_Web_Site/pdf/Biogas/BioenergyReport.pdf. Accessed 14th Nov 2016
61. Kambo HS, Dutta A (2015) A comparative review of biochar and hydrochar in terms of production, physico-chemical properties and applications. Renew Sustain Energy Rev 45:359–378. https://doi.org/10.1016/j.rser.2015.01.050
62. Kan Y, Yue Q, Li D, Wu Y, Gao B (2017) Preparation and characterization of activated carbons from waste tea by $H_3PO_4$ activation in different atmospheres for oxytetracycline removal. J Taiwan Inst Chem Eng 71:494–500. https://doi.org/10.1016/j.jtice.2016.12.012
63. Kasikamphaiboon P, Chungsiriporn J, Bunyakan C, Wiyaratn W (2013) Simultaneous removal of $CO_2$ and $H_2S$ using MEA solution in a packed column absorber for biogas upgrading. Songklanakarin J Sci Technol 35:683–691. https://doi.org/10.1016/j.energy.2010.04.014
64. Kelleher Robins (2013) Canadian biogas study technical document. https://www.biogasassociation.ca/bioExp/images/uploads/documents/2014/biogas_study/Canadian_Biogas_Study_Technical_Document_Dec_2013.pdf. Accessed 12th Dec 2016
65. Kim TH (2013) Pretreatment of lignocellulosic biomass. In: Yang ST, El-Enshasy HA, Thongchul N (eds), Bioprocessing technologies in biorefinery for sustainable production of fuels, chemicals, and polymers. Wiley, pp 91–109
66. Knaebel KS, Hill FB (1985) Pressure swing adsorption: development of an equilibrium theory for gas separations. J Chem Eng Sci 40:2351–2360. https://doi.org/10.1016/0009-2509(85)85139-3
67. Knaebel KS (2008) Adsorption. In: Albright LF (ed) Albright's chemical engineering handbook. CRC Press, Boca Raton, pp 1119–1172
68. Korres NE (2013) Bioenergy production by anaerobic digestion : using agricultural biomass and organic wastes. In: O'Kiely P, Benzie JAH, West J (eds), New York, pp 137–258
69. Kwaśny J, Balcerzak W (2016) Sorbents used for biogas desulfurization in the adsorption process. Pol J Environ Stud 25:37–43. https://doi.org/10.15244/pjoes/60259
70. Lau LC, MohamadNor N, Lee KT, Mohamed AR (2016) Adsorption isotherm, kinetic, thermodynamic and breakthrough curve models of $H_2S$ removal using $CeO_2$/NaOH/PSAC. Int J Petrochem Sci Eng 1:1–10. https://doi.org/10.15406/ipcse.2016.01.00009
71. Leonzio G (2016) Upgrading of biogas to bio-methane with chemical absorption process: simulation and environmental impact. J Clean Prod 131:364–375. https://doi.org/10.1016/j.jclepro.2016.05.020
72. Lestinsky P, Vecer M, Navratil P, Stehlik P (2015) The removal of $CO_2$ from biogas using a laboratory PSA unit: design using breakthrough curves. Clean Technol Environ Policy 17:1281–1289. https://doi.org/10.1007/s10098-015-0912-2
73. Li H, Larsson E, Thorin E, Dahlquist E, Yu X (2015) Feasibility study on combining anaerobic digestion and biomass gasification to increase the production of biomethane. Energy Convers Manag 100:212–219. https://doi.org/10.1016/j.enconman.2015.05.007
74. Li W, Li Q, Zheng L, Wang Y, Zhang J, Yu Z, Zhang Y (2015) Potential biodiesel and biogas production from corncob by anaerobic fermentation and black soldier fly. Bioresour Technol 194:276–282. https://doi.org/10.1016/j.biortech.2015.06.112
75. Li S, Han K, Li J, Li M, Lu C (2017) Preparation and characterization of super activated carbon produced from gulfweed by KOH activation. Microporous Mesoporous Mater 243:291–300. https://doi.org/10.1016/j.micromeso.2017.02.052

76. Lijó L, González-García S, Bacenetti J, Fiala M, Feijoo G, Moreira MT (2014) Assuring the sustainable production of biogas from anaerobic mono-digestion. J Clean Prod 72:23–34. https://doi.org/10.1016/j.jclepro.2014.03.022
77. Liu J, Reddy KP, Gadkaree PK (2014) Method for making alkali activated carbon. US201408809231B2
78. Llewellyn PL, Hamon L, Devic T, Ghoufi A, Clet G, Driver G, Beek WV, Jolimaı E (2009) Co-adsorption and separation of $CO_2$-$CH_4$ mixtures in the highly flexible MIL-53 (Cr) MOF. J Am Chem Soc 131:17490–17499. https://doi.org/10.1021/ja907556q
79. Lozano-castell D, Lozar JPM, Falco C, Titirici M-M, Amoros DC (2013) Porous biomass-derived carbons : activated carbons. In: Titirici MM (Ed), Sustainable carbon materials from hydrothermal processes. Wiley, pp 75–100
80. Luo L (2013) Intensification of adsorption process in porous media. In: Heat and mass transfer intensification and shape optimization: a multi-scale approach. pp 19–43. https://doi.org/10.1007/978-1-4471-4742-8
81. Machnikowski J, Kierzek K, Lis K, Machnikowska H, Czepirski L (2010) Tailoring porosity development in monolithic adsorbents made of KOH-activated pitch coke and furfuryl alcohol binder for methane storage. Energy Fuels 24:3410–3414. https://doi.org/10.1021/ef901536y
82. Mahmood T, Ali R, Naeem A, Hamayun M, Aslam M (2017) Potential of used Camellia sinensis leaves as precursor for activated carbon preparation by chemical activation with $H_3PO_4$; optimization using response surface methodology. Process Saf Environ Prot 109:548–563. https://doi.org/10.1016/j.psep.2017.04.024
83. Mamun MRA, Torii S (2017) Enhancement of methane concentration by removing contaminants from biogas mixtures using combined method of absorption and adsorption. Int J Chem Eng 2017:1–9. https://doi.org/10.1155/2017/7906859
84. Marzbali MH, Esmaieli M, Abolghasemi H, Marzbali MH (2016) Tetracycline adsorption by $H_3PO_4$-activated carbon produced from apricot nut shells: a batch study. Process Saf Environ Prot 102:700–709. https://doi.org/10.1016/j.psep.2016.05.025
85. Meijden CM Van Der, Drift A Van der, Vreugdenhil BJ, Smit R, Rabou LPLM (2011) Large scale production of Biomethane from wood. In: International gas union research conference. Energy Research Center of the Netherlands, Seoul, pp 1–17
86. Micale C (2015) Bio-methane generation from biogas upgrading by semi-permeable membranes: an experimental, numerical and economic analysis. Energy Proc 82:971–977. https://doi.org/10.1016/j.egypro.2015.11.854
87. Michel R, Mischler N, Azambre B, Finqueneisel G, Machnikowski J, Rutkowski P, Zimny T, Weber JV (2006) Miscanthus × Giganteus straw and pellets as sustainable fuels and raw material for activated carbon. Environ Chem Lett 4:185–189. https://doi.org/10.1007/s10311-006-0043-4
88. Miltner M, Makaruk A, Harasek M (2017) Review on available biogas upgrading technologies and innovations towards advanced solutions. J Clean Prod 161:1329–1337. https://doi.org/10.1016/j.jclepro.2017.06.045
89. Molino A, Nanna F, Iovane P (2015) Low pressure biomethane production by anaerobic digestion (AD) for the smart grid injection. Fuel 154:319–325. https://doi.org/10.1016/j.fuel.2015.03.054
90. Mopoung S, Moonsri P, Palas W, Khumpai S (2015) Characterization and properties of activated carbon prepared from tamarind seeds by KOH activation for Fe (III) adsorption from aqueous solution. Sci World J 2015:1–9. https://doi.org/10.1155/2015/415961
91. Nahar G, Mote D, Dupont V (2017) Hydrogen production from reforming of biogas: review of technological advances and an Indian perspective. Renew Sustain Energy Rev 76:1032–1052. https://doi.org/10.1016/j.rser.2017.02.031
92. Niesner J, Jecha D, Stehlík P (2013) Biogas upgrading technologies: state of art review in European region. Chem Eng Trans 35:517–522. https://doi.org/10.3303/CET1335086
93. Nock WJ, Walker M, Kapoor R, Heaven S (2014) Modeling the water scrubbing process and energy requirements for $CO_2$ capture to upgrade biogas to biomethane. Ind Eng Chem Res 53:12783–12792. https://doi.org/10.1021/ie501280p

94. Panjičko M, Zupančič GD, Fanedl L, Logar RM, Tišma M, Zelić B (2017) Biogas production from brewery spent grain as a mono-substrate in a two-stage process composed of solid-state anaerobic digestion and granular biomass reactors. J Clean Prod 166:519–529. https://doi.org/10.1016/j.jclepro.2017.07.197
95. Pari G, Darmawan S, Prihandoko B (2014) Porous carbon spheres from hydrothermal carbonization and KOH activation on cassava and tapioca flour raw material. Proc Environ Sci 20:342–351. https://doi.org/10.1016/j.proenv.2014.03.043
96. Petersson A, Wellinger A (2006) Biogas upgrading to vehicle fuel standards and grid, 1st edn. IEA Bioenergy, Stockholm
97. Petersson A, Wellinger A (2009) Biogas upgrading technologies–developments and innovations, 1st ed. IEA Bioenergy
98. Plaza-Recobert M, Trautwein G, Pérez-Cadenas M, Alcañiz-Monge J (2017) Preparation of binderless activated carbon monoliths from cocoa bean husk. Microporous Mesoporous Mater 243:28–38. https://doi.org/10.1016/j.micromeso.2017.02.015
99. Pobeheim H, Munk B, Lindorfer H, Guebitz GM (2011) Impact of nickel and cobalt on biogas production and process stability during semi-continuous anaerobic fermentation of a model substrate for maize silage. Water Res 45:781–787. https://doi.org/10.1016/j.watres.2010.09.001
100. Rashed M, Mamun A, Torii S (2015) Removal of $H_2S$ and $H_2O$ by chemical treatment to upgrade methane of biogas generated from anaerobic co-digestion of organic biomass waste. IPASJ Int J Mech Eng 3:42–52
101. Rasi S (2009) Biogas composition and upgrading to biomethane (Doctoral Dissertation, University of Jyvaskyla, Jyvaskyla, Finland). Retrieved from https://urn.fi/URN:ISBN:978-951-39-3618-1
102. Rauch R, Hrbek J, Hofbauer H (2014) Biomass gasification for synthesis gas production and applications of the syngas. Wiley Interdiscip. Rev Energy Environ 3:343–362. https://doi.org/10.1002/wene.97
103. Ray NHS, Swain PR, Mohanty MK, Mohanty RC (2014) A case study of biogas upgradation by water scrubbing: laboratory model for purification and effective storage in LPG cylinders for cooking applications. Int J Adv Eng Res Sci 1:68–75
104. Rezaei F, Webley P (2010) Structured adsorbents in gas separation processes. Sep Purif Technol 70:243–256. https://doi.org/10.1016/j.seppur.2009.10.004
105. Ryckebosch E, Drouillon M, Vervaeren H (2011) Techniques for transformation of biogas to biomethane. Biomass Bioenerg 35:1633–1645. https://doi.org/10.1016/j.biombioe.2011.02.033
106. Sato A, Terashima H, Takei Y (2014) Monolith adsorbent and method and apparatus for adsorbing samples with the same. US201408795410B2
107. Saxena R, Singh VK, Kumar EA (2014) Carbon dioxide capture and sequestration by adsorption on activated carbon. Energy Proc 54:320–329. https://doi.org/10.1016/j.egypro.2014.07.275
108. Sevilla M, Fuertes AB, Demir-cakan R (2013) Applications of hydrothermal carbon in modern nanotechnology. In: Titirici MM (ed), Sustainable carbon materials from hydrothermal processes. Wiley, pp 213–294
109. Shah IK, Pre P, Alappat BJ (2013) Steam regeneration of adsorbents: an experimental and technical review. Chem Sci Trans 2:1078–1088. https://doi.org/10.7598/cst2013.545
110. Shahin O, Saka C (2013) Preparation and characterization of activated carbon from acorn shell by physical activation with $H_2O$-$CO_2$ in two-step pretreatment. Bioresour Technol 136:163–168. https://doi.org/10.1016/j.biortech.2013.02.074
111. Shao P, Dal-cin M, Kumar A, Li H, Paul D (2012) Design and economics of a hybrid membrane—temperature swing adsorption process for upgrading biogas. J Memb Sci 413–414:17–28. https://doi.org/10.1016/j.memsci.2012.02.040
112. Shen C, Yu J, Li P, Grande CA, Rodrigues AE (2011) Capture of $CO_2$ from flue gas by vacuum pressure swing adsorption using activated carbon beads. Adsorption 17:179–188. https://doi.org/10.1007/s10450-010-9298-y

113. Shen Y, Linville JL, Urgun-demirtas M, Mintz MM, Snyder SW (2015) An overview of biogas production and utilization at full-scale wastewater treatment plants (WWTPs ) in the United States: challenges and opportunities towards energy-neutral WWTPs. Renew Sustain Energy Rev 50:346–362. https://doi.org/10.1016/j.rser.2015.04.129
114. Silva LA da, Borges SMS, Paulino PN, Fraga MA, Oliva ST de, Marchetti SG, Rangel M do C (2017) Methylene blue oxidation over iron oxide supported on activated carbon derived from peanut hulls. Catal Today 289:237–248. https://doi.org/10.1016/j.cattod.2016.11.036
115. Singhal S, Agarwal S, Arora S, Sharma P, Singhal N (2017) Upgrading techniques for transformation of biogas to bio-CNG : a review. Int J Energy Res. https://doi.org/10.1002/er.3719
116. Song M, Jin B, Xiao R, Yang L, Wu Y, Zhong Z, Huang Y (2013) The comparison of two activation techniques to prepare activated carbon from corn cob. Biomass Bioenerg 48:250–256. https://doi.org/10.1016/j.biombioe.2012.11.007
117. Sricharoenchaikul V, Pechyen C, Aht-ong D (2008) Preparation and characterization of activated carbon from the pyrolysis of physic nut (Jatropha curcas L.) waste. Energy Fuels 22:31–37. https://doi.org/10.1021/ef700285u
118. Suhas, Gupta VK, Carrott PJM, Singh R, Chaudhary M, Kushwaha S (2016) Cellulose: a review as natural, modified and activated carbon adsorbent. Bioresour Technol 216:1066–1076. Doi: https://doi.org/10.1016/j.biortech.2016.05.106
119. Sun Q, Li H, Yan J, Liu L, Yu Z, Yu X (2015) Selection of appropriate biogas upgrading technology-a review of biogas cleaning, upgrading and utilisation. Renew Sustain Energy Rev 51:521–532. https://doi.org/10.1016/j.rser.2015.06.029
120. Sun Y, Li H, Li G, Gao B, Yue Q, Li X (2016) Characterization and ciprofloxacin adsorption properties of activated carbons prepared from biomass wastes by $H_3PO_4$ activation. Bioresour Technol 217:239–244. https://doi.org/10.1016/j.biortech.2016.03.047
121. Techsci Research (2016) Global Activated Carbon Market Competition Forecast and Opportunities, 2011–2021. https://www.techsciresearch.com/report/global-activated-carbon-market-by-type-powdered-activated-carbon-granular-activated-carbon-and-others-by-raw-material-wood-coconut-shells-coal-and-others-by-application-competition-forecast-and-opportunities-2011-202. Accessed 21st Aug 2017
122. Thakur GSRS, Kaistha N (2011) Process intensification in PSA processes for upgrading synthetic landfill and lean natural gases. Adsorption 17:121–133. https://doi.org/10.1007/s10450-010-9302-6
123. Tounsadi H, Khalidi A, Machrouhi A, Farnane M, Elmoubarki R, Elhalil A, Sadiq M, Barka N (2016) Highly efficient activated carbon from Glebionis coronaria L. biomass: Optimization of preparation conditions and heavy metals removal using experimental design approach. J Environ Chem Eng 4:4549–4564. https://doi.org/10.1016/j.jece.2016.10.020
124. TransCanada Pipelines, Canadian Pipelines, Canada Alliance, Canadian Mainline, North Baja, Mainline Tariff (2016) Gas Quality Specifications TransCanada and other pipelines, pp 1–2
125. Tromly K (2001) Renewable energy: an overview https://www.nrel.gov/docs/fy01osti/27955.pdf. Accessed 24th Dec 2016
126. Turco M (2016) Processes of biogas production : anaerobic digestion and thermal gasification. In: Treatment of biogas for feeding high temperature fuel cells. pp 1–29. Doi: https://doi.org/10.1007/978-3-319-03215-3
127. TUV (2012) Biogas To Biomethane Technology Review, 1st edn. Intelligent Energy Europe, Vienna
128. US Energy (2017a) AEO 2017 with projections to 2050. Annu Energy Outlook 2017 1–127
129. US Energy (2017b) Renewable natural gas (biomethane) production. https://www.afdc.energy.gov/fuels/natural_gas_renewable.html. Accessed 24th Jul 2017
130. Vemula RR, Kothare MV, Sircar S (2015) Anatomy of a rapid pressure swing adsorption process performance. AIChE 61:2008–2015. https://doi.org/10.1002/aic.14779
131. Venkatesan S (2013) Adsorption. In: Ramaswamy S, Huang HJ, Ramarao BV (eds), Separation and purification technologies in biorefineries. Shell Global Solutions International, pp 103–148

132. Wang D, Ai P, Yu L, Tan Z, Zhang Y (2015) Comparing the hydrolysis and biogas production performance of alkali and acid pretreatments of rice straw using two-stage anaerobic fermentation. Biosyst Eng 132:47–55. https://doi.org/10.1016/j.biosystemseng.2015.02.007
133. Wang S, Ruan Y, Zhou W, Li Z, Wu J, Liu D (2018) Net energy analysis of small-scale biogas self-supply anaerobic digestion system operated at psychrophilic to thermophilic conditions. J Clean Prod 174:226–236. https://doi.org/10.1016/j.jclepro.2017.10.186
134. West Virginia University (2015) Production and application of highly porous activated carbon (HPAC) materials from bioenergy crops https://portal.nifa.usda.gov/web/crisprojectpages/1007044-production-and-application-of-highly-porous-activated-carbon-hpac-materials-from-bioenergy-crops.html. Accessed 22nd Aug 2017
135. Wheeldon I, Caners C, Karan K, Peppley B (2007) Utilization of biogas generated from ontario wastewater treatment plants in solid oxide fuel Cell systems: a process modeling study. Int J Green Energy 4:221–231. https://doi.org/10.1080/15435070601015585
136. Xebec Adsorption Inc. (2015) Xebec Methane Upgrading https://www.xebecinc.com/pdf/BGX_biogas-brochure_2015_12_single.pdf. Accessed 6th Oct 2016
137. Xu J, Chen L, Qu H, Jiao Y, Xie J, Xing G (2014) Preparation and characterization of activated carbon from reedy grass leaves by chemical activation with $H_3PO_4$. Appl Surf Sci 320:674–680. https://doi.org/10.1016/j.apsusc.2014.08.178
138. Yang RT (2003) Adsorbents: fundamentals and applications, 1st edn. Wiley, New Jersey
139. Yorgun S, Yıldız D (2015) Preparation and characterization of activated carbons from Paulownia wood by chemical activation with $H_3PO_4$. J Taiwan Inst Chem Eng 53:122–131. https://doi.org/10.1016/j.jtice.2015.02.032
140. Yousef AMI, Eldrainy YA, El-maghlany WM, Attia A (2016) Upgrading biogas by a low-temperature $CO_2$ removal technique. Alexandria Eng J 55:1143–1150. https://doi.org/10.1016/j.aej.2016.03.026
141. Zhang Z, Yan Y, Zhang L, Chen Y, Ran J, Pu G, Qin C (2014) Theoretical study on $CO_2$ absorption from biogas by membrane contactors: effect of operating parameters. Ind Eng Chem Res 53:14075–14083. https://doi.org/10.1021/ie502830k

# $CO_2$ Sequestration Through Phytoremediation Techniques with Special Emphasis on Urban Forestry to Mitigate Climate Change Impact

**M. Govindaraju, J. Fowmitha Banu, S. Senthamil selvi, and Malti Goel**

**Abstract** Concentration of carbon dioxide in the Earth's atmosphere reached high, according to the World Meteorological Organization (WMO) Report and became parts per million in December 2018. The main causes of global warming are emission of $CO_2$ and other Green House Gases (GHG), and it is due to anthropogenic activities. To reduce the impact, sequestrating $CO_2$ through photosynthetic process is more effective. This process known as phytoremediation has been used to clean up greenhouse gases, heavy metals, pesticides, xenobiotics, organic compounds, and toxic aromatic pollutants. It is a cost-effective technique to mitigate global warming. Plants capture atmospheric $CO_2$ and convert it into organic compounds. The $CO_2$ sequestration of plants through the leaf and extensive root systems is useful for reduction of $CO_2$ by the potential assimilating capacity plants. Different plant species have different capacity of carbon sequestration potential. Plants that are identified and grown in and around the urban and industrial areas can reduce the air pollutants. They act as a wind breaker to minimize the transportation of air pollutants; the article has emphasis on the development of greenbelt model in urban forestry by planting more $CO_2$ assimilating plants. The benefits of developing this model will not only sequester $CO_2$, but also improve micro-climate of the urban area and increase urban biodiversity as best practices for sustainable environment and urban development. Three case studies are presented.

**Keywords** Phytoremediation · Carbon dioxide sequestration · Urban forestry · Climate change · Assimilating plants

M. Govindaraju (✉) · J. Fowmitha Banu
Department of Environmental Biotechnology, School of Environmental Science, Centre for Climate Change Research, Bharathidasan University, Tiruchirappalli, Tamil Nadu 620024, India
e-mail: mgrasu@bdu.ac.in

S. Senthamil selvi
Department of Botany, School of Life Science, Bharathidasan University, Tiruchirappalli, Tamil Nadu 620024, India

M. Goel
Science and Technology Solution for Sustainable Energy Future, Climate Change Research Institute, New Delhi, India

 
M. Goel et al. (eds.), *Climate Change and Green Chemistry of $CO_2$ Sequestration*, Green Energy and Technology, https://doi.org/10.1007/978-981-16-0029-6_15

# 1 Introduction

Anthropogenic activities, particularly the burning of fossil fuels such as coal, oil, and gas, have caused a rapid increase in the concentration of $CO_2$ in the atmosphere. Carbon dioxide circulates and accumulates in the atmosphere, oceans, and land; these $CO_2$ pools are a natural part of the carbon cycle. But, the natural land and ocean pools are unable to absorb all of the anthropogenic $CO_2$ currently being emitted. According to IPCC [8], the concentration of different greenhouse gases especially carbon dioxide has increased from 1990s. Keeping this atmospheric $CO_2$ concentration in a desired limit is real big goal to humanity. Technically and economically feasible strategies are needed to mitigate the consequences of increased atmospheric $CO_2$. There are two main possible approaches for mitigating these greenhouse gases, first is to reduce emissions, and second is to capture $CO_2$ and store it, i.e., sequestration [2]. The term $CO_2$ sequestration is the process of removing $CO_2$ from the atmosphere and depositing it into underground reservoirs. Plants can also play important roles as $CO_2$ sinks. By capturing atmospheric $CO_2$ through photosynthesis, plants store large amounts of $CO_2$ in above- and belowground biomass. In urban forestry approach, the amount of carbon stored in soils and plants is increased. It helps to restore degraded soils, enhances biomass production, and purifies surface and groundwaters. By improvement in techniques, plants can play important roles as $CO_2$ sinks called as phytoremediation also called phyto C-sequestration.

# 2 Mode of $CO_2$ Sequestration by Plants

By capturing atmospheric $CO_2$ through photosynthesis, plants store large amounts of $CO_2$ [12] in above- and belowground biomass [7]. Carbon is the major component of all cellular life forms,trees utilize carbon as a building material with which to form trunks, roots, stems, branches, and leaves. Trees remove (sequester carbon from the atmosphere through photosynthesis [4]. The process of photosynthesis permits green plants to uptake $CO_2$ from the atmosphere and convert it into organic carbon as they grow and in turn organic carbon is converted back to $CO_2$ when it is eaten or decomposed known as the process of respiration. During photosynthesis, atmospheric $CO_2$ enters the leaf through surface pores, combines with water, and is converted into cellulose, sugars, and other materials in a chemical reaction catalyzed by sunlight. Most of these materials become fixed as wood, although some are respired back to $CO_2$ or used to make leaves that are eventually shed by the tree [17].

Stomata are pores present in the plant epidermis and function as gateway by linking the intercellular gas to the external environment. Two guard cells surround each stomatal pore, and changes in turgor pressure of the guard cells regulate the size of the pore aperture. An increase in guard cell turgor results in stomatal opening, whereas a reduction in turgor leads to stomatal closure. Stomata can control $CO_2$ assimilation and limit excessive water loss. The $CO_2$ concentration inside the leaves changes as a result of photosynthesis and transpiration; low $CO_2$ induces stomatal opening while high $CO_2$ induces stomatal closure. Furthermore, the continuing rise in atmospheric $CO_2$ is predicted to interfere with the regulation of stomatal conductance and to have diverse effects on the productivity of the plant [13]. Trees play a significant role in the estimation of tree biomass and carbon of a forest as the trees which are greater in diameter (mature trees) tend to accumulate more biomass than the trees which have fewer diameters (Young trees). Almost 74% of the total aboveground carbon was sequestered in woody biomass of trees with diameter between 20 and 49.99 cm, and the rate of sequestration was higher in trees of diameter classes (20–29.99) cm and (10–19.99) cm with values 0.95 tons/ha/year and 0.62 tons/ha/year, respectively; accounting 71% of the annual sequestration rate of the forest and aboveground biomass was greater in the 40–60 cm diameter class in plantation forest and in 60–80 cm dbh class in natural forest [14].

## 3 Model Greenbelt in Urban Ecosystem

Greenbelt (GB) development is having multiplicity of objectives encompassing the micro-level air pollutant abatement to enhancement of socioeconomic value of the region. The plants play an important role in reducing air pollution and climate change. The idea of greenbelt development in industrial and urban areas has acquired much attention recently. The concept has been developed as pollution mitigation tool. Plants act as filters to reduce air pollution by absorption, adsorption, detoxification, accumulation, or metabolization without damage or decline in growth. [9]. Identification of suitable plant species for greenbelts is very important. Greenbelt needs a suitable scientific design for powerful reduction in air pollution. Some of the design parameters of greenbelts are tree height, width of greenbelt, distance of greenbelt from the source, visibility, etc. The species used in greenbelts must be pollutant tolerant. Capacity of air pollution tolerance varies from species to species. Since the suitability of a species to be used in greenbelt may be site specific, some of the characteristics of an ideal plant for greenbelt are fast growth rate for quick development of a canopy,strong branches to combat storms; large leaf size for enhanced holding of pollutants; dense foliage for better entrapment; relatively good tolerance to insects and diseases, tolerance to soil compaction and nutrient stress, and long life span for extended longevity of the greenbelt itself. Therefore, greenbelt model would prove more suitable for urban forestry which is very much essential of these days in the effort of sustainable environment [6, [20]. The Green India Mission (GIM), one of the missions under India's National Action Plan on Climate Change (NAPCC), hopes to sequester 4.3 crore tons of greenhouse gases each year by 2020 and absorbs 6.35% of annual emissions. This action plan mentions about enhancing tree cover in over 2 lakh ha in urban and peri-urban areas. It has been estimated that about 2.21% of the carbon stocks are stored in through urban tree cover, and the mission will promote greening forest patches threatened by expanding urban development, diffused planting and planting on institutional lands [1].

## 4 Significance of Urban Forestry and Phyto C-Sequestration

'Urban forest' is the collection of trees, shrubs and green space that is present within the urban forests and can be defined as networks or systems comprising all woodlands, groups of trees, and individual trees located in urban and peri-urban areas; they include, therefore, forests, street trees, trees in parks and gardens, and trees in derelict corners [5]. Urban forests are the backbone of the green infrastructure, linking rural and urban areas, and enrich a city's environmental quality. The biggest challenge for urban development is to maintain the green cover. Urban greening needs best management practices where it needs more opportunities for regeneration and enhancement of adapting best practices like greenbelts development. There is a difference in rural and urban infrastructure. Because of higher tree densities, rural forests sequester about twice as much $CO_2$ as urban forests per unit land area. The result of Wang et al. [22] showed that the weighted mean carbon concentration by biomass was approximately 48.15%. In this study, the carbon concentration of aboveground tree organs is ranked with descending order from living branch, bark, foliage, dead branch, and stem; and in the belowground, it is ranked as large roots, stumps, thick roots, medium root, and small roots.

The net save in carbon emissions that can be achieved by urban planting can be up to 18 kg $CO_2$/year per tree, and this benefit corresponds to that provided by 3–5 forest trees of similar size and health [4]. Beckett et al. [3] found that roadside trees capture more large-size particulate matter than trees not near the road. These effects have implications for air quality standards. The $CO_2$ is fixed as organic carbon accumulating in the form of biomass. The 73.59 tons of $CO_2$ are removed by trees planted on road sides of Vadodara city which represents 22% of the city's estimated total $CO_2$ production [10]. The study in Pune city estimated and figured that of carbon sequestered is 15,000 tons per year [21]. *Olea ferruginea* species has sequestrated 34.52 tons of carbon in its standing biomass, followed by *Pinus roxburghii* (20.7 tons) in Jammu and Kashmir [7]. *Ficus benghalensis* has sequestered 1333.44 kg/tree of $CO_2$ which is highest compared to other tree species. Total $CO_2$ sequestered from all trees in the Aurangabad City is 4161.203 tons [15]. In Chennai city, *Albizia saman* absorbed 204.13 Mg $CO_2$ for carbon sequestration and $CO_2$ absorption ranged from 385.46 to 3009.29 kg for one hectare/year [19]. Furthermore, roadside trees have high emotional and aesthetic value to residents and high ecological value to urban areas as part of our green infrastructure. However, in urban environments, trees face an array of man-made and natural stresses that may lead to degradation of urban forest and reduce their life spans compared to trees in rural areas or natural stands.

## 5 Benefits of Urban Forestry

The multitude of urban forestry advantages has been described such as social benefits, aesthetic and architectural benefits, climatic and physical benefits, and ecological and economic benefits. Furthermore, greenbelt model increases the benefits of urban forestry extensively [5], 19. The main ecological benefits are as follows:

- Improve soil conditions and prevent erosion
- Increased biodiversity will improve the microclimate
- Reduction of air pollutants and buffer noise
- $CO_2$ sequester carbon and mitigate climate change
- Mitigate local climate and build resilience
- Save energy through shading and cooling and grow wood for fuel
- Reduces the temperature in the built environment through a large canopy and evapotranspiration.
- Provide more natural and green space which is useful for public
- Improve the physical and mental health of the public
- Mitigate storm water runoff and reduce flooding and erosion
- Provide opportunities for recreation and environmental education
- Provide shelter
- Enable infiltration and the reuse of wastewater/groundwater recharge
- Provide recreation places like parks for outdoor interaction.

Some of the case studies on identification of high $CO_2$ assimilating plant species are described as follows:

## 6 Case Study I

### *6.1 Study Area: Neyveli Lignite Corporation Campus (NLC), Cuddalore, Tamil Nadu*

It is an integrated approach to develop a specialized greenbelt model to minimize $CO_2$ emission for Neyveli Lignite Corporation Campus (NLC), Cuddalore, Tamil Nadu, India. Neyveli is an industrial area consists of thermal power stations I and II. There are 30 different species having identical age which were selected. The plants involved in more $CO_2$ assimilation are identified by various characteristics such as tree height, diameter at breast height, stomatal density, conductance, total chlorophyll, carotenoid, ascorbic acid, relative water content, pH of leaf, leaf structure, leaf size and Rubisco content, and sulfur of the leaf samples were tested to assess the air pollution tolerance index (APTI) and anticipated performance index (API) of the tree species. Another important estimation was Rubisco protein analysis. It is the most important enzyme in photosynthesis. Plants can trap the dust particle from the wind to reduce velocity for which they should have compound leaves with thick epidermis and waxy cuticle. *Polyalthia longifolia* Sonn., *Casuarina equisetifolia, Bambusa arundinacea,* and *Eucalyptus* hybrid species were added as wind breakers, *Bougunvillea spectabillis* species can be grown within the greenbelt to maintain the soil moisture and air quality. Among those species which were involved for analysis, *Mangifera indica*is the most tolerant to grow around the thermal power station followed by *Mimusops elengi, Ficus benghalensis, Anacardium Occidentale, Manilkara zapota, Ficus religiosa and Artocarpus heterophyllus* is assessed to be very good performer. The plant species like *Mangifera indica and Azardirachta indica* were selected for the present study based on APTI and API value. The result of this study *Mangifera indica* has been identified as pollution-tolerant species and has high potential for $CO_2$ assimilation with excellent and judged as a keystone species. *Azardirachta indica* has been coming under the rate of 694 $\mu$ mol$^{-2}$ s$^{-1}$. Hence, those species which were suggested for the proposed greenbelt will be effective $CO_2$ sink in the urban and industrial areas.

# 7 Case Study II

## *7.1 Study Area: Pachamalai Reserve Forest, Tamil Nadu*

This study is on the identification and evaluation of novel plant species for $CO_2$ sequestration in different altitude ranges of Pachamalai Reserve Forest. It is located in the Tiruchirapalli, Salem and Perambalur district of Tamil Nadu. The elevation is 1060 m above the mean sea level. The density of trees in different altitude was ranged between 62 and 115 trees. The maximum number of trees was identified in the altitude 630 and 530 m. If the height increases, the number of trees decreases. The amount of carbon dioxide is stored in trees estimated by two ways: destruction method and non-destruction method. In non-destruction method, carbon storage is estimated by allometric equations that use several parameters to calculate tree biomass: diameter at breast height, tree height, tree condition, wood density, and moisture content. The amount of $CO_2$ storage is estimated among the tree species which are having the above said characteristics. The total biomass estimated in Pachamalai Reserve Forest is 1679.29 tons and the total amount of $CO_2$ sequestration is 3081.41 tons. The plant species like *Ficus religiosa, Macaranga peltata, Anacadium occidentale, Artocarpus heterophyllus,Celtis Phillipensis, Ficus benghalensis, Mangnifera indica, Margaritaria indica, Pterocarpus masrsupium, Santalum album, Sterculiaurens, Tectonagrandis, and Wrightia tintoria* has more potential to sequestrate more $CO_2$ which is present in the altitude of 580–830 m. The rate of carbon dioxide sequestration is different between trees based on several factors such as age of tree, soil nutrition, speed of growth, floral diversity composition, type of soil, type of forest, forest elevation, size of tree and root leaf abscission, climatic factor, etc. By planting above said trees in urban areas will improve the quality of air and combat sequestrate more $CO_2$ to the climate change impact.

# 8 Case Study III

## *8.1 Study Area: Delhi Ridge, Delhi State*

Delhi the capital city of India faces high level of air pollution during most months of the year. Major sources of air pollution in Delhi are automobiles; construction; thermal power plants, and industry, with automobiles having the largest share. Changing land use pattern and biomass burning around the city is also intensifying air pollution. The mega city is divided in two main regions as Yamuna flood plains and Delhi Ridge, and the ridge constitutes 6% of Delhi's geographical area and acts as the green lungs of Delhi. With forest area of 178 $km^2$, there are fourteen major forests and in addition 14,500 big and small parks. Tree cover in avenues is estimated to be 98 $km^2$. From a comprehensive study of biomass cover in Delhi, total carbon pool in forests, gardens, parks, and avenues is estimated as 3.1 million tons in a year [16]. Urban forest has a critical role in helping combat increasing levels of atmospheric $CO_2$, as it sequesters $CO_2$ as well as affects the emissions of $CO_2$ from urban areas. Over the years, shrinking of Delhi Ridge area is seen owing to illegal encroachment for construction of buildings and roads. A study of land use changes in Delhi Ridge area estimated the potential for release of as much as 91.24 tons $CO_2$ into atmosphere by conversion of one hectare of protected forest to unprotected forest [11].

Augmentation of green cover was first envisaged in National Forest Policy 1988. 'Delhi Preservation of Trees Act' was enacted in 1994 for preservation of trees. To coordinate the management of parks/gardens in NCT, the Delhi State Government has formed a society, namely Delhi Parks and Gardens Society under the Department of Environment of Delhi Government. From a field mapping of 36 parks in East Delhi [18], it was concluded that 18 species of trees have significant potential for carbon sequestration. The species for plantation in Delhi Ridge for carbon sequestration have been identified as follows: *Acacia leucophloea, Acacia, senegal, Azadirachta indica, Balanites aegyptiaca, Butea, monosperma, Holoptelea integrifolia, Flacourtia indica, Maytenus senegalensis, Prosopis cineraria, Cappais decidua, Ziziphus nummularia, Calotropis procera, Grewia tenax, Aristida, Crotolaria burhia, Eragrostis, and heteropogon contortus.* Indian Government's concern to address climate change is reflected in New Forest Policy of India 2019 which emphasizes further promotion of trees outside forests and urban greens as natural sink of carbon dioxide. Six biodiversity parks have been created in ridge area for improvement of ecosystem. Urban forestry can be a double-edged tool to enhance $CO_2$ sequestration and ameliorate air pollution in cities for clean air. There is a need to have a dynamic urban forestry model in areas that are already exceeding air quality index in Delhi and work toward protection of Delhi Ridge by massive plantation.

## 9 Summary

Urban areas are generally warmer than their surrounding rural environments due to high proportion of imperious surface and high levels of fossil fuels combustion. Therefore, impacts of heat may be exacerbated in urban areas. Urban forests can play an important role mitigating the impacts of climate change by reducing $CO_2$ in urban areas. Urban forests can reduce the levels of atmospheric $CO_2$ through sequestration and reducing $CO_2$ emission by conserving energy used for heating and cooling. Tree growth may be subjected to environmental gradients associated with elevation, providing convenient scenarios to investigate the potential response of forest growth. The carbon dioxide sequestration in trees is achieved by several ways. *Mangifera indica and Mimus opselengi* have been shown to be good performer in $CO_2$ assimilation based on the case study I. The amount of carbon dioxide storage is estimated among ten species which are having the above said characteristics. *Ficus religiosa, Macaranga peltata, and anacadium occidentale* etc. are the major plants, and case study II has more capacity to sequestrate more $CO_2$. These plants in greenbelt will provide numerous advantages. The benefits of developing this greenbelt will improve micro-climate of the urban area increasing urban biodiversity and more carbon dioxide sequestration. Strategic planning and framework of execution would facilitate successful implementation of urban forestry concept, strong collaboration is must from local government and administration, and NGOs, urban planners, agronomist, institutions, business sector, and local communities would improve the environmental quality in urban ecosystem.

**Acknowledgements** The authors are thankful to the UGC-SAP, DST-FIST, and MoEF & CC, New Delhi, for their financial support to carry out the work.

## References

1. Arya A, Negi SS, Kathota JC, Patel AN, Kalubarme MH, Garg JK (2017) Carbon sequestration analysis of dominant tree species using geo-informatics technology in Gujarat State (INDIA). Int J Environ Geoinformatics 4(2):79–93
2. Asha A, Juwarkar AA, Varghese AO, Singh SK, Aher VV, Thawale PR (2011) Carbon sequestration potential in above ground biomass of natural reserve forest of Central India,(2011). Int J Agric Res Rev 1(2):80–86
3. Beckett KP, Freer Smith PH, Taylor G (2000) Effective tree species for local air quality management. Arboricultural J 26(1):12–19
4. Escobedo FJ, Kroeger T, Wagner JE (2011) Urban forests and pollution mitigation: Analyzing ecosystem services and disservices. Environ pollut 159(8–9):2078–2087
5. FAO (2016) Guidelines on urban and peri-urban forestry
6. Govindaraju M, Ganeshkumar RS, Muthukumaran VR, Visvanathan P (2012) Identification and evaluation of air-pollution-tolerant plants around lignite-based thermal power station for greenbelt development. Environ Sci Pollut Res 19(4):1210–1223
7. Gupta B, Sharma S (2014) Estimation of biomass and carbon sequestration of trees in informally protected areas of Rajouri, J & K. Int Res J Environ Sci 3(6):56–61

8. IPCC (2001) Intergovermental panel for climate. In: Climate change: the scientific basis. Cambridge University Press, Cambridge, UK
9. Kapoor CS, Chittora AK (2016) Efficient control of air pollution through plants a cost effective alternatives. J Climatol Weather Forecast 4(184):2
10. Kiran GS, Kinnary S (2011) Carbon sequestration by urban trees on roadsides of Vadodara city. Int J Eng Sci Technol 3(4):3066–3070
11. Kumari P, Nema AK (2017) Impact of urbanization on climate change in Delhi NCR due to land use changes. In: ASCE India conference 2017. Doi: https://doi.org/10.1061/978078448 2032.033
12. Maiti R, Rodriguez HG, Kumari ChA (2015) Trees and shrubs with high carbon fixation/concentration. For Res S 1:2
13. Negi J, Hashimoto-Sugimoto M, Kusumi K, Iba K (2013) New approaches to the biology of stomatal guard cells. Plant Cell Physiol 55 (2):241–250
14. Pant H, Tewari A (2013) Carbon sequestration potential of Chir Pine (*Pinusroxburghii. Sarg*) forest on two contrasting aspects in Kumaun Central Himalaya between 1650 and 1860 m elevation. Appl Ecol Environ Sci 1(6):110–112
15. Potdar Vishnu R, Patil Satish S (2016) Carbon Storage and Sequestration by trees in and around University Campus of Aurangabad City, Maharashtra. Int J Innov Res Sci Eng Technol 5(4):5459–5468
16. Sharma M, Dikshit O (2016) Comprehensive Study on Green House Gases (GHGs) in Delhi, Report submitted to DPCC. www.indiaenvironmentportal.org.in
17. Toochi EC (2018) Carbon sequestration: how much can forestry sequester $CO_2$ Forest Res EngInt 2(3):60
18. Tripathi M, Joshi H (2015) Carbon flow in Delhi urban forest ecosystems. Ann Biol Res 6(8):13–17
19. Udayakumar M, Selvam A, Sekar T (2018) Aboveground biomass stockpile and carbon sequestration potential of *Albiziasaman* in Chennai Metropolitan City, India. Plant 6(3):60
20. Urban Greening Guidelines (2014) Town and country planning organisation government of India ministry of urban development
21. Waran A, Patwardhan A (2001) Urban carbon burden of Pune City: a case study from India (Doctoral dissertation, Master thesis submitted to University of Pune
22. Wang X, Fu Y, Sun Y (2013) Carbon concentration variability of Larixolgensis in North-Eastern China. Adv J Food Sci Technol 5(5):627–632

# Effect of Anthropogenic Activities on the Rate of Soil $CO_2$ Flux in the Dipterocarpus Forest Ecosystem of Northeast India

**P. S. Yadava and Amrabati Thokchom**

**Abstract** The aim of the present study is to study the effect of anthropogenic activities, i.e. burning and logging on the rate of soil $CO_2$ flux and its relationship with the abiotic and the biotic factor in the Dipterocarpus forest of Northeast India. Rates of soil $CO_2$ flux were found to be the highest in burnt and the lowest in logged forest site. Seasonally soil $CO_2$ flux rate was found to be maximum in rainy season and minimum in winter season in all the study sites. Simple linear regression shows there is a strong positive relationship between soil $CO_2$ flux and soil moisture, temperature, and soil organic carbon. Soil $CO_2$ flux rate can be altered through different forest management practices such as harvesting, thinning, and burning to mitigate climate change.

**Keywords** Soil $CO_2$ flux · Control · Burnt · Logged · Forest management

## 1 Introduction

Soil is a major biosphere reservoir for carbon containing globally twice as much as the atmosphere and three times as much as vegetation [1]. With the increase in atmospheric greenhouse gases and global climate change, soil can either be a net source or a net sink of $CO_2$ in the future. This depends on which flux prevails the input of carbon into the soil due to plant growth or the output of carbon through soil $CO_2$ flux.

Soil $CO_2$ flux is a major source of carbon loss from an ecosystems and influences storage and cycling. Carbon flows through the forest, entering by photosynthesis and exiting to the atmosphere as $CO_2$ via $CO_2$ flux. The balance between carbon inputs (photosynthesis and biomass production) and losses (autotrophic $CO_2$ flux from trees and heterotrophic $CO_2$ flux from soil organisms) is directly related to its ability to function as a C sink for atmospheric $CO_2$.

P. S. Yadava (✉) · A. Thokchom
Waikhom Mani Girls' College, Thoubal 795138, Manipur, India
e-mail: yadava.ps1@gmail.com

 
M. Goel et al. (eds.), *Climate Change and Green Chemistry of $CO_2$ Sequestration*, Green Energy and Technology, https://doi.org/10.1007/978-981-16-0029-6_16

Soil $CO_2$ flux rates reflect the soil's capacity to sustain flora and fauna. It indicates soil health and can be thought of as an ecosystem's metabolism or "pulse" because measurements of soil $CO_2$ flux provide information about a forest's soil organic matter content and the rates that crucial nutrients become available for use by trees. Soil $CO_2$ flux rates increase with the increase of aboveground and belowground biomass as well as of as microbial activity.

Soil $CO_2$ flux is defined as the production of $CO_2$ by an organism and the plant parts in the soil. It is a composite process integrating two major carbon fluxes, i.e. autotrophic $CO_2$ flux of plants root and heterotrophic $CO_2$ flux through the soil microbial activities and is influenced by multitudes of environmental factors [2]. It provides the main carbon efflux from terrestrial ecosystems to the atmosphere and is therefore an important component of the global carbon cycle balance [3].

Small changes in soil $CO_2$ flux across large areas can produce a great effect on $CO_2$ atmospheric concentration and provide a potential positive feedback between increasing temperature and enhanced soil $CO_2$ flux that may ultimately accelerate global warming [3].

It is considered likely that global warming will increase soil $CO_2$ flux, releasing more $CO_2$ that will further exacerbate warming [4–6]. It is therefore critical to develop a better understanding of the controls on soil $CO_2$ flux and its components, particularly the decomposition of soil organic matter (SOM). This should promote a better assessment of the rate and direction of change of soil carbon, which is an essential component of developing and implementing policies and measures to mitigate climate change. Therefore, detailed information on soil $CO_2$ flux and its controlling factors is critical for constraining the ecosystem C-budget and for understanding the response of soils to changing land use and global climate change [1].

Soil $CO_2$ flux for a specific ecosystem can be characterized by its magnitude and its temporal and spatial availability. The rate of soil $CO_2$ flux is controlled primarily by the rate of $CO_2$ production by biota within the soil, but regulated by environmental factors influencing the $CO_2$ movement out of the soil [7]. Generally, soil temperature and soil moisture are considered the most influential environmental factors controlling soil $CO_2$ flux [2, 8–13]. These factors interact to affect the productivity of terrestrial ecosystems and the decomposition rate of soil organic matter, thereby driving the temporal variation of soil $CO_2$ flux [14]. The heterogeneity of vegetation coverage, root distribution, major environmental factors, and soil organic carbon contributes to the spatial variation of soil $CO_2$ flux in the different ecosystems.

Slash and burning are an old practice in the north-eastern region of the country in order to clear certain area in the forest for the cultivation of agriculture crops by local inhabitants to sustain their livelihood. Burning and logging practices influence both autotrophic and heterotrophic soil $CO_2$ flux through biomass removal and the opening of the forest canopy which alter the microclimatic properties of the soil as well as influence the quality and quantity of the substrate input into the soil. It affects the soil carbon dynamics by changing the forest floor energy balance, soil moisture content, nutrient availability, and biomass production.

Therefore, it is necessary to understand the effect of logging and slash and burnt treatments in the forest on the sustainability of the soil quality and to develop management strategies for preventing soil degradation for sustainable production. The major aim of the present study is to understand the effect of anthropogenic activities, i.e. burning and logging on the rate of soil $CO_2$ flux in the Dipterocarpus forest ecosystem of Northeast India and its interrelationships with abiotic and biotic variables.

## 2 Study Site

The study is located at 24° 22′ N latitude and 94° 29′ E longitude near Myanmar border in the Chandel District of Manipur state about 110 km from Imphal city. The forest vegetation is dominated by *Dipterocarpus tuberculatus*. Three experimental sites were earmarked randomly in the forest study site. Site I is a control protected forest site, and site II is recently logged one month before the onset of the experimental work. Site III is a slash and burnt area where Dipterocarpus is cut down and burnt one year before the onset of experiment. Thereafter, the experiment work was conducted from September 2017 in all the study sites.

Climate of the area is monsoonic with warm moist summer and cool dry winter. The mean maximum temperature varied from 24.15 °C (January) to 35.9 °C (May), and the main minimum temperature ranged from 4.5 °C (January) to 21.1 °C (August). The mean monthly rainfall ranged from 4.5 mm (January) to 196.0 mm (July). The mean annual rainfall is 1245 mm. The relative humidity of air varied between 61.5% (February) and 82.8% (July).

## 3 Material and Methods

Soil samples were collected from 0 to 10 cm soil depth at monthly interval from September 2017 to August 2018 from all the study sites. Three replicates were taken for analysis and measurement in the present study.

Subsamples of the soil were air-dried and used for determination of texture by international pipette method, moisture by gravimetric method, temperature by soil thermometer, and pH in 1:5 (soil:water) suspension by a pH metre, and soil organic carbon was estimated by Walkley Black Method [15]. Total soil nitrogen and total soil phosphorous were determined by FIAStar 5000 (flow injection analyser) (Foss Tecator AB, Sweden).

Soil $CO_2$ flux rate was measured by alkali absorption method [16]. Open-ended aluminium cylinders 13 cm diameters and 25 cm tall were inserted into the soil upto 15 cm depth. The surface area enclosed in each experimental cylinder was 132.7 $cm^2$. Five cylinders each of same size were used in each of the Dipterocarpus forest site I, II, and III out of which one cylinder each was used as blank in the forest sites. A 50 ml of 0.25 N NaOH solution was kept in 100 ml plastic vials, and the cylinder was made airtight with anchor grip and placed for 24 hrs to absorb $CO_2$ released. The carbon dioxide absorbed was then determined by titrating the NaOH solution with 0.25 N standard dilute HCl solution using phenolphthalein as an indicator. Carbon dioxide absorbed from the soil was calculated by using the formula.

$$\mathrm{mgCO_2} = \mathrm{V} \times \mathrm{N} \times 22$$

where V = volume of the acid; N = normality of the acid.

The data on the rate of soil $CO_2$ flux were analysed statistically using ANOVA and simple linear regression.

# 4 Result and Discussion

## *4.1 Physico-Chemical Characteristic of the Soil*

Soil texture, soil pH, soil moisture, soil temperature, organic carbon, total soil nitrogen, and phosphorous were determined on monthly basis from September 2017 to August 2018 and are set in Table 1.

The soil is laterite, sandy loam, and acidic in nature with a pH range from 5.7 to 6.4. The pH is slightly higher in burnt site III than that of site I and II which may be due to the addition of ash content in the soil on burning treatment. Losses of cations through rain water could be reason of acidic nature of the soil as region falls under high rainfall zone. The pH was comparatively higher in the slash and burnt site followed by logged and control Dipterocarpus forest site. Higher pH in slash and burnt and logged site may be due to the accumulation of higher organic matter in the soil and the addition of bases from the ash. [17] also have reported that application

**Table 1** Physico-chemical characteristic of soil in the Dipterocarpus forest sites (mean ± SE)

| Parameter | | Control site I | Logged sitec II | Slash and burnt site III |
|---|---|---|---|---|
| Soil texture | Sand | 70.26 ± 0.82 | 70.90 ± 0.57 | 70.00 ± 0.71 |
| | Silt | 17.20 ± 0.23 | 17.20 ± 0.27 | 18.20 ± 0.43 |
| | Clay | 12.54 ± 0.59 | 1190 ± 0.32 | 12.00 ± 0.28 |
| Soil Ph | | 5.62 ± 0.04 | 5.84 ± 0.03 | 6.50 ± 0.03 |
| Soil moisture (%) | | 13.85 ± 5.45 | 12.64 ± 4.27 | 13.63 ± 5.17 |
| Soil temperature (°C) | | 25.07 ± 4.52 | 27.67 ± 3.21 | 28.22 ± 5.95 |
| Bulk density (g $cm^{-3}$) | | 1.22 ± 0.01 | 1.20 ± 0.01 | 1.21 ± 0.01 |
| Soil organic carbon ($mgg^{-1}$) | | 6.42 ± 2.63 | 6.12 ± 2.62 | 8.13 ± 2.78 |
| Total soil nitrogen (%) | | 0.92 ± 0.06 | 1.25 ± 0.06 | 1.80 ± 0.08 |
| Total soil phosphorous (%) | | 0.39 ± 0.02 | 0.49 ± 0.03 | 0.58 ± 0.03 |

of wood ash raised the soil pH and degree of base saturation to the same extent in Myrtilhes type of forest and clear cut and burnt forest of Finland. The mean annual soil temperature was found to be the highest in the slash and burnt site as compared to the other two study sites. This may be due to the direct exposure of soil surface to the sunlight and also of the burning of the forest. Soil moisture was comparatively high in the control Dipterocarpus forest followed by the slash and burnt site and the logged sites depending upon the intact vegetation in the former and expose directly to the sunlight owing to slash and burn and deforested treatment in the later. Mac Donald and Huffman [18] mentioned the persistence of fire-induced soil water repellence, hence leading to loss of moisture content. Soil organic carbon was found to be the highest in the slash and burnt site followed by control forest and logged which may be due to addition of residual ash content in the soil on burning treatment. Total soil nitrogen and total soil phosphorous were also found to be the highest in burnt site followed by logged and control Dipterocarpus forest.

## 4.2 *Soil $CO_2$ Flux Rate in Control, Logged and Burnt Forest Sites ($mgCO_2m^{-2}\ hr^{-1}$)*

The rate of soil $CO_2$ flux ranged from 263.5 to 609.9 $mgCO_2m^{-2}\ hr^{-1}$ in control site I, 233.4 to 594.8 $mgCO_2m^{-2}\ hr^{-1}$ in logged site II and 308.7 to 700.2 $mgCO_2m^{-2}\ hr^{-1}$ in burnt forest site III in different months throughout the year. The soil $CO_2$ flux was recorded to be the highest in burnt site and the lowest in the logged site (Fig. 1).

In the present study, the high rate of soil $CO_2$ flux was recorded in burnt site III which shows that burning promotes the overall activities of microbes owing to partially burnt material on burning treatment. After the fire, large amounts of incompletely burnt organic material were available for rapid microbial decomposition that leads to the increase of soil $CO_2$ flux. Burning also leads to the increase input of nutrient into the soil by ash and relatively higher temperature in the soil which strongly enhanced the rate of decomposition of aboveground and belowground parts. Despite the loss of nutrients in fires, nutrient availability to plants typically increased

**Fig. 1** Soil $CO_2$ flux rate ($mgCO_2m^{-2}\ hr^{-1}$) under different forest sites in different months of the year, Manipur Northeast India

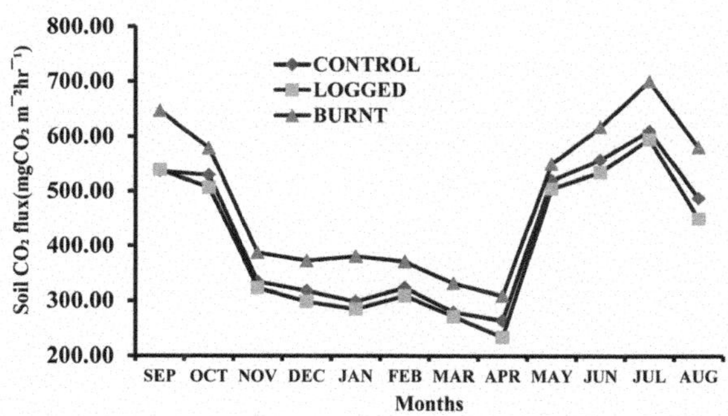

after fire as a result of heat-induced release of nutrients, reduced competition among plants, and perhaps sustained changes in soil conditions (such as temperature and water content). Forest fires leave behind large amounts of ash, and the concentrations of nutrients are unusually high in ash. Some of the nutrients in ash are water soluble or readily released by microbial activity, and this should provide a large supply of available nutrients for the recovery of the vegetation. The effect of fires on soil $CO_2$ flux changes with plant regeneration and soil microbial community. The post-fire increases in temperature, and substrate quality was reported to increase decomposition of humic materials. Forest fires typically destroy forest communities, reduce their growth, remove nutrients, and increase decomposition of soil organic matter from forests in the short term. Effects of forest fire on soil $CO_2$ flux (by changing root respiration and soil microbial respiration) are attracting increasing interest since soil $CO_2$ flux is highly related to the global carbon dynamics. The temporal variations in soil $CO_2$ flux result from variations in environmental conditions after fires. Moreover, following forest fires, the soil $CO_2$ flux on forest grounds becomes more sensitive to fluctuations in surface moisture conditions. Carbon loss may be accentuated if microbial activity increases after fires, owing to elevated soil temperatures and soil moisture.

Soil $CO_2$ flux was found to be minimum in the logging forest site which may be due to the removal of aboveground biomass on logging restricting the continual supply of litter to the soil and thus resulted in the decrease of the soil organic carbon which leads to the slower rate of soil $CO_2$ fluxes. When a tree was logged, the root respiration of the logged tree stopped, and then decomposition of the root began to take place. Therefore, the soil $CO_2$ flux rate decreased shortly after logging and eventually it may increase. The decrease in soil $CO_2$ flux rate in the logged forest may be due to decrease in heterotrophic respiration in the logged forest. Environmental conditions (i.e. soil compaction and reduced canopy cover) in logged forests changed as a result of logging disturbances. In the present study site, soil disturbance (e.g. removal of litter layer and extracted timbers) was observed after the logging operations. Soil compaction and decrease in the litter amount are known to cause reductions in the rate of heterotrophic respiration [19]. Second, the removal of aboveground biomass will lower or eliminate the flux of tree photosynthates to soils [20] and thus reduce the associated microbial respiration. Third, changes in the spatial and temporal variability in soil temperature and moisture may also affect plant and soil microbial activities and thus alter soil $CO_2$ flux [21]. Fourth, clear-cutting may also affect soil $CO_2$ flux by accelerating turnover rates of detrital and soil carbon pools, including roots, litter, forest floor organic matter, and mineral soil organic matter [22]. Root respiration usually ceases a short time after clear-cutting [20].

In our present study, we find that low soil $CO_2$ flux rate in the logged forest may be due to little organic material or soil microbial activity, limited biological activity and organic decomposition, limited nutrient recycling available to feed plants and microorganisms, and reduced root activity.

In the control forest site, optimal rate of soil $CO_2$ flux may be due to active microbial communities and productive root systems resulted from steady cycling of

**Table 2** Soil $CO_2$ flux rate in control site I, logged site II and burnt site III in Dipterocarpus forest ecosystem of Northeast India ($mgCO_2m^{-2}\ hr^{-1} \pm$ S.E.)

| Season | Control site-I | Logged site-II | Burnt site-III |
|---|---|---|---|
| Summer | 248.8 ± 2.07 | 228.2 ± 2.2 | 293.8 ± 1.66 |
| Rainy | 472.02 ± 1.96 | 444.1 ± 1.8 | 539.7 ± 2.58 |
| Winter | 215.5 ± 2.03 | 198.7 ± 1.9 | 271.9 ± 1.24 |
| Annual | 330.74 ± 2.16 | 308.36 ± 2.06 | 388.97 ± 1.96 |

nutrients availability for plant and microbial uptake. It may also be owing to longer-term balance between annual litter and carbon inputs and soil C loss in control site.

In the burning site, high rate of soil $CO_2$ flux may be due to short-term disturbance changes in factors affecting respiration, excess or imbalance of a specific nutrient, potential for brief depletion of site nutrients, result in leaching losses of nutrients, and reduce overall productivity over the long-term and rapid loss of soil organic matter.

There is a strong seasonality in the rate of soil $CO_2$ flux in all the study sites, and it was estimated to be the highest in the rainy season and the lowest in winter season in all the study sites (Table 2). Minimum rate of soil $CO_2$ flux is in winter season in all the study sites may be due to low moisture content of soil and temperature thereby inhibiting the microbial activity and rate of decomposition and thus leading to low $CO_2$ emission from the soil. However, the maximum rate of soil $CO_2$ flux was exhibited in the rainy season in all the study sites which may be due to burst mineralization of labile soil organic matter that has been accumulated during the dry period and which is available to microorganism after rewetting of the soil on the set of monsoon rains. The moist soil condition enhanced the physiological activities of soil microbes, and the root $CO_2$ flux resulted in the increase of the soil $CO_2$ flux rate. The similar seasonal pattern of soil $CO_2$ flux rates high in rainy and low rates during the winter cold and dry season has also been reported by [23] in temperate forest, [8] in subtropical forest.

## 5 Relationship Between Soil $CO_2$ Flux and Biotic and Abiotic Variables

The two important components of soil microclimate, i.e. soil temperature and moisture, also determine the magnitude of soil $CO_2$ flux, and therefore, simple linear regression was used to find out the relationship between biophysical variables and soil $CO_2$ flux (Figs. 2, 3 and 4).

The rate of soil $CO_2$ flux displayed a positive correlation with soil temperature in all the study sites (Fig. 2). 77–82% of variability in the soil $CO_2$ flux was due to the variability in the soil temperature in all the forest study sites. Significant difference was observed in the soil temperature between the forest types ($F = 13.7; p < 0.01$).

Soil $CO_2$ flux generally increases as soil temperature increases until it reaches a maximum temperature of 95–104 °F, at which point it begins to decline. On a smaller

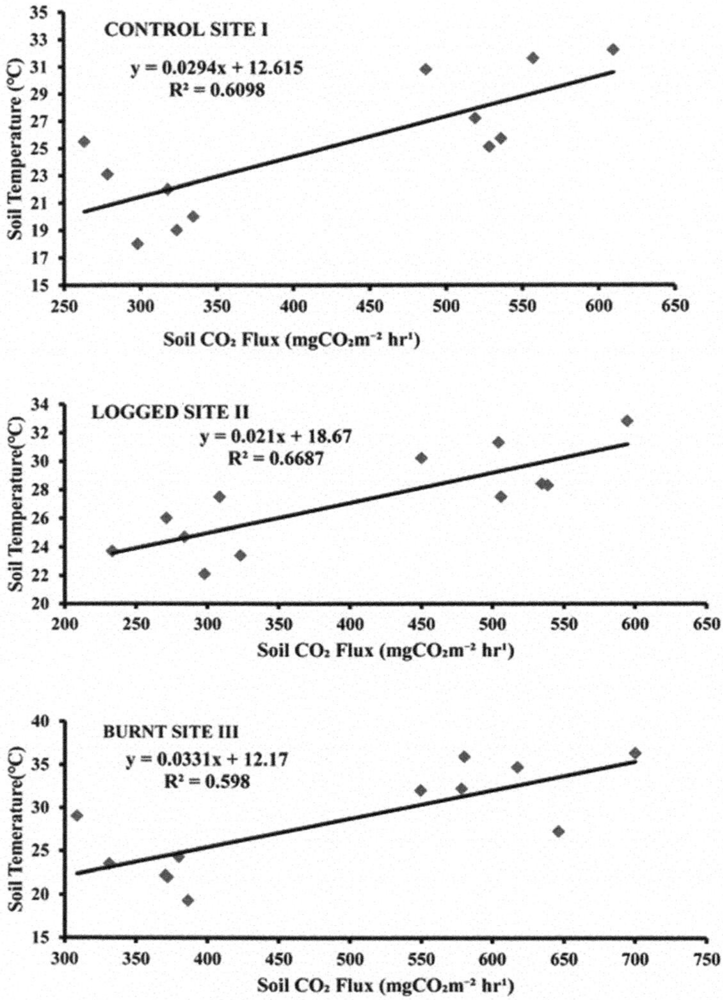


**Fig. 2** Relation of the soil temperature (°C) with the soil $CO_2$ flux ($mgCO_2m^{-2}$ $hr^{-1}$) in all the forest sites

scale, soil temperature may also influence soil $CO_2$ flux indirectly via its effects on dissolved organic matter and oxygen transport in the soil. These transport and diffusion are important for microbial metabolism and are influenced by water content of the soil. Degradation processes are rather influenced by increasing temperature than composition processes in the soil [5, 24]. Increasing degradation processes lead to increasing soil $CO_2$ flux, which results in decreasing organic matter content of the soil and consequently promotes the deterioration of the habitat. Several studies

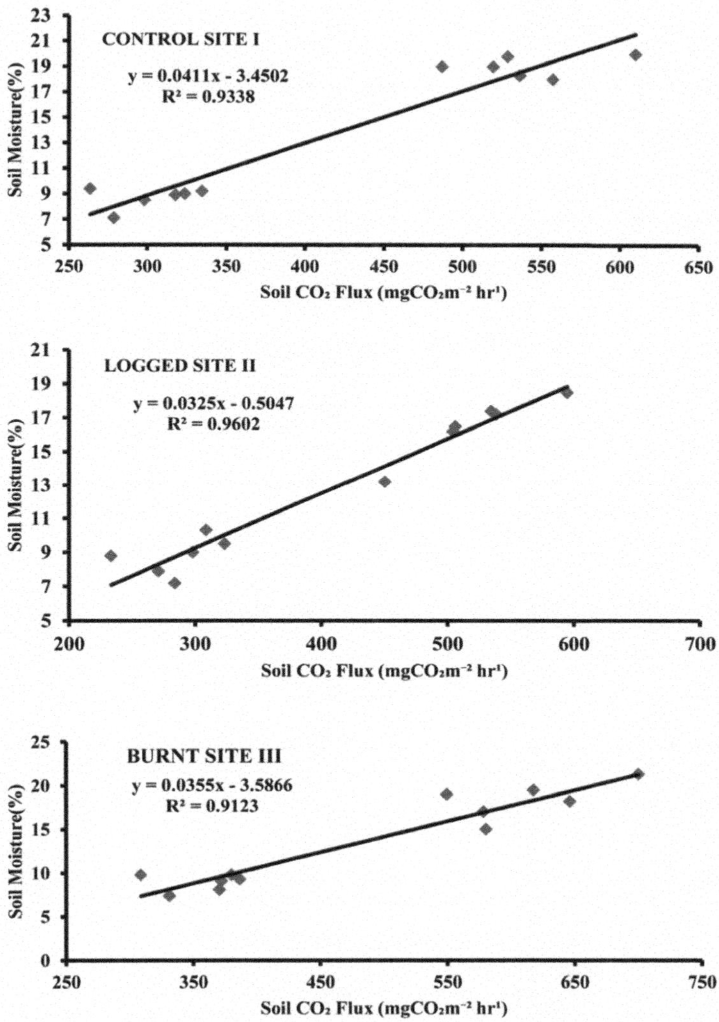


**Fig. 3** Relation of the soil moisture (%) with the soil $CO_2$ flux ($mgCO_2m^{-2}\ hr^{-1}$) in all forest sites

have shown a positive correlation between soil $CO_2$ flux and temperature forest ecosystems [25, 26].

There is a strong positive relationship between soil moisture and soil $CO_2$ flux in all the study sites (Fig. 3). Soil moisture caused a variation of 95–98% in the soil $CO_2$ flux. Significant difference was observed in the different forest sites on soil moisture content to a depth of 10 cm ($F = 12.67$; $p < 0.01$). In general, the highest rate of soil $CO_2$ flux during the rainy season in all the forest study sites is the effect of soil moisture on rewetting of the soil after dry periods of summer which causes a short

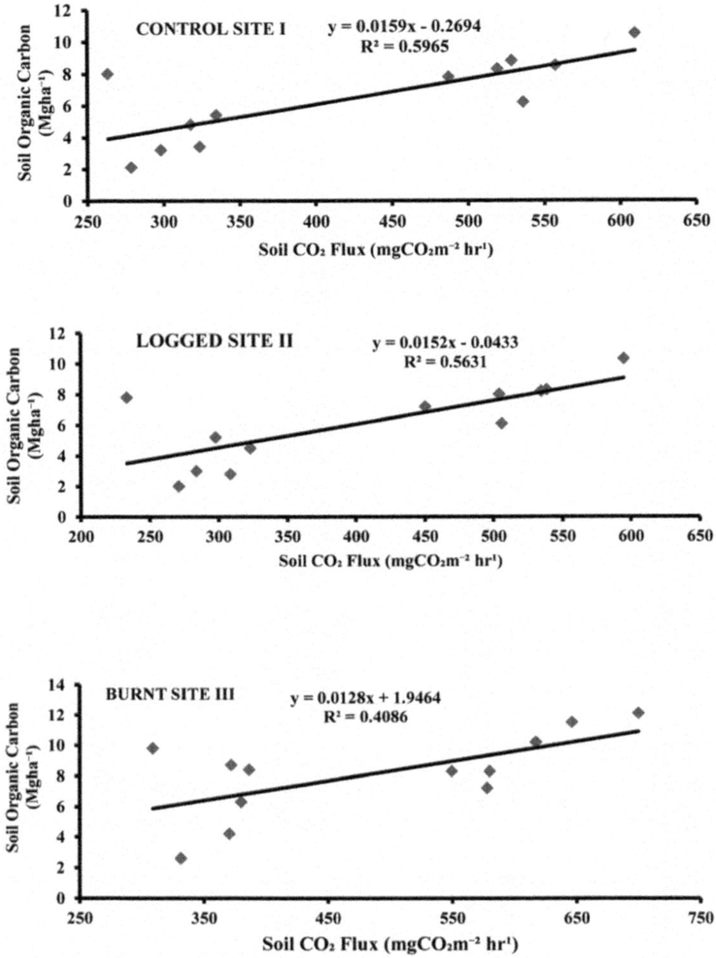


**Fig. 4** Relation of the soil organic carbon ($Mgha^{-1}$) with the soil $CO_2$ flux ($mgCO_2m^{-2}\ hr^{-1}$) in all the forest sites

lived but strong increase in soil $CO_2$ flux. Further it may led to burst mineralization of labile soil organic matter that has been accumulated during the dry period and is available to microorganism after rewetting of the soil. Sometime oxygen deficit as a result of too high soil moisture decreases activity of plant root and heterotrophic decomposition of soil organic matter [27]. Thus, soil moisture is another important component influencing soil $CO_2$ flux. In general, soil $CO_2$ flux is low under dry conditions, reaches a maximum in intermediate soil moisture levels, and decreases at high soil moisture levels when oxygen becomes limiting and anaerobic conditions dominate. The optimum water content is usually somewhere near field capacity, because the macropore spaces are mostly air-filled, thus facilitating oxygen diffusion, and the micropore spaces are mostly water-filled, enabling diffusion of dissolved organic matter. When soil water content exceeds optimum conditions, soil $CO_2$ flux slows due to oxygen deficits.

Strong positive relationship between the rate of soil $CO_2$ flux and soil organic carbon shows that 64–77% variability of soil $CO_2$ flux (Fig. 4) is due to the variability in soil organic carbon stock to the soil. Low concentration of soil carbon in the forest may be attributed to less litter production. Soil organic carbon concentration is strongly affected by litter thickness and litter accumulation in the soil and rate of litter decomposition. The concentrations of carbon in the soil are highly linked with micro-environment, litter dynamics, altitude, and type of vegetation. Release of nutrients through litter decomposition process is recognized as an important part of carbon released in the soil.

Soil organic carbon content is also influenced by human activities, since organic matter decomposition and impending soil erosion can be enhanced by timbering and soil disturbance. Values of soil $CO_2$ flux intensity provide input data for large-scale researches of carbon circulation, contribute to the revelation of interactions between the biosphere and atmosphere, and promote the drawing up of carbon balances from associations to global level. These values are also needed for making predictions and for modelling the global climatic change.

Thus, our study reveals that the rates of soil $CO_2$ flux in all the study sites are highly influenced by soil temperature, soil moisture, and soil organic carbon in all the study sites as shown by simple regression. The significant and positive correlation between soil $CO_2$ flux and soil moisture and soil temperature was reported in the forest ecosystems by several workers [8, 12]. The significant positive correlation between soil $CO_2$ flux and soil organic carbon in all Dipterocarpus forest sites shows that soil $CO_2$ flux rate is also influenced by the organic carbon present in the soil which was also reported by several workers [27, 28] and [2].

Thus, our finding in the present study shows that the rate of soil $CO_2$ flux was strongly influenced by the anthropogenic activities exhibiting the higher rate of $CO_2$ flux on burning treatment and low on logging forest. Besides this rate of soil $CO_2$ flux was also influenced seasonally with the maximum in the rainy season and minimum in the winter season. Thus, it also shows that the rate of soil $CO_2$ flux is also basically controlled by soil moisture, soil temperature, and soil organic carbon in the present Dipterocarpus forest ecosystem.

## 6 Soil $CO_2$ Flux and Climate Change

The most recent report of the Intergovernmental Panel on Climate Change (IPCC) indicates that the average global temperature will probably rise between 1.1 and 6.4 °C by 2090–2099, as compared to 1980–1999 temperatures, with the most likely rise being between 1.8 and 4.0 °C [29]. In the last century, considerable changes took place in the gas composition of the atmosphere due to natural processes and human activities. This has led to a rise in global temperature and high spatial and temporal variability. The idea that the Earth's climate is changing is now almost universally accepted in the scientific community [30], and even many scientists who dispute that climate change is anthropogenic are in agreement that it is happening [31]. With climate change happening, there will be effects on the environment, including the soil. Soil is a naturally occurring body that has been evolved owing to combined influence of climate and organisms acting on parent material as conditioned by topography over a period of time. Climate change and its hydrological consequences may result in the significant modification of soil conditions. Climate change impacts soil chemical, physical, and biological functions through a range of predicted global change drivers such as $CO_2$, N deposition, temperature, and rainfall.

The disruption of the global carbon cycle by human activity in both developed and developing countries is one of the key environmental issues facing human populations as we move into the twenty-first century. More recently, anthropogenic activities such as deforestation, agricultural practices, and the burning of fossil fuels have resulted in large shifts among carbon pools, particularly since the beginning of the Industrial Revolution [32]. Estimates of the current global carbon budget indicate atmospheric sources of ~1.6 Pg C $yr^{-1}$ via tropical deforestation and other land use changes and ~5.4 Pg C $yr^{-1}$via burning of fossils fuels [33]. These atmospheric sources are balanced by a ~2.0 Pg C $yr^{-1}$ oceanic sink and a~3.3 Pg C $yr^{-1}$ atmospheric sink and ~1.7 Pg C $yr^{-1}$ in terrestrial ecosystems sink [34].

Atmospheric $CO_2$ concentrations have been increasing in response to these disruptions to the carbon cycle [32]. This has raised concerns about potential global warming and, perhaps more seriously, about possible positive feedback effects that warming could have on further release of $CO_2$ from terrestrial carbon pools, particularly soils [35]. This is more than two times the amount of carbon in the atmosphere. Increased storage of carbon in world soils could help offset further anthropogenic emissions of $CO_2$, whereas a release could significantly exacerbate the atmospheric increases. The flux of carbon from soils to the atmosphere occurs primarily in the form of $CO_2$ and is the result of "soil $CO_2$ flux". Soil $CO_2$ flux represents the combined respiration of roots and soil micro and macroorganisms. Even a small change in soil $CO_2$ flux could thus equal or exceed the annual input of $CO_2$ to the atmosphere via land use changes and/or fossil fuels combustion and could significantly exacerbate—or mitigate—atmospheric increases of $CO_2$, with consequent feedbacks to climate change.

## 7 Soil $CO_2$ Flux and Forest Management Consideration

Soil $CO_2$ flux rates can be altered through silviculture and proactive forest management and aim for an optimal range for a given forest. High soil $CO_2$ flux rates do not necessarily mean healthier and more productive forests. Instead, high rates may indicate an unstable system and loss of organic matter due to site disturbances. Other factors that may degrade forest or soil health include a crowded understory, overstocked stand, or residual effects from fertilization. On the other hand, low $CO_2$ flux rates may be an indicative of poor soil structure and health or saturated soils.

Forest management practices can optimize above- and belowground productivity and C storage by increasing inputs and controlling losses of C in the forest. The extent of which soil $CO_2$ flux occurs affects soil health, soil carbon storage, and nutrient availability, ultimately impacts forest growth. Forest management actions can influence (increase or reduce) the rate of soil $CO_2$ flux, impacting soil health and forest productivity. Therefore, prudent forest management practices can help sustain healthy forest soil conditions and keep forests productive.

Forest management activities can alter soil conditions such as soil temperature, moisture, organic matter, chemical elements, and soil structure. As soil conditions change, populations of different microbes rise and fall as well. Because the health, vigour, and resiliency of the forest trees depend heavily upon a healthy, active, and diverse community of soil organisms, forest management strategies should carefully consider the impact of various treatments on soil health when developing management prescriptions. Forest can be managed by human activities of harvesting, thinning, and fire.

**Harvesting**: Forest harvesting can have a dramatic impact on soil physical and chemical properties and soil $CO_2$ flux due to tree removal, but also from the harvesting equipment itself which can disturb the site and compact the soil. Harvest methods, forest type, speed of regeneration (or replanting), and climate conditions will influence how much soil $CO_2$ flux is affected. Biomass removal as a result of harvesting generally increases soil temperature and reduces soil water content by increasing water evaporation at the soil surface. A large amount of forest litter and dying tree roots that decompose easily are typically left after a harvest, which can increase soil organic matter content. Proactive organic matter management can result in optimum soil $CO_2$ flux rates and increase soil carbon storage.

**Thinning**: Forest thinning alters many different site factors like soil temperature, moisture, and substrate availability, which influences respiration. Thinning partially removes trees from a stand to reduce competition, improve crop tree productivity, and reduce wildfire risk. Like forest harvesting, thinning decreases stand density and leaf area, increases light and nutrient availability, and alters soil temperature and moisture regimes. In addition, mechanical thinning may cause soil compaction from machinery, which decreases soil aeration and restricts root growth and microbial activities. Generally, when implemented under appropriate conditions, forest thinning results in healthier, more productive forests that balance soil nutrient cycling and soil $CO_2$ soil rates with aboveground productivity.

**Prescribed Fire**: Frequently, low intensity fires are a natural part of the forest landscape. Fire can result in rapid changes in soil $CO_2$ flux and soil organic matter, but can have beneficial long-term effects depending on fire frequency and intensity. Burning aboveground biomass creates charcoal. Charcoal is essentially SOM that is resistant to decomposition and persists in soils for decades to millennia, thus contributing to long-term soil C stores. Charcoal can favour microbial communities and can improve soil structure thereby enhancing or maintaining healthy respiration rates. Avoid intense fires as they rapidly consume soil organic matter which then affects soil nutrition, soil moisture holding capacity, and soil temperature.

Healthy soils are the foundation for healthy, productive, and resilient forests. Soil $CO_2$ flux represents yet another important indicator of forest and soil health. Management practices that optimize soil health are those that contribute biomass and organic matter to the soil, cause minimal soil disturbance, minimize soil erosion, conserve water, improve soil structure, enhance activity and species diversity of soil fauna, and improve nutrient cycling. Increasing plant carbon inputs and minimizing soil disturbances in order to regulate decomposition rates are the key to optimal soil $CO_2$ flux rates. Ultimately, soil health and ideal soil $CO_2$ rates can be achieved, and soil carbon can be built through adjustments to forest management practices.

## 8 Conclusions and Outlook

From our study, we find that the soil $CO_2$ flux was found to be the highest in burnt site and the lowest in the logged site. It shows that burning promotes the overall activities of microbes owing to partially burnt material on burning, whereas in logged forest, the root $CO_2$ flux of the logged forest stopped and then decomposition of the rhizome began to take place. Thus, it resulted in the decrease of the soil $CO_2$ flux rate. Seasonally, soil $CO_2$ flux was estimated to be the highest during rainy season followed by summer and the lowest winter in irrespective of human activities of burning and logging in all the forest sites. The rate of soil $CO_2$ flux has significant positive correlation with soil moisture, soil temperature, and soil organic carbon in all the study sites, and thereby it indicates that the soil moisture, soil temperature, and soil organic carbon are the main controlling factors on the rate of soil $CO_2$ flux in the present forest. Anthropogenic disturbance of forest ecosystems can induce large net of $CO_2$ emissions, both by directly releasing biomass carbon and by indirect $CO_2$ emissions from the accelerated decomposition of tree debris and soil organic matter. From our study, we can conclude that disturbances can create a large effect on the rate of soil $CO_2$ flux by increasing or decreasing which can ultimately lead to accelerate global warming. Thus, investigating how human activities, in particular, logging and fire affect $CO_2$ emissions from soils is important to help us understand how soil processes may respond to abrupt environmental changes. Thus, the information generated on the rate of soil $CO_2$ emission into the atmosphere could be utilized in the computation of overall carbon budget and cycling in the tropical forest ecosystems at national and global level to mitigate climate change.

## References


1. Han G, Zhou G, Xu Z, Yang Y, Liu J, Shi K (2007) Biotic and abiotic factors controlling the spatial and temporal variation of soil $CO_2$ flux in an agricultural ecosystem. Soil Bio Biochem 39:418–425
2. Zhou Z, Zhang Z, Zha T, Luo Z, Zheng J, Sun OJ (2013) Predicting soil $CO_2$ flux using carbon stock in roots, litter and soil organic matter in forests of Loess Plateau in China. Soil Biol Biochem 57:135–143
3. Schlesinger WH, Andrews JA (2000) Soil $CO_2$ flux and the global carbon cycle. Biogeochem 48:7–20
4. Raich JW, Schlesinger WH (1992) The global carbon dioxide flux in soil $CO_2$ flux and its relationship to vegetation and climate. Tellus 44:81–89
5. Schimel DS, Braswell BH, Holland EA, McKeown R, Ojima DS, Painter TH, Parton WJ, Townsend AR (1994) Climatic, edaphic, and biotic controls over storage and turnover of carbon in soils. Glob Biog Cyc 8:279–293
6. Townsend AR, Vitousek PM, Holland EA (1992) Tropical soils could dominate the shortterm carbon cycle feedbacks to increased global temperatures. Climatic Change 22:293–303
7. Tufekcioglu A, Raich JW, Isenhart TM, Schultz RC (2001) Soil $CO_2$ flux within riparian buffers and adjacent crop fields. Plant Soil 229:117–124
8. Devi NB, Yadava PS (2009) Emission of $CO_2$ from the soil and immobilization of carbon in microbes in a sub-tropical mixed oak forest ecosystem, Manipur, N.E., India. Cur Sc 96:1630–1637
9. Maier M, Schack-Kirchner H, Hildebrand EE, Schindler D (2011) Soil $CO_2$ efflux vs. soil $CO_2$ flux: implications for flux models. Agri For Meteor 151:1723–1730
10. Qi Y, Xu M (2001) Separating the effects of moisture and temperature on soil $CO_2$ efflux in a coniferous forest in the Sierra Nevada mountains. Plant Soil 237:15–23
11. Yohannes Y, Shibistova O, Abate A, Fetine M, Guggenberger G (2011) Soil $CO_2$ efflux in an Afromontane forest of Ethiopia as driven by seasonality and tree species. For Ecol Manag 261:1090–1098
12. Chen W, Jia X, Zha T, Wu B, Zhang Y, Li C, Wang X, He G, Yu H, Chen G (2013) Soil $CO_2$ flux in a mixed urban forest in China in relation to soil temperature and water content. Eur J Soil Biol 54:63–68
13. Thokchom A, Yadava PS (2014) Soil $CO_2$ flux in different ecosystem of North-East India. Curr Sci 107(1):99–105
14. Wiseman PE, Seiler JR (2004) Soil $CO_2$ flux across four age classes of plantation loblolly pine (Pinus taeda L.) on the Virginia Piedmont. For Ecol Manag 192:297–331
15. Anderson JM, Ingrame SI (1993) Tropical soil biology and fertility (TSBF), Methods handbook internation union of biological sciences, 77
16. Singh JS, Gupta SR (1977) Plant decomposition and soil $CO_2$ flux in terrestrial ecosystems. Bot Rev 43:449–528
17. Fritze H, Smolender A, Levula T, Kitunen V, Malkoner E (1994) Wood ash fertilization and fire treatments in a scots pine forest stand: effects on the organic layer, microbial biomass and microbial activity. Biol Fert Soils 17:57–63
18. Mac Donald LH, Huffman EL (2004) Post fire soil water repellency: persistance and soil moisture thresholds. Soil Sc Soc Amer J 68:1729–1734
19. Zimmermann M, Meir P, Bird M, Malhi Y, Ccahuana A (2009) Litter contribution to diurnal and annual soil $CO_2$ flux in tropical montane cloud forest. Soil Biol Biochem 41:1338–1340
20. Högberg P, Nordgren A, Buchmann N et al (2001) Large-scale forest girdling shows that current photosynthesis drives soil $CO_2$ flux. Nature 411:789–792
21. Flerchinger GN, Pierson FB (1997) Modelling plant canopy effects on variability of soil temperature and water: model calibration and validation. J Arid Environ 35:641–653
22. Lytle DE, Cronan CS (1998) Comparative soil $CO_2$ evolution, litter decay, and root dynamics in clearcut and uncut spruce-fir forest. For Ecol Manage 103:121–128

23. Laishram ID, Yadava PS, Kakati LN (2002) Soil $CO_2$ flux in a mixed oak forest ecosystem at shiroy hills, Manipur in North Eastern India. Inter J Ecol Envt Sc 28:133–137
24. Kirschbaum MUF (1995) The temperature dependence of soil organic matter decomposition, and the effect of global warming on soil organic C storage. Soil Bio Biochem 27:753–760
25. Bond-Lamberty B, Thomson A (2010) Temperature-associated increases in the global soil $CO_2$ flux record. Nature 464:579–582
26. Ryan MG, Linder S, Vose JM, Hubbard RM (1994) Dark respiration of pines (Copenhagen). Ecol Bul 43:50–63
27. Adachi M, Bekku YS, Rashidah W, Okuda T, Koizumi H (2006) Difference in soil $CO_2$ flux between different tropical ecosystems. Appl Soil Ecol 34:258–265
28. Li LJ, You MY, Shi HA, Ding XL, Qiao YF, Han XZ (2013) Soil $CO_2$ emissions from a cultivated mollisol: effects of organic amendments, soil temperature and moisture. Eur J Soil Biol 55:83–90
29. IPCC (2007) Climate change 2007 synthesis report. In: Pachauri RK, Reisinger A (eds) Working group contributions to the fourth assessment report of the intergovernmental panel on climate change. Cambridge University Press, Cambridge, UK, pp 23–73
30. Cooney CM (2010) The perception factor: climate change gets personal. Environ Health Pers 118(11):A484–A489
31. Kutílek M (2011) Soils and climate change. Soil Tillage Res. 117:1–7
32. IPCC (1995) Scientific assessments of climate change. The policymaker's summary of working group 1 to the intergovernmental panel on climate change, WMO/UNEP
33. Sundquist ET (1993) The global carbon budget. Science 259:934–941
34. Ciais P, Tans PP, Trolier M, White JWC, Francey RJ (1995) A large northern hemisphere terrestrial $CO_2$ sink indicated by the 13C/12C ratio of atmospheric $CO_2$. Science 269:1098–1102
35. Houghton JT, Filho LGM, Callander BA, Harris N, Kattenburg A, Maskell K (1996) Climate change 1995: the science of climate change: contribution of working group I to the second assessment report of the intergovernmental panel on climate change. Cambridge University Press, Cambridge, U.K.

# Estimation of Carbon Dioxide Fluxes Between Land, Ocean and Atmosphere During 2006–2011 with a 4D Variational Assimilation Scheme and Special Reference to Asia

**P. S. Swathi, N. K. Indira, and M. Ramonet**

**Abstract** The rates of exchange of carbon dioxide between atmosphere, land and ocean is a key component of the global carbon budget. It is extremely important to quantify these both on a spatial and temporal basis as it can identify sources and sinks of anthropogenic carbon which are important in setting goals for country-wise emissions and uptake. This chapter presents results from a 4D variational assimilation scheme of ingesting carbon dioxide data between 2006 and 2011 into the adjoint of the LMDZ transport model to obtain fluxes between land, ocean and atmosphere. Carbon dioxide data from three stations from India (Hanle, Pondicherry and Port Blair) along with one hundred global stations have been assimilated into the model, and this is expected to increase the confidence in the results for the tropical and temperate Asian regions. The grid point-wise results are aggregated into larger regions representing India, China, USA, Western Europe and the Rest of the World to reflect the quantification of sources and sinks in these regions. In addition, aggregation in terms of Transcom regions is also presented.

**Keywords** Carbon fluxes Asia · 4D variational assimilation · LMDZ model · Transcom

## Abbreviations

| | |
|---|---|
| CDIAC | Carbon dioxide Information and Analysis Centre |
| COP | Conference of Parties |
| CSIR | Council of Scientific and Industrial Research |
| DST | Department of Science and Technology |
| EDGAR | Emission Database for Global Atmospheric Research |

P. S. Swathi (✉) · N. K. Indira
CSIR Fourth Paradigm Institute, NAL Belur Campus, Bangalore 560041, India
e-mail: pananjady.swathi@gmail.com

M. Ramonet
LSCE, Gif sur Yvette, France

 
M. Goel et al. (eds.), *Climate Change and Green Chemistry of $CO_2$ Sequestration*, Green Energy and Technology, https://doi.org/10.1007/978-981-16-0029-6_17

| | |
|---|---|
| ESRL | Earth System Research Laboratories |
| FCO | Foreign and Commonwealth Office |
| FF | Fossil Fuel |
| GAW | Global Atmospheric Watch |
| GMD | Global Monitoring Division |
| GR | Growth Rate |
| GTC | Gigatonne of Carbon |
| HLE | Hanle |
| ICOS | Integrated Carbon Observing System |
| IFCPAR | Indo-French Centre for the Promotion of Advanced Research |
| LMDZ | Laboratoire de Meteorologie Dynamique Zoom |
| LSCE | Le Laboratoire des Sciences du Climate et de l'Environnment |
| NIOT | National Institute of Ocean Technology |
| PBL | Port Blair |
| PON | Pondicherry |
| ROW | Rest of the World |
| WMO | World Meteorological Organisation |

## 1 Introduction

In view of the commitments and compliance of Intended Nationally Determined Contributions (INDCs) under the COP 21 Paris accord, there is an urgency to quantify sources and sinks of carbon at a national level. Although a mid-term review will take place only in 2023, national efforts toward meeting the targets can be assessed by estimating sources and sinks as precisely as possible in the interim.

There are essentially two ways to estimate carbon fluxes between the three compartments (land, ocean and atmosphere) of the earth system and closing the carbon budget. The first is a bottom-up approach where aggregation of flux measurements can be extrapolated and consolidated for larger regions. The difficulty here is the heterogeneity of the land compartment which encompasses many types of vegetation and compromises the effort in arriving at a robust estimate. Undersampling of the oceanic compartment also raises important concerns about the reliability of the extrapolation. Biomass burning is also a major source of uncertainty. The fossil fuel component is probably the best known among the sources, basically on account of information about trade.

The second way to estimate fluxes of carbon is a top-down approach where a Bayesian framework is employed. Prior information about the fluxes along with their uncertainty is assumed. This is used in a transport model to estimate $CO_2$ concentrations at all grid points. The deviations between observed and modelled concentrations are used to correct the assumed prior fluxes. How this is done in 3D and 4D variational assimilation schemes is described later. The current state of the art in reconciling the bottom-up and top-down approaches is described by Kondo et al.

[1]. The top-down approach is still under vigorous development, and the expectation is that it may be mature enough to contribute to assessment in the present decade.

The early attempts at assimilating data into transport models were based on a 3D variational scheme, where an explicit linear relationship between sources and concentrations was constructed, often by running transport models for obtaining Green's functions relating concentrations with unit impulses from large geographic areas. These efforts culminated with the IGBP sponsored Transcom experiment [2, 3]. The earth was divided into 22 regions (11 land and 11 ocean) as shown in Fig. 1, and 1 Gigatonne of carbon (GTC) source per year (spatially distributed according to the net primary production in each region) was applied in each region for a month, and the resulting CO2 distribution was tracked for 3 years with a transport model.

The results from these simulations were used to construct a simple model $GS = D$ where $S$ is a vector of sources, $G$ is a Green's function counterpart and $D$ is the data. The matrix $G$ could be constructed in such a way that it could represent either annual, cyclostationary or interannual conditions. Most of the computing effort is spent toward generating the Green's functions. Once it is done, the rest of the procedure is relatively straightforward. If $S_0$ and $C(S_0)$ are the prior solution and its covariance, respectively, and $C(D)$ the covariance of the data, then the posterior solution and its covariance are given by

$$S = S_0 + \left(G^T C(D)^{-1} G + C(S_0)^{-1}\right)^{-1} G^T C(D)^{-1}(D - GS_0) \tag{1}$$

$$C(S) = \left(G^T C(D)^{-1} G + C(S_0)^{-1}\right)^{-1} \tag{2}$$

The above equations are the results of minimising a cost function, $J$ given by

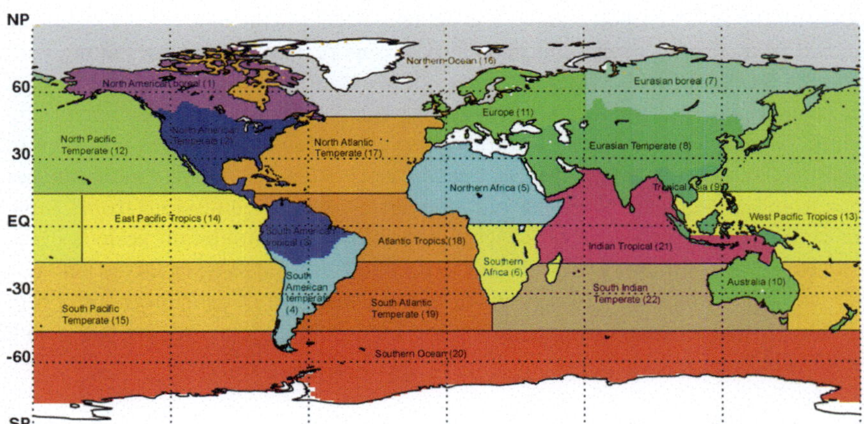


**Fig. 1** Transcom regions

$$J = (S - S_0)^T C(S_0)^{-1}(S - S_0) + (GS - D)^T C(D)^{-1}(GS - D) \quad (3)$$

In contrast, in 4D variational data assimilation, the nonlinear transport model $D = M(S)$ is allowed to evolve in time and can be locally linearised to $GS = D$. Instead of a direct procedure as given in Eq. 1, we use an iterative procedure to minimise $J$, i.e. solving $\nabla J = 0$, with respect to $S$. The corrections to the priors are made iteratively as the adjoint of the transport model is integrated backwards in time to determine $\nabla J$. A minimisation scheme based on a conjugate gradient approach is used to correct $S$, and the procedure is repeated for a set number of times [4–6].

A synthesis of the terrestrial carbon budgets from several sources for South and Southeast Asia was performed by Cervarich et al. [7] where they computed the net biome productivity based on bottom-up models. Based on their calculations, only Bhutan and Laos were net uptake countries, while all others were net emitters. A multi-model land uptake assessment by Rao et al. [8] over 1900–2010 showed that India had a NEP of -0.138 GTC/year with a positive trend of 2% per year over this period but had a lot of interannual variability on account of precipitation. Gahlot et al. [9], estimated that net biome production by Indian forests to be 27 TgC/year in the 1980s, 34 TgC/year in the 1990s and 23 TgC/yr in the 2000s.

Maksyutov et al. [10] was the first study to look at Asian fluxes in detail after ingesting data from several sites in Japan and a Sydney-Tokyo aircraft profile into Transcom inversions. A multi-model study estimated a sink of 0.5 GTC/year over the Eurasian temperate region with a smaller error band compared with the standard inversion of Gurney et al. [2]. Another study applying 3DVar inversion with flask data from 81 sites including Hanle showed the presence of a large terrestrial sink in the Eurasian temperate region [11]. The present study is a refinement incorporating more data and employing a 4DVar procedure to get better spatial and temporal estimate of fluxes. Thompson et al. [12] made a careful comparison of results from six inverse models for the Asian carbon budget. The data ingested into the model were primarily obtained from aircraft campaigns and ships of opportunity programmes along with data from NOAA ESRL's GMD database. They identified a major land sink in East Asia (comprising of China, Korea and Japan) while South Asia remained essentially neutral. There was a considerable spread between different models as was seen in the earlier Transcom experiments. Saeki and Patra [13] have attributed increased net fluxes from East Asia due to the use of higher erroneous CO2 emissions from China. Kim et al. [14] used the Carbon Tracker with different assimilation periods (3, 5, 7 and 10 weeks) and ensembles (100, 150 and 300) to study Asian fluxes between 2007 and 2009. They used 14 stations in the Asian region in their inversion. From their suite of models, they estimated that the Eurasian temperate region was a sink of 0.45 GTC/year.

## 2 Methodology

The procedure for 4DVar, called PYVAR, has been outlined by Chevallier in a series of papers [4–6], but we summarise the method again for clarity. PYVAR is written in Python with separate modules for LMDZ and its adjoint and tangent linear codes (both written in Fortran) along with modules for processing data, the implementation of the assimilation scheme and the conjugate gradient scheme for minimisation, which are all written in Python. The transport model has a resolution of $3.75^\circ * 2.5^\circ$ (longitude-latitude) with 19 vertical levels.

The control variables which are optimised in the 4DVar include surface fluxes at every 8-day intervals with day and night values separated. Besides these, surface values of $CO_2$ at the start of the simulation are also optimised. The system is run for the period 1 January 2006 to 31 December 2011.

The priors used in the system come from a variety of sources. Fossil fuel emissions [FF] are taken from EDGAR 4.2 Database [15] scaled with annual global totals from CDIAC (https://cdiac.ess-dive.lbl.gov/ftp/ndp030/global.1751_2013.ems). The slight differences between CDIAC and global totals in Table 1 are due to interpolation effects. They are yearly values, and they are interpolated to the model grid. The global total along with a regional break up is given in Table 1.

The reduction in 2009 was due to the worldwide economic slump. The sudden increase in 2008 in China is on account of the Olympics. In the five years, global emissions jumped by 9.2%, India by 18%, China by 24%, Rest of the World (ROW) by 8.8%. USA and W. Europe have held steady. W. Europe in this Table is smaller than the Europe region in Transcom (Fig. 1). The emissions for 2011 are assumed to be the same as 2010. For emissions in carbon dioxide units, multiply all values by 3.67.

The fossil fuel distribution for 2010 is shown in Fig. 2. Note the high emissions from Eastern China, US and W. Europe.

Air–sea exchange is prescribed from the Takahashi database [16] of monthly means. The annual global air–sea flux is -1.508 GTC (sink). The biospheric fluxes are provided by the ORCHIDEE model [17]. The model can provide fluxes ranging from diurnal variability to long-term variations. The GFEDv2 database [18] provides biomass burning fluxes where emissions originate at hotspots, but they are returned evenly back to the earth throughout the year.

**Table 1** Fossil fuel emission totals in Gigatonnes of carbon (GTC)

| Year | Global, GTC | India | China | USA | W Europe | ROW |
|---|---|---|---|---|---|---|
| 2006 | 8.388 | 0.406 | 1.693 | 1.681 | 1.424 | 3.184 |
| 2007 | 8.581 | 0.431 | 1.796 | 1.687 | 1.408 | 3.260 |
| 2008 | 8.763 | 0.459 | 2.007 | 1.583 | 1.407 | 3.308 |
| 2009 | 8.662 | 0.453 | 1.983 | 1.564 | 1.391 | 3.270 |
| 2010 | 9.180 | 0.481 | 2.102 | 1.658 | 1.474 | 3.466 |

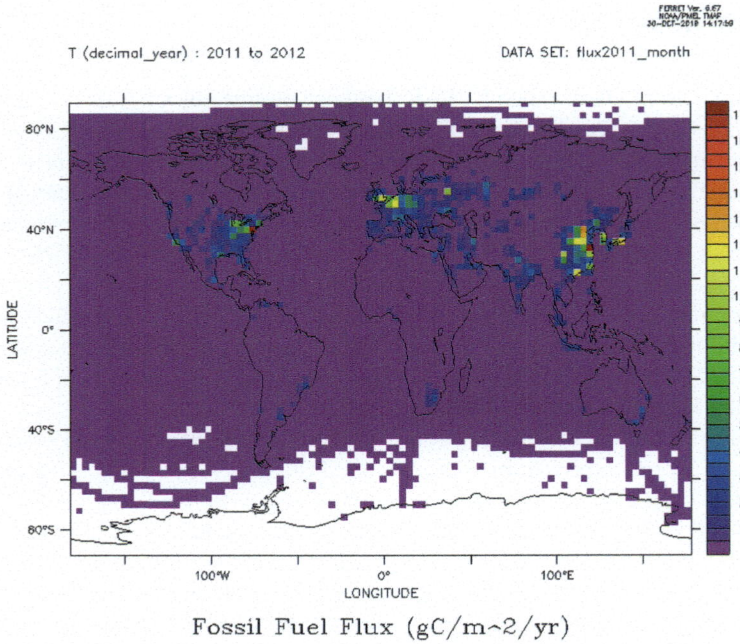


**Fig. 2** Annually averaged fossil fuel flux for 2011

The uncertainties associated with surface fluxes are a little difficult to quantify. Land surface flux variability is assumed to be similar, but scaled, to heterotrophic respiration. Based on a comparison of net ecosystem exchange from flux towers at a few sites and the ORCHIDEE model, Chevallier [19], provided a prescription to calculate $C(S_0)$ over land. Over the ocean, the uncertainty is set to a constant value of 0.2 gC/m$^2$/d. The uncertainties are also spatially and temporally correlated. The land values have an e-folding measure of 500 km spatially and 1 month temporally, while the oceans have a value of 1000 km and 1 month, respectively. Unlike 3DVar where the posterior covariance matrix can be explicitly computed, 4DVar does not permit this evaluation. Based on a Monte Carlo simulation, Chevallier [5] has estimated the posterior errors.

## 3 Observations

We have used data from 114 observations from 103 sites from all over the world as shown in Table 2. It is a mix of continuous and discrete measurements over land and ocean stations.

In case there are more than one set of measurements at a site they are given in separate rows. Of the stations listed above, three (Hanle, Port Blair and Pondicherry)

**Table 2** Stations considered in this study

| Station | Latitude | Longitude | Height (m) | Cont/Disc | No of days | Error (ppm) |
|---|---|---|---|---|---|---|
| Alert, Canada | 82.5 N | 62.5 W | 210 | C, D | 308<br>2024 | 1.151<br>1.125 |
| Amsterdam Island | 37.8S | 77.6E | 0 | D | 1457 | 0.558 |
| Argyle, Maine, USA | 45.0 N | 66.7 W | 50 | C | 1771 | 4.015 |
| Anmeyon-do, Korea | 36.53 N | 126.3R | 47 | C | 2071 | 6.937 |
| Ascension Island | 7.9S | 14.4 W | 54 | D | 511 | 0.824 |
| Assekrem, Algeria | 23.6 N | 5.6 N | 2710 | D | 274 | 0.663 |
| Azores | 38.7 N | 27.3 W | 19 | D | 155 | 1.808 |
| Baltic Sea | 55.3 N | 17.2E | 3 | D | 542 | 5.178 |
| Begur, Spain | 41.9 N | 3.2E | 0 | D | 219 | 3.657 |
| Baring Head, NZ | 41.4S | 171.8E | 85 | C, D | 99<br>194 | 1.056<br>0.379 |
| Bukit, Indonesia | 0.2S | 100.3E | 845 | D | 182 | 3.315 |
| Bermuda East | 32.3 N | 64.6 W | 12 | D | 49 | 1.270 |
| Bermuda West | 32.3 N | 64.6 W | 30 | D | 185 | 1.831 |
| Barrow, Alaska | 72.3 N | 156.6 W | 11 | C | 2128 | 1.643 |
| Cold Bay, Alaska | 55.2 N | 165.7 W | 21 | D | 404 | 2.165 |
| Cabauw, Netherlands | 52 N | 4.9E | 0 | C | 1048 | 6.950 |
| Candle Lake, Canada | 53.9 N | 104.7 W | 489 | C | 1811 | 2.745 |
| Cape Ferguson, Aus | 19.3S | 147.1E | 2 | D | 121 | 1.465 |
| Cape Grim, Aus | 40.7S | 144.7E | 97 | D | 187 | 0.294 |
| Christmas Island | 1.7 N | 157.2 W | 3 | D | 227 | 0.522 |
| Cape Town | 34.4S | 18.5E | 230 | C | 2023 | 1.367 |
| Crozet Island, France | 46.4S | 51.8E | 197 | D | 272 | 0.317 |
| Casey Station, Aus | 66.3S | 110.5E | 60 | D | 124 | 0.338 |
| Drake Passage | 59S | 64.7 W | 0 | D | 119 | 0.340 |
| Egbert, Canada | 44.2 N | 79.8 W | 255 | C | 2080 | 5.162 |
| Easter Island, Chile | 27.2S | 109.4 W | 0 | D | 200 | 2.4 |
| Estevan Point, Canada | 49.4 N | 126.5 W | 7 | C, D | 1074 | 2.335 |
| Finokalia, Crete | 35.3 N | 25.7E | 150 | D | 123 | 10.053 |
| Fraserdale, Canada | 49.8 N | 81.5 W | 210 | C | 1952 | 3.036 |
| Mariana Island, Guam | 13.4 N | 144.7E | 0 | D | 330 | 3.191 |
| Halley Station, Antarctica | 75.6S | 26.2 W | 30 | D | 257 | 0.255 |
| Hanle, India | 32.8 N | 79E | 4500 | C, D | 1417 | 0.715 |
| Hohenpeissenberg, Germany | 47.8 N | 11E | 936 | D | 255 | 6.611 |

(continued)

**Table 2** (continued)

| Station | Latitude | Longitude | Height (m) | Cont/Disc | No of days | Error (ppm) |
|---|---|---|---|---|---|---|
| Hegyhutsal, Hungary | 47 N | 16.7E | 248 | C, D | 1265<br>289 | 5.82<br>6.01 |
| Storhofdi, Iceland | 63.4 N | 20.3 W | 118 | D | 296 | 1.754 |
| Izana, Canary Islands | 28.3 N | 16.5 W | 2373 | C, D | 261<br>2120 | 1.207<br>1.007 |
| Jubany, Antarctica | 62.2S | 58.7 W | 15 | C | 1410 | 0.509 |
| Jungfrauhoch, Switzerland | 46.5 N | 8E | 3580 | C, D | 260<br>1968 | 2.337<br>2.382 |
| Kasprowy Wierch, Poland | 49.1 N | 20E | 1987 | C | 94 | 4.467 |
| Key Biscayne, Florida | 25.7 N | 80.2 W | 0 | D | 218 | 4.067 |
| Cape Kumukahi, Hawaii | 19.7S | 155 W | 0 | D | 307 | 2.115 |
| Sary Taukum, Kazhakistan | 44 N | 76.6E | 0 | D | 159 | 1.882 |
| Park falls, Wisconsin | 45.9 N | 90.3 W | 472 | C | 1927 | 3.640 |
| Lulin, Taiwan | 23.5 N | 120.9E | 2862 | D | 185 | 3.759 |
| Lampedusa, Italy | 35.5 N | 12.6E | 45 | D | 241 | 2.019 |
| Ile Grande, France | 48.8 N | 3.6 W | 20 | D | 166 | 5.828 |
| Mawson, Antarctice | 67.6S | 62.9E | 20 | D | 101 | 0.348 |
| Mace Head, Ireland | 52.3 N | 9.9 W | 5 | C, D | 262<br>1504 | 3.308<br>2.939 |
| MidWay Island | 28.2 N | 177 W | 11 | D | 286 | 1.408 |
| Mt. Kenya | 0 | 37.3E | 3644 | D | 84 | 1.784 |
| Mauna Loa, Hawaii | 19.5 N | 155.6 W | 3397 | C | 2121 | 0.774 |
| Minamitorishima, Japan | 24.3 N | 154E | 7 | C | 2036 | 1.103 |
| Macquarie Island, Antarctica | 54,5S | 158.9E | 6 | D | 139 | 0.434 |
| Gobabeb, Namibia | 23.6S | 15 N | 456 | D | 171 | 1.070 |
| Niwot Ridge, Colorado | 40 N | 105.6 W | 3523 | D | 297 | 1.833 |
| Oschenkopf, Germany | 50.3 N | 11.8E | 1022 | D | 196 | 12.421 |
| Pallas, Finland | 68 N | 24.1E | 565 | C, D | 275<br>1063 | 3.0334<br>2.926 |
| Port Blair, India | 11.6 N | 92.7E | 30 | D | 49 | 2.060 |
| Pic du Midi, France | 43 N | 0.1E | 2877 | D | 183 | 2.538 |
| Pacific Ocean, POC000 | 0 | 155 W | 10 | D | 79 | 0.505 |

(continued)

**Table 2** (continued)

| Station | Latitude | Longitude | Height (m) | Cont/Disc | No of days | Error (ppm) |
|---|---|---|---|---|---|---|
| POCN05 | 5 N | 151 W | 10 | D | 77 | 0.569 |
| POCN10 | 10 N | 149 W | 10 | D | 77 | 0.645 |
| POCN15 | 15 N | 145 W | 10 | D | 78 | 0.606 |
| POCN20 | 20 N | 141 W | 10 | D | 78 | 0.838 |
| POCN25 | 25 N | 139 W | 10 | D | 70 | 0.872 |
| POCN30 | 30 N | 135 W | 10 | D | 65 | 1.833 |
| POCS05 | 5S | 159 W | 10 | D | 76 | 0.605 |
| POCS10 | 10S | 161 W | 10 | D | 77 | 0.483 |
| POCS15 | 15S | 164 W | 10 | D | 73 | 0.549 |
| POCS20 | 20S | 167 W | 10 | D | 80 | 0.496 |
| POCS25 | 25S | 171 W | 10 | D | 75 | 0.469 |
| POCS30 | 30S | 176 W | 10 | D | 75 | 0.426 |
| Pondicherry, India | 11.9 N | 79.8E | 30 | D | 149 | 5.080 |
| Plateau Rosa, Italy | 46 N | 8E | 3480 | C | 476 | 1.778 |
| Palmer Station, Antarctica | 64.9S | 64 W | 10 | D | 307 | 0.343 |
| Point Arena, California | 39 N | 123.7 W | 17 | D | 238 | 5.997 |
| Puy Dome, France | 45.8 N | 3E | 1465 | D | 223 | 4.607 |
| Ragged Point, Barbados | 13.2 N | 59.4 W | 15 | D | 288 | 0.793 |
| Ryori, Japan | 39 N | 141.9E | 260 | C | 2090 | 3.688 |
| Schauinsland, Germany | 48 N | 8E | 1205 | C | 360 | 3.803 |
| Seychelles | 4.7S | 55.7E | 0 | D | 269 | 1.007 |
| Southern Great Plains, Oklahoma | 363.6 N | 97.5 W | 314 | D | 289 | 5.486 |
| Shemya Island, Alaska | 52.7 N | 174.1E | 23 | D | 210 | 2.640 |
| American Samoa | 14.2S | 170.6 W | 42 | C | 1983 | 0.621 |
| Sonnblick, Austria | 47 N | 13E | 3106 | C | 1417 | 2.423 |
| South Pole, Antarctica | 90S | 24.8 W | 2810 | C | 2158 | 0.224 |
| Ocean Station M, Norway | 66 N | 2E | 0 | D | 372 | 2.425 |
| Summit, Greenland | 72.6 N | 32.4 W | 3209 | D | 306 | 1.212 |
| Syowa Station, Antarctica | 69S | 39.6E | 14 | D | 141 | 0.244 |

(continued)

**Table 2** (continued)

| Station | Latitude | Longitude | Height (m) | Cont/Disc | No of days | Error (ppm) |
|---|---|---|---|---|---|---|
| Tae-Ahn Peninsula, Korea | 36.7 N | 126.1E | 16 | D | 244 | 7.296 |
| Tierra Del Fuego, Argentina | 54.8S | 68.3 W | 0 | D | 157 | 1.537 |
| Trinidad Head, California | 41 N | 124 W | 107 | D | 298 | 5.177 |
| Trainou, France | 48 N | 2.1E | 311 | D | 184 | 6.707 |
| Wendover, Utah | 39.9 N | 113.7 W | 1327 | D | 273 | 2.408 |
| Ulaan Uul, Mongolia | 44.5 N | 111.1E | 1007 | D | 291 | 2.598 |
| Weizman Institute, Israel | 30 N | 35 N | 151 | D | 304 | 2.303 |
| Moody, Texas | 31.3 N | 97.3 W | 251 | C | 1802 | 2.932 |
| Mt. Waliguan, China | 36.3 N | 100.9E | 3810 | D | 298 | 2.178 |
| Western Pacific Cruise | 0 | 151 W | 0 | D | 76 | 2.104 |
| Sable Island, Canada | 43.9 N | 60 W | 2 | C | 1884 | 2.953 |
| Yonagunijima, Japan | 24.5 N | 123E | 30 | C | 2101 | 2.553 |
| Svalbard, Norway | 78.9 N | 11.9E | 474 | C, D | 722<br>276 | 1.196<br>1.503 |

are maintained by CSIR Fourth Paradigm Institute with support from LSCE, France, Indian Institute of Astrophysics, Pondicherry University and NIOT, Port Blair. At Hanle (HLE), both continuous and discrete flask measurements are carried out. The continuous measurements are based on a LICOR-based system called Caribou (Ramonet et al. paper in preparation), and hourly averages from the measurements are used, while the flask analysis at all the stations is made according to ICOS protocols [20, 21].

The data is first pre-processed before it is ingested into the system. First, the data is detrended and deseasonalised according to the procedure described by Masarie and Tans [22]. The daily means at each station are first computed. In case of continuous data, only data between local times of 12:00 to 20:00 h are used for continental stations, while 01:00 to 06:00 was used for stations higher than 1000 m. At Hanle, all 24-h data were used as the site has very little vegetation and is not affected by the daytime updraft from the plains. A least squares fit of the daily mean data with a function involving a constant, linear and quadratic terms and four annual harmonics is performed, and residuals of the daily means from this fit are computed. The standard deviation of the residuals is computed, and half of this value is taken as a first-cut measure of $C(D)$. This time-invariant standard deviation of the residuals and the number of days on which measurements are available are listed in the seventh and sixth columns in Table 2. However, an additional screening was performed to assess the ability of the model to simulate the data. Concentrations at the data points were computed using a first guess model which spreads the increase in annual loading obtained from increase in $CO_2$ annual mean concentrations from well-mixed stations such as Mauna Loa and South Pole. Chevallier [6] called this a poor man's guess. A 1 ppm increase in the annual mean $CO_2$ translates to a loading of 2.12 GTC [6]. This is spread over land in proportion to $C(S_0)$. If deviations of this model from observed data were more than three times the residual computed in the step above, the data were rejected. Finally, for continuous data, the residuals were multiplied by the square root of the number of valid hourly data in a day. No cross-correlations between data errors are considered. Due to all the additional procedures, the errors given in the seventh column of Table 2 are only indicative. The actual $C(D)$ is higher and time-dependent. But notice the small values of $C(D)$ in several high altitude sites like Hanle, Mauna Loa, etc., remote Antarctic and Arctic sites and several oceanic island sites. These data will have a larger influence on the inversion by providing a tighter constraint.

The conjugate gradient algorithm for minimisation of the cost function $J$ brought down its value from 13,733,305 to 203,288, a reduction of 2 orders of magnitude over 16 iterations. Further iteration is not expected to reduce this much.

## 4 Results and Discussion

The annual average of the land uptake for 2011 is shown in Fig. 3 at each grid point. However, this is of very limited use as many grid points are not properly resolved due to lack of data, and the results have to be aggregated into larger regions to gain confidence. However, Fig. 3 still has the broad outlines of the uptake regions which are concurrent with large vegetated regions such as the Amazon rain forests, Eurasian Boreal and the Indo-China regions.

Figure 4 shows the uptake from the ocean for 2011. The outgassing from the equatorial regions and the uptake in the northern Atlantic and the southern oceans are clearly visible. Again, it must be emphasised that the grid-level values are not very reliable, and we need to aggregate over larger areas.

The aggregated uptakes over different regions of the world are given in Table 3.

Units are Gigatonnes of carbon. Negative indicates sink. The Atm (9th column) is the residual after land and ocean sinks are subtracted from global FF emissions. The second-last column is annual atmospheric growth rate based on the atmospheric loading. (2.12 GTC atmospheric load results in a 1 ppm increase when it is completely mixed in the atmosphere.) The global atmospheric growth rate in the last column is provided by NOAA (https://www.esrl.noaa.gov/gmd/ccgg/trends/gl_gr.html) which is based on marine surface data.

From Table 3, we see that aggregated uptake is highly variable over the years. The sign of the uptake changes for consecutive years for USA and W. Europe. But we see that the global growth rate agrees reasonably well with independent NOAA estimates, especially in years where there is a large increase (2007 and 2010). However, this agreement is slightly deceptive as the atmospheric loading is already prescribed crudely with a poor man's guess. Chevallier [6] demonstrates clearly that

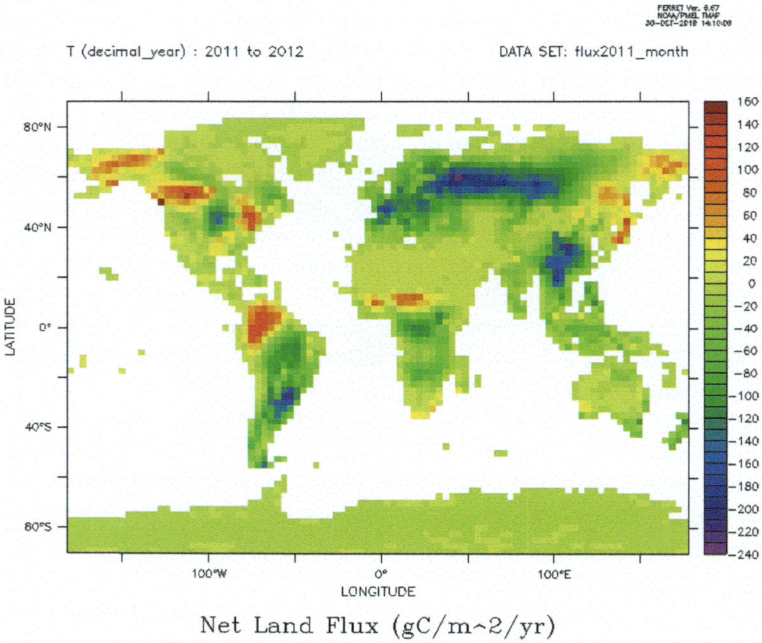


**Fig. 3** Annually averaged residual land flux

in a larger view, the inversion improves the data fit at many stations and contributes significantly to the information content of the inversion. The oceanic total has shown not too much of a change from the prior value of -1.508 GTC except in 2011. The changes in atmospheric loading are mainly on account of changes in land uptake. The land uptake by India and China is quite significant, around 20% of the global land uptake.

## 5 Transcom Aggregation

The regions for this aggregation are given in Fig. 1, and these are quite different from the ones used in Tables 1 and 2, especially for the European and W. European regions. Figure 5 shows the fluxes for all the regions seen in Fig. 1. The reduction in error when there is a preponderance of data is striking as can be seen in North American Boreal, North American temperate and European land regions. The major land sinks are in North American temperate, Eurasian Boreal, Eurasian temperate and Europe, while the major ocean sinks are in North Pacific temperate, Northern Ocean, North Atlantic temperate, South Pacific temperate, South Atlantic temperate, South Indian temperate and the Southern Ocean. When the global data is aggregated

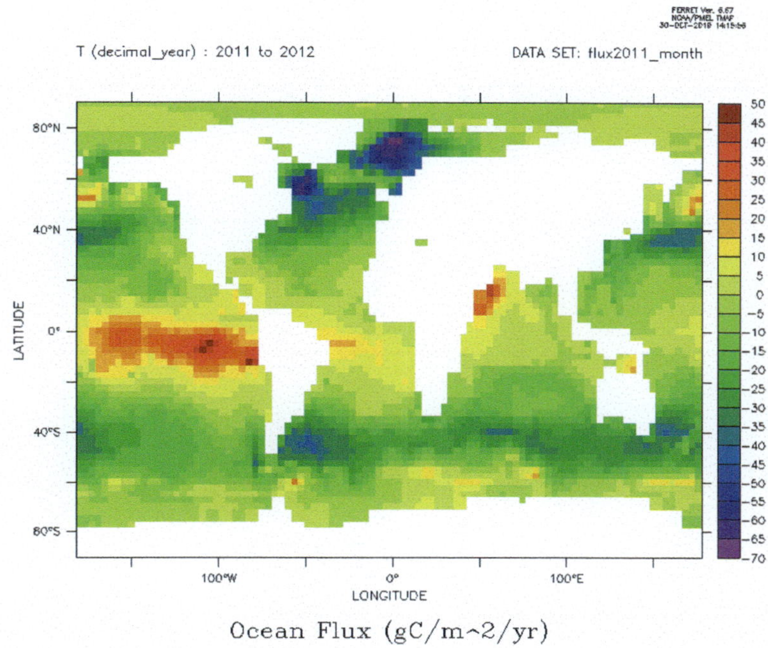


**Fig. 4** Annually averaged ocean flux for 2011

to cover the total land and ocean areas, the uncertainty reduction on land is far more significant than that of the ocean.

## 6 Fit to Data with Inversion

Due to the large amount of data ingested into the inversion, it is difficult to compare the fit of the final inversion at each site. As the main focus of the paper is the effect of the inversion due to the addition of three Indian sites, namely Hanle, Pondicherry and Port Blair, we focus on these sites for the fit. The fit for Hanle is shown in Fig. 6. The agreement between the two is striking. The difference between the two is mostly within 1 ppm. The seasonal cycle is reproduced very well in the analysis. This site is one of the better constrained ones as can be seen in Table 2 along with the fact that hourly values are assimilated. The fits at Pondicherry (Fig. 7) and Port Blair (Fig. 8) to the weekly flask data are not as good as Hanle, but this is to be expected in view of the large $C(D)$ values assigned to them.

**Table 3** Aggregated fluxes over continental regions

| Year | Global FF GTC | India | China | USA | W Europe | ROW | Ocean | Atm | Atm GR ppm | Atm GR NOAA |
|---|---|---|---|---|---|---|---|---|---|---|
| 2006 | 8.388 | −0.274 | −0.357 | 0.203 | −0.045 | −2.471 | −1.575 | 3.868 | 1.83 | 1.75 |
| 2007 | 8.581 | −0.258 | −0.317 | −0.252 | 0.160 | −1.373 | −1.415 | 5.126 | 2.42 | 2.09 |
| 2008 | 8.763 | −0.261 | −0.587 | −0.340 | −0.190 | −2.314 | −1.602 | 3.469 | 1.64 | 1.78 |
| 2009 | 8.662 | −0.232 | −0.598 | −0.122 | −0.334 | −2.060 | −1.517 | 3.798 | 1.79 | 1.61 |
| 2010 | 9.180 | −0.148 | −0.204 | −0.059 | −0.729 | −0.861 | −1.664 | 5.517 | 2.60 | 2.43 |
| 2011 | 9.180 | −0.078 | −0.449 | 0.017 | −0.639 | −2.595 | −2.015 | 3.421 | 1.61 | 1.70 |

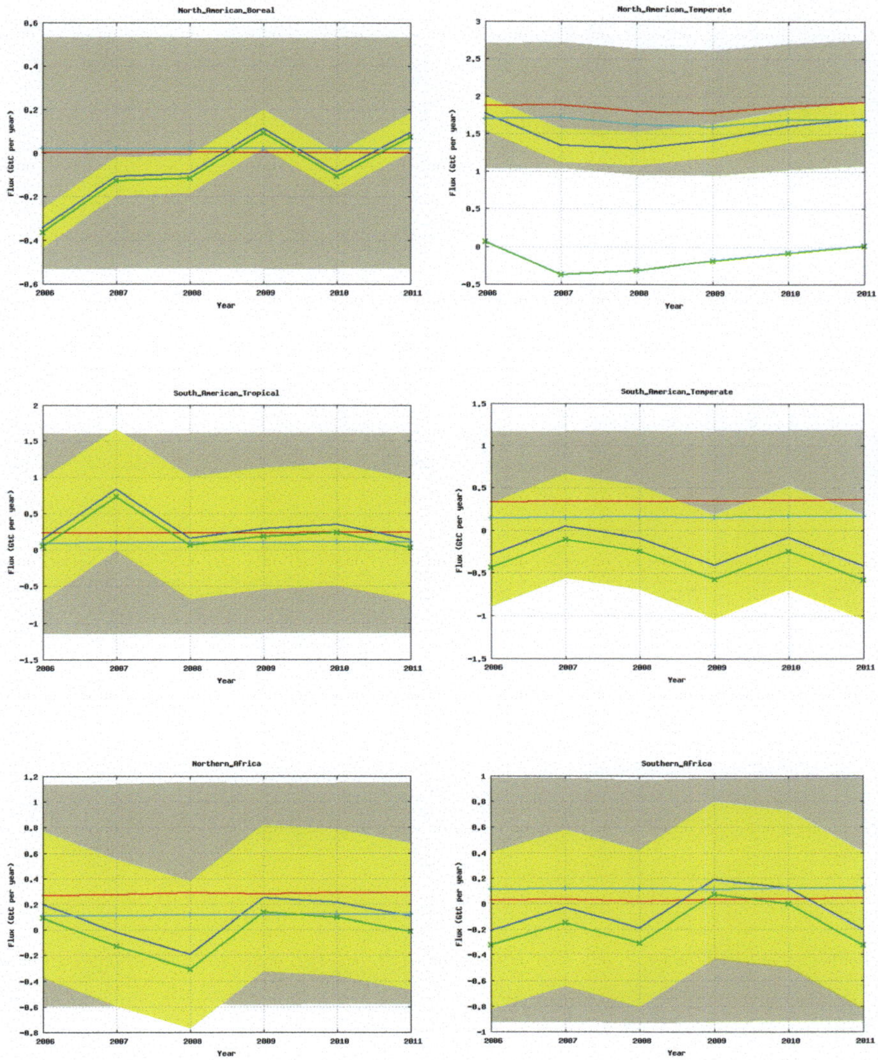


**Fig. 5** Inversion results for Transcom regions. Aggregation over total land and ocean is also given. The red line is the prior flux which includes fossil fuel emissions and land or ocean priors. The beige shaded region is the prior uncertainty. The yellow shaded region is the posterior uncertainty computed with a Monte Carlo approach described in Chevallier [5]. The blue line is the posterior total flux which includes fossil fuel emissions, and the green line is the same without fossil fuel. The cyan line is the FF emission. The difference between the red line (prior flux with FF) and the cyan line (FF) is the assumed prior flux. Positive values denote sources, and negative values indicate sinks. Note that graphs have different scales for different Transcom regions

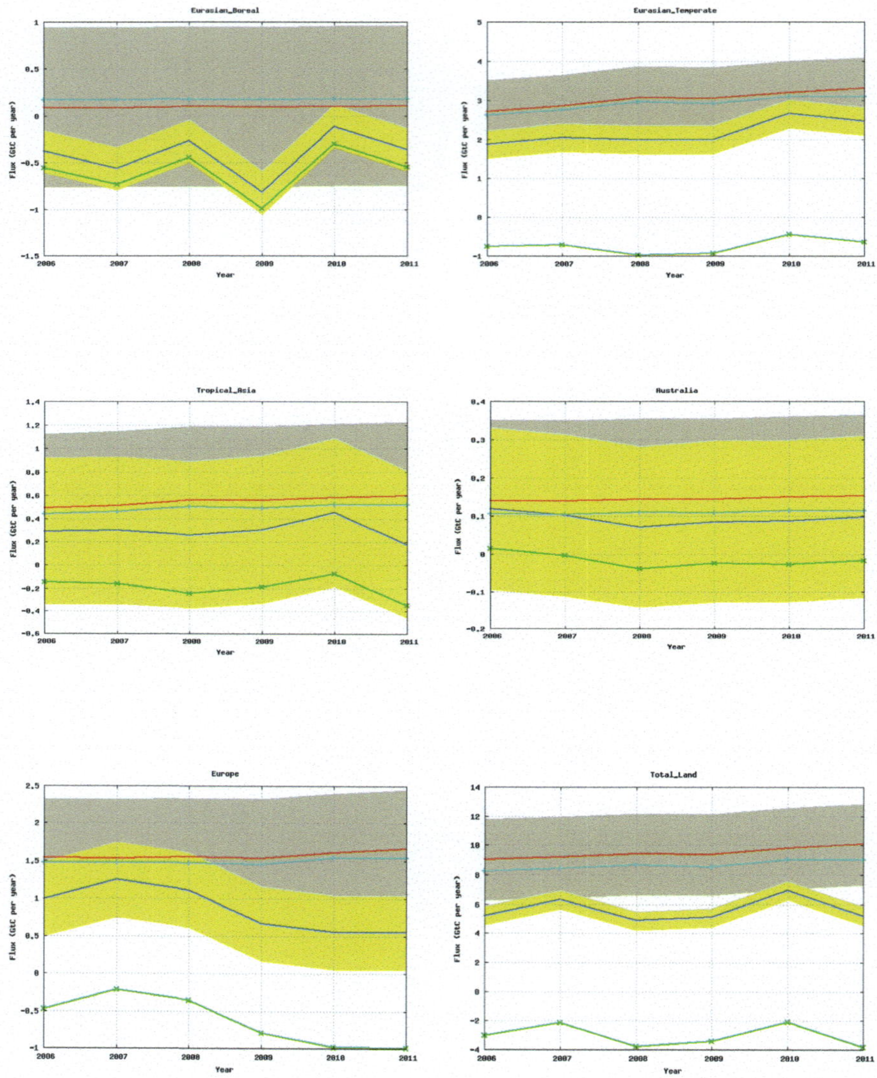


**Fig. 5** (continued)

## 7 Conclusion and Outlook

Carbon dioxide data from three Indian stations along with 100 other global sites have been incorporated into PYVAR, a variational data assimilation system for the 2006–2011 period. The grid-level inverted fluxes have been aggregated to regions representing India, China, USA, W. Europe and the Rest of the World. In addition,

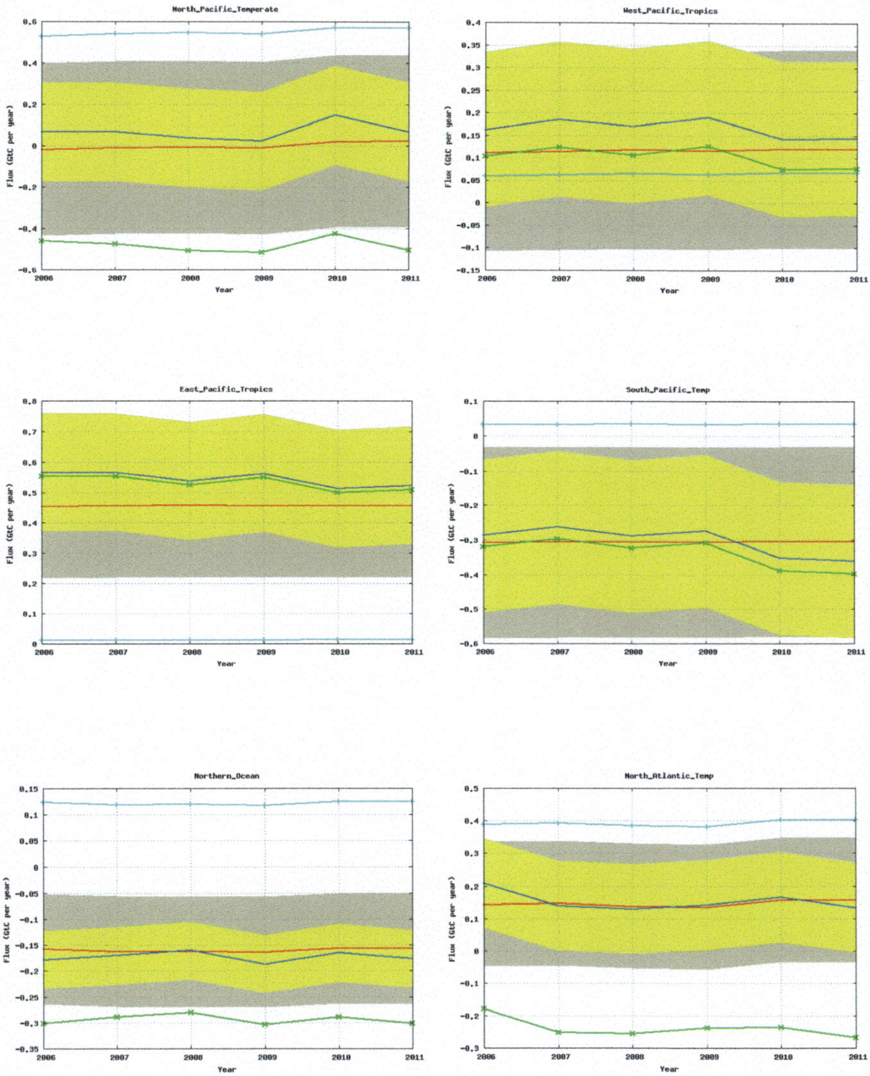


**Fig. 5** (continued)

aggregation according to Transcom specifications is also made. Our results show that India and China are major land sinks but with large interannual variability. The ocean sinks were reasonably constant in 2006–2010 but increased by 30% in 2011. The atmospheric loading and hence the atmospheric growth rates inferred from inversion match very well with independent estimates of ESRL, NOAA. Finally, the concentration data computed from the final inversion show a very good fit with the observed data at Hanle, lending support to the inversion process.

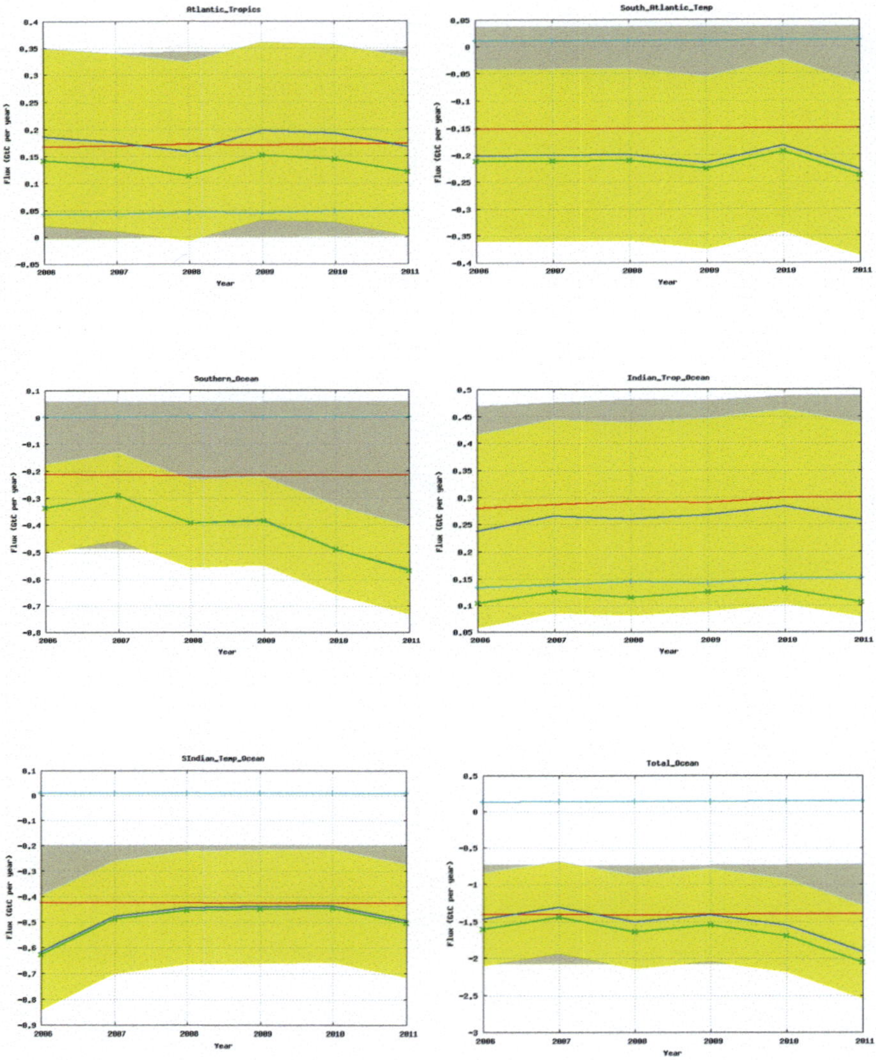


**Fig. 5** (continued)

The posteriori error estimates are still on the higher side to make definitive statements about the magnitude of the sources and sinks. We can improve the estimates if more data of WMO standard quality become available. It is recommended that at least half a dozen such stations be established in India in the near future.

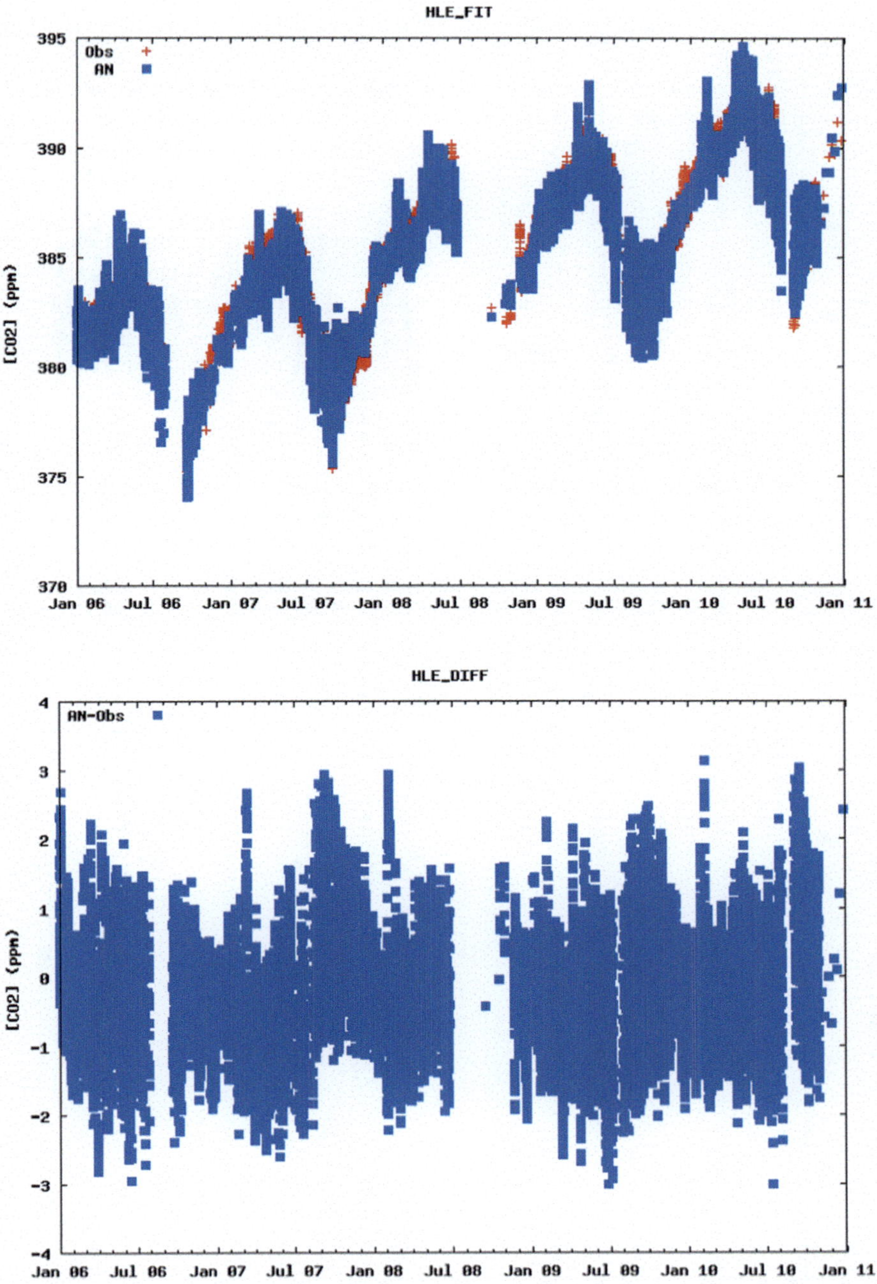


**Fig. 6** Observations and analysis data for Hanle (Top). The difference is given in the bottom

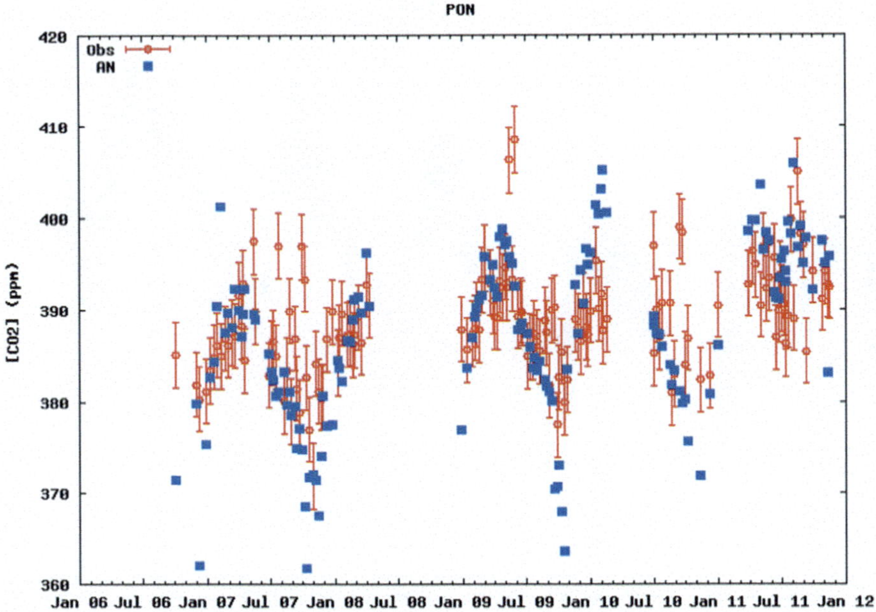


**Fig. 7** Observations and analysis data for Pondicherry

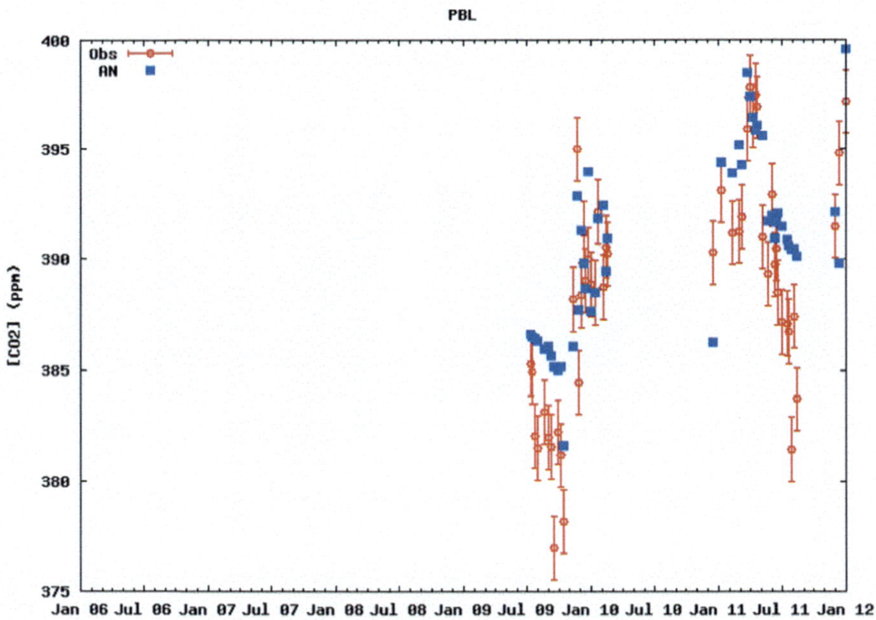


**Fig. 8** Observations and analysis for Port Blair

**Acknowledgements** We are deeply grateful to Frederic Chevallier, LSCE, for providing the PYVAR code and setting up the model configuration. We also thank ESRL, Colorado, GAW, WMO and RAMCES, LSCE, for the data that have been used in the inversion. We thank all the data providers who contributed to the database. We would also like to acknowledge Bhuvan Bhatt, Dorje Angchuk, T. Dorjai, Sonam Jorphail, Indian Institute of Astrophysics, Bangalore, and Hanle, Swati Patnaik, Vikram Reddy and S. Balakrishnan, Pondicherry University, Mehmuna, Dileep Kumar, Vinith Kumar and Kirubagaran NIOT, Port Blair, for their logistical support and cooperation in running GHG stations. We would also like to thank Martina Schmidt and Marc Delmotte, LSCE, for flask analysis. We thank Vinod Gaur for his valuable and informed discussions on inverse problems and Head, CSIR-4PI for institutional and computational resources. We thank CSIR, DST, IFCPAR and FCO, British High Commission for funding the projects under which this work was carried out.

## References

1. Kondo M, Patra PK, Stich S, Friedlingstein P, Poulter B, Chevallier F, Ciais P, Canadell JG, Bastos A, Lauerwald R, Calle L, Ichii K, Anthoni P, Arneth A, Haverd V, Jain AK, Kato E, Kautz M, Law RM, Lienert S, Lombardozzi D, Maki T, Nakamura T, Peylin P, Roddenbeck C, Zhuravlev R, Saeki T, Tian HQ, Zhu D, Ziehn T (2019) State of the science in reconciling top-down and bottom-up aroaches for terrestrial $CO_2$ budget. Glob Change Biol. https://doi.org/10.1111/gcb.14917
2. Gurney KR, Law RM, Denning AS, Rayner PJ, Baker D, Bousquet P, Bruhwiler L, Chen YH, Ciais P, Fan S, Fung IY, Gloor M, Heimann M, Higuchi K, John J, Maki T, Maksyutov S, Masarie K, Peylin P, Prather M, Pak BC, Randerson J, Sarmiento J, Taguchi S, Takahashi T, Yuen CW (2002) Towards robust regional estimates of $CO_2$ sources and sinks using atmospheric transport models. Nature 415(6872):626–630. https://doi.org/10.1038/415626a
3. Baker DF, Law RM, Gurney KR, Rayner P, Peylin P, Denning AS, Bousquet P, Bruhwiler L, Chen YH, Ciais P, Fung IY, Heimann M, John J, Maki T, Maksyutov S, Masarie K, Prather M, Pak B, Taguchi S, Zhu Z (2006) TransCom 3 inversion intercomparison: impact of transport model error on the interannual variability of regional $CO_2$ fluxes, 1988–2003. Glob Biogeochem Cycles 20:1. https://doi.org/10.1029/2004GB002439
4. Chevallier F, Fisher M, Peylin P, Serrar S, Bousquet P, Breon F-M, Chedin A, Ciasis P (2005) Inferrung $CO_2$ sources and sinks from satellite observations: method and application to TOVS data. J Geophys Res 110:D24309. https://doi.org/10.1029/2005JD006390
5. Chevallier F, Breon F-M, Rayner PJ (2007) The contribution of the ocean carbon observatory to the estimation of $CO_2$ sources and sinks: theoretical study in a variational data assimilation framework. J Geophys Res 112:D09307. https://doi.org/10.1029/2006JD007375
6. Chevallier F, Ciais P, Conway TJ, Aalto T, Anderson BE, Bousquet P, Brunke EG, Ciattaglia L, Esaki Y, Frohlich M, Gomez A, Gomez-Palaez AJ, Haszpra L, Krummel PB, Langenfelds RL, Lauenberger M, Machida T, Maignan F, Matsueda H, Morgui JA, Mikai H, Nakazawa T, Peylin P, Ramonet M, Rivier L, Sawa Y, Schmidt M, Steele LP, Vay SA, Vermeulen AT, Wofsy S, Worthy D (2010) $CO_2$ surface fluxes at grid point scale estimated from a global 21 year reanalysis of atmospheric measurements. J Geophys Res 115:D21307. https://doi.org/10.1029/2010JD013887
7. Cervarich M, Shu SJ, Jain AK, Arneth A, Canadell J, Friedlingstein P, Houghton RA, Kato E, Koven C, Patra P, Poulter B, Stich S, Stocker B, Viovy N, Wiltshire A, Zeng N (2016) The terrestrial carbon budget of South and Southeast Asia. Env Res Lett 11:10. https://doi.org/10.1088/1748-9326/11/10/105006
8. Rao AS, Bala G, Ravindranath HN, Nemani R (2019) Multi-model assessment of trends, variability and drivers of terrestrial carbon uptake in India. J Earth Syst Sci 128:99. https://doi.org/10.1007/s-12040-019-120-y

9. Gahlot S, Shu S, Jain AK, Roy SB (2017) Estimating trends and variation of net biome productivity in India for 1980–2012 using a land surface model. Geophys Res Lett 44:11573–11579. https://doi.org/10.1002/2017GL075777
10. Matsyukov S, Machida Y, Mukai S, Patra PK, Nazakawa T (2003) Inoue G and transcom-3 modellers Effect of recent observation on Asia $CO_2$ flux estimates by transport model inversions. Tellus 55B:522–529
11. Swathi PS, Indira NK, Rayner PJ, Jagadeesha D, Bhatt BC, Gaur VK (2013) Robust inversion of carbon dioxide fluxes over temperate Eurasia in 2006–2008. Curr Sci 105:201–208
12. Thompson RL, Patra PK, Chevallier F, Matsyukov S, Law RM, Ziehn T, vad der Luijkx IT, Peters W, Ganshin A, Zhuralev R, Maki T, Nakamura T, Shirai T, Ishizawa M, Saeki T, Machida T, Poulter B, Canadell JG, Ciais P (2016) Top-down assessment of Asian carbon budget since mid 1990s. Nat Comm 7:10724. https://doi.org/10.1038/ncomms10724
13. Saeki T, Patra PK (2017) Implications of overestimated anthropogenic $CO_2$ emissions on East Asian and global land $CO_2$ flux inversion. Geosci Lett 4. https://doi.org/10.1186/s40562-017-0074-7
14. Kim H, Kim HM, Kim J, Cho CH (2018) Effect of data assimilation parameters on the optimized surface flux in Asia. Asia-Pac J Atmos Sci 54:1–17. https://doi.org/10.1007/s13143-017-0049-9
15. Janssens-Maenhout G, Crippa M, Guizzardi D, Muntean M, Schaaf E, Dentener F, Bergamaschi P, Pagliari V, Olivier JGJ, Peters JAHW, van Aardenne JA, Monni S, Doering U, Roxana Petrescu AM, Solazzo E, Oreggioni GD (2019) EDGAR v4.3.2 global atlas of the three major greenhouse gas emissions for the period 1970–2012. Earth Syst Sci Data 11:959–1002. https://doi.org/10.5194/essd-11-959-2019
16. Takahasi T, Sutherland SC, Wanninkhof R, Sweeney C, Feely RA, Chipman DW, Hales B, Friederich G, Chavez F, Watson A, Bakker DCE, Schuster U, Metzl N, Yoshikawa-Inoue H, Ishii M, Midorikawa T, Nojiri Y, Sabine C, Olafsson J, Arnarson TS, Tilbrook B, Johannessen T, Olsen A, Bellerby R, Körtzinger A, Steinhoff T, Hoppema M, de Baar HJW, Wong CS, Delille B, Bates NR (2009) Climatological mean and decadal changes in surface ocean $pCO_2$, and net sea-air $CO_2$ flux over the global oceans. Deep-Sea Res II 56:554–577
17. Kinner G, Viovy N, de Noblet-Ducoudre N, Ogee J, Polcher J, Friedlingstein P, Ciais P, Stich S, Prentice IC (2005) A dynamic global vegetation model for the studies of the coupled atmosphere-biosphere system. Glob Biogeochem Cycles 19. https://doi.org/10.1029/2003GB002199
18. Randerson JT, van der Werf GR, Giglio L, Collatz GJ, Kasibhatla PS (2013) Global fire emission database, version 2.1. Oak Ridge Natl Labs. https://doi.org/10.3334/ORNLDAAC/849
19. Chevallier F, Viovy N, Reichstein M, Ciais P (2006) On the assignment of prior errors in bayesian inversion of $CO_2$ fluxes. Geophys Res Lett 33:L13802. https://doi.org/10.1029/2006GL026496
20. Lin X, Indira NK, Ramonet M, Delmotte M, Ciais P, Bhatt BC, Reddy MV, Angchuk D, Balakrishnan S, Jorphail S, Dorjai T, Mahey TT, Patnaik S, Begum M, Brenninkmeijer C, Durairaj S, Kirubagaran R, Schmidt M, Swathi PS, Vinithkumar NV, Yver Kwok C, Gaur VK (2015) Long-lived atmospheric trace gases measurements in flask samples from three stations in India. Atmos Chem Phys 15:9819–9849. https://doi.org/10.5194/acp-15-9819-2015
21. Laurent O (2017) ICOS atmospheric station specifications, version 1.3. https://www.icos-ri.eu/sites/default/files/cmis/ICOS%20Atmospheric%20Station%20specifications%20Version%201.3%20-%20November%202017.pdf
22. Masarie KA, Tans PP (1995) Extension and integration of atmospheric carbon dioxide data into a globally consistent measurement record. J Geophys Res 100:11593–11601. https://doi.org/10.1027/95JD00859

The manufacturer's authorised representative in the EU is Springer Nature Customer Service Centre GmbH, Europaplatz 3, 69115 Heidelberg, Germany. If you have any concerns regarding our products, please contact ProductSafety@springernature.com

Printed and bound by CPI Group (UK) Ltd, Croydon, CR0 4YY
30/09/2026
02225247-0006